NUMERICAL ANALYSIS

NUMERICAL ANALYSIS

Rama B. Bhat
Ashok Kaushal

Alpha Science International Ltd.
Oxford, U.K.

Numerical Analysis
386 pgs.

Rama B. Bhat
Ashok Kaushal
Department of Mechanical Engineering
Concordia University
Montreal, Canada

Copyright © 2017

ALPHA SCIENCE INTERNATIONAL LTD.

7200 The Quorum, Oxford Business Park North
Garsington Road, Oxford OX4 2JZ, U.K.

www.alphasci.com

ISBN 978-1-78332-346-3

Printed from the camera-ready copy provided by the Authors.

PREFACE

Numerical Methods as a subject has not changed very much over the years. However, this book is different from the others available at present in the market because our approach in the presentation of the material is simple and crisp. This will serve as a very good textbook for any undergraduate course on numerical methods.

The first ten chapters of this book originally evolved out of the course notes developed over a period of several years at Concordia University, Montreal, Quebec, Canada. For this edition of the book, MATLAB® solutions to problems have been added.

In preparing the book, we have taken into account the needs of other courses in the undergraduate engineering programs such as circuit analysis, electricity, magnetism, fluid mechanics, heat transfer, structural analysis, acoustics, network analysis, mechanical vibrations, hydraulics in preparing this book. Example problems illustrating the methods give good insight into methods discussed.

This book also contains some MATLAB® programs with an introduction to the use of MATLAB® programming given in the appendix. This was done with the intention to assist undergraduate students in learning the basics of programming in general and programming MATLAB® in particular. The codes are written in a way that is easy to follow and require little prior programming knowledge, however, if some users want a more comprehensive introduction, there are many other resources available. The MATLAB codes provided might not be the shortest, fastest or the most efficient but can be easily understood and help the students advance their knowledge of the software. If some users have prior knowledge of MATLAB® they are encouraged to write programs that might be more efficient and compare their results with those in this book. The main focus is to cover the basics of how to program an algorithm for the various numerical methods.

We acknowledge the helpful comments from our colleagues as well as the students. Thanks are particularly due to Mr. Bijoy Kothari who helped in preparing the manuscript in the final form.

Rama B. Bhat
Ashok Kaushal

CONTENTS

Chapter 1

SOLUTION OF EQUATIONS FOR ENGINEERING DESIGN AND ANALYSIS

1.1 INTRODUCTION

In order to implement any physical systems that engineers conceive in their minds, they should verify whether such physical systems would do their intended operations satisfactorily in the operating environments. This is normally done by developing a mathematical model for the physical system, subjecting them to external loads that they would be subjected to during their operation and ensuring the systems would be stable and perform their intended tasks in the operating environments. The conditions of stable operation under operating environments may be expressed in the form of system of simple equations, system of differential equations or system of integral equations. When it is difficult, cumbersome or sometimes impossible to solve these equations in closed form engineers resort to numerical techniques to solve them.

With the advent of electronic computers, the numerical solutions to engineering problems cast in the form of mathematical equations have become simpler. The accuracy of such numerical solutions also can be improved by systematic applications of the numerical techniques or algorithm, in an iterative form.

1.2 TAYLOR'S SERIES EXPANSION OF FUNCTIONS

For satisfactory operation under operating environments, physical systems satisfy equilibrium conditions, which can be expressed in the form of simple equations, differential equations or integral equations. In any of these forms, a function can be expressed in its neighborhood, if we know the function value and their higher derivatives at a given point, using Taylor's series expression about the point.

The Taylor's series is the foundation of numerical methods. Many of the numerical techniques are derived directly from these series.

Taylor's series obtains the function in the form of a polynomial in the neighborhood of the known point. The expansion is given by

$$f(x) = f(a) + (x-a)f'(a) + \frac{(x-a)^2}{2!}f''(a) + \cdots + \frac{(x-a)^n}{n!}f^{(n)}(a) + \cdots$$

If the estimation of the function at a point b which is fairly close to a is desired, it is given by

$$f(b) = f(a) + (b-a)f'(a) + \frac{(b-a)^2}{2!}f''(a) + \cdots + \frac{(b-a)^n}{n!}f^{(n)}(a) + \cdots$$

These are infinite series. In practical applications the series can be terminated after a few terms, depending on the distance of point b from a. If b is very close to *a*, only a few terms will give a good estimation.

The error in Taylor's series for $f(x)$ when the series is terminated after the term containing $(x-a)^n$ is not greater than

$$\left| f^{(n+1)} \right|_{max} \cdot \frac{\left(|x-a| \right)^{n+1}}{(n+1)!}$$

where *max* denotes the maximum magnitude of the derivative in the interval a to x. However, $f^{(n+1)}$ itself is not known if the function $f(x)$ is not known. This frustrating state of affairs is common in numerical analysis. If the expansion is used to obtain the estimate of the function for very small values of $(x-a)$, then the error also comes down. If the series is truncated after n terms, we can say that $f(x)$ is accurate to $0(x-a)^n$.

EXAMPLE 1.1
Determine a) $\sinh(0.9)$ and b) $\cosh(0.9)$ to $0(0.9)^4$.
Solution

$$\sinh(x) = \frac{e^x - e^{-x}}{2}; \quad \cosh(x) = \frac{e^x - e^{-x}}{2};$$

$$\sinh(0) = 0 \qquad\qquad \cosh(0) = 1$$

$$\sinh'(0) = 1 \qquad\qquad \cosh'(0) = 0$$

$$\sinh''(0) = 0 \qquad\qquad \cosh''(0) = 1$$

$$\sinh'''(0) = 1 \qquad\qquad \cosh'''(0) = 0$$

a)
$$\sinh(0.9) = \sinh(0) + 0.9\sinh'(0) + \frac{0.9^2}{2!}\sinh''(0) + \frac{0.9^3}{3!}\sinh'''(0)$$

$$= 0 + 0.9 + 0 + \frac{0.9^3}{6!} = 1.0215$$

(exact: 1.0265....)

b)
$$\cosh(0.9) = \cosh(0) + 0.9\cosh'(0) + \frac{0.9^2}{2!}\cosh''(0) + \frac{0.9^3}{3!}\cosh'''(0)$$

$$= 1 + \frac{0.9^2}{2} = 1.405$$

(exact: 1.4331....)

EXAMPLE 1.2
Consider one more non-zero term in the above expansions:
Solution

a)
$$\sinh(0.9) = 0 + 0.9 + 0 + \frac{0.9^3}{6} + 0 + \frac{0.9^5}{5!} = 1.0264$$

(exact: 1.0265....)

b)
$$\cosh(0.9) = 1 + \frac{0.9^2}{2} + 0 + \frac{0.9^4}{4!} = 1.4320$$

(exact: 1.4331...)

EXAMPLE 1.3
Given $f(1.8) = -1.1664$ and $f'(1.8) = 3.888$. Find out when $f(x) = 0$.

Solution

$$f(x) = f(a) + f'(a)(x - a)$$
$$0 = -1.1664 + 3.888(x - 1.8)$$

Hence, $x = 1.8 + \dfrac{1.1664}{3.888} = 2.1$

(The example is based on the function $f(x) = x^4 - 2x^3$, which has a root at $x = 2$)

EXAMPLE 1.4
Given the differential equation $\dfrac{df}{dx} = 2x$ with $f(1) = 1$ obtain $f(1.2)$.

Solution

$$f(x) = f(1) + f'(1)(x - 1) = f(1) + 2(x - 1) + 2(x - 1)^2/2!$$
$$f(1.2) = 1 + 2(1.2 - 1) + 2(1.2 - 1)^2/2 = 1.44$$

1.3 DIGITAL COMPUTERS

There exist basically two types of computers: 1) digital and 2) analog. The analog computer generates a continuous output signal when it is given a continuous input signal. This type of computer is popular in control systems; however, it has little value when number crunching is the main concern as in numerical methods. The digital computer deals with numbers in the form of finite number of digits. A sequence of simple binary operations, such as on or off, assigned with digits 0 and 1, respectively, can represent numbers to any number of digits of accuracy. The digital computer, like the calculator, can perform arithmetic operations; however, unlike a normal non-programmable calculator, it can perform several operations using data from previous operations if necessary, without help from the user. In other words, the digital computer has the ability to make decisions based on how it is programmed, *e.g.* if the computer is instructed by the program to compare two numbers A and B, then depending on whether A is less than, greater than, or equal to B, the computer will take one of three entirely different paths for subsequent operations. The computer also has the ability to store previous calculations for future use if they are needed.

The most important characteristic of the digital computer is its high speed. By employing simple routines, calculations that previously took days and weeks, can now be performed in seconds or minutes.

The major contribution which has revolutionized the use of computers is the development of a memory unit. The memory unit is capable of storing a few hundred to several thousand numbers. Commands can also be stored in the unit in a similar fashion.

A computer system consists of five major components: 1) input device, 2) a memory unit, 3) a control unit, 4) an arithmetic logic unit, and 5) output devices. These five components are either combined into one unit or operate as individual units. (see Fig. 1.1)

<u>**Computer operating system**</u>

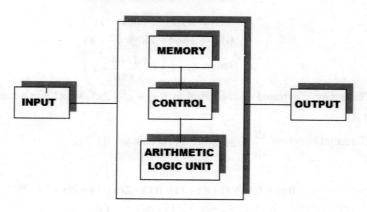

Figure 1.1: Computer operating system

The input unit serves as a device to feed data and commands into the memory unit. This is achieved by components such as keyboards, punch cards, magnetic disk, or magnetic tape. Magnetic disk is the fastest of the above mentioned.

The memory unit, control unit, and arithmetic logic unit are referred to as the Central Processing Unit or C.P.U. This is the unit which will, through the use of a compiler program, open and close various electronic switches or gates, allowing data to pass from one part of the machine to another. The C.P.U. is also responsible for making logical decisions and performing arithmetic operations.

The output devices, such as printers, plotters, magnetic disks, video screens, or magnetic tape units, serve to display the computed data outside the machine.

The units described so far are known as hardware. Hardware are the actual physical components of the computer, whereas software are the programs or instructions fed to the computer. Notice that software has no physical substance, it is just a set of instructions. This course will deal only with software.

1.4 NUMBER REPRESENTATION: FLOATING POINT AND FIXED POINT

When doing calculations on paper, the decimal point can be kept track of quite easily after each calculation. However, when using a computer, where several operations are performed internally before a final result is displayed, there are two techniques in which the decimal point is taken care of.

The first method is the fixed-point method which is seldom used. This is due to the inconvenience of the user having to keep track of decimal places in the input and output. The second and much more accepted method employed in the computer hardware is the floating point method.

In this method the computer represents a number in the following form.

$$\text{Number} = S \times M \times B^k$$

S = sign (+ or −)
M = mantissa, a value which lies between 0.1 and 1
B = base (B = 10 for decimal computers; B = 2 for binary computers)
k = exponent
To illustrate the above formula, it is convenient to consider a simple example.

EXAMPLE 1.5
Consider a familiar number such as Young's modulus:
$$E = 30,000,000 \text{ psi}$$
$$N = + 0.3 \times 10^8 \text{ psi}$$
Notice that the mantissa is always between 0.1 and 1. Also notice that this would be a decimal computer since the number is represented with base 10.

If an operation is carried out between two numbers in this form, zeros might appear before the decimal point. This is taken care of by the computer by normalizing the result:

EXAMPLE 1.6

$$0.5789875 \times 10^7$$
$$-0.5778764 \times 10^7$$
$$\overline{0.0011111 \times 10^7}$$

Normalizing this for the next operation would be to put it into the form:

$$0.11111 \times 10^5$$

So far the way in which a number is handled by the decimal based computer has been shown. For simplicity assume that we have a decimal computer with a storage space available for 10 digits plus a sign. In such a case the number:

$$N = -31.579643 \text{ would be represented by:}$$
$$N = -0.31579643 \times 10^2 \text{ which would be stored as}$$

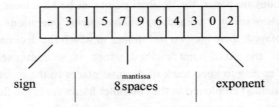

| − | 3 | 1 | 5 | 7 | 9 | 6 | 4 | 3 | 0 | 2 |

sign   mantissa   exponent
                  8 spaces

The computer handles a negative exponent by adding 50 to it;

$$N = 0.31579643 \times 10^{-4} \text{ this would be stored as;}$$

| + | 3 | 1 | 5 | 7 | 9 | 6 | 4 | 3 | 4 | 6 |

The computer knows that 46 represents an exponent of − 4.

At this point it must be noted that the available amount of storage space is 8 and therefore the precision of the machine is also 8. Suppose the mantissa has 12 digits and the computer precision is only 8 as above, or two numbers with 8 digits are multiplied producing a number 15 digits long. Some questions that would arise are how accurate are the results after 100 such multiplication and how does the computer store the result in its limited memory space.

The machine precision dictates the amount of accuracy of the result.

1.5 ALGORITHMS AND FLOWCHARTS

Algorithm is a set of step-by-step logical instructions that are to be followed to reach a goal efficiently. Some examples of an algorithm would be a recipe for cooking a dish or instructions for building a house.

An algorithm is the step-by-step thought process that is used to solve a problem. Algorithms should possess the following qualities:

(1) They should be precise: This means that if an iterative process is being performed that permits a relative error of 0.001, the operation should continue until this condition is satisfied, and not go farther.

(2) They should be finite: The process should terminate after a finite number of steps.

(3) They should be effective: This means that the problem should be solved efficiently. For instance computer time is expensive when running a large problem. If the program is well structured it will be effective.

Algorithms consist of three major parts, some input, a process, and resulting output.

A flowchart is a pictorial representation of an algorithm. For example the flowchart representing the algorithm to machine a component is shown below:

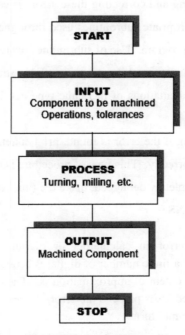

Figure 1.2: Logical Process

A logical approach to solving a typical engineering problem in real life is outlined below:

STEP 1 Formulate a mathematical model for the given problem. From this model certain mathematical equations can be developed.

STEP 2 Select a numerical method suitable for solving the mathematical equations. An algorithm should be developed at this stage.

STEP 3 Draw a detailed flowchart. Each block of the flowchart should represent a major section of the program. Make the flow chart as detailed as possible.

STEP 4 Write the codes, which is another name for computer program, in a suitable computer language. Each block of the flowchart should be represented by its own section of code. Often subroutines are employed to show the interdependence between two blocks of the flowchart. Subroutines or subprograms are an intricate part of structured programming.

STEP 5 Run the program on the digital computer. The first time the program is run, it is bound to have errors, whether it be in the actual logic, or just typing errors (referred to as syntax errors). The process of recognizing and correcting these errors is called debugging. If the program is well structured, with appropriate subroutines used throughout the program, its is usually easy to locate the error in the program. Use of subroutines makes it much easier for someone other than the programmer to understand the program. Subroutines can also be called at any point in the program so that calculations of the same type can be performed without writing new codes.

Once the general location of the error is found, print statements can be used to indicate what the program is doing incorrectly. This type of procedure is called tracing.

STEP 6 After the program is completely debugged, make the final run.

1.6 ERROR CONSIDERATIONS

There is an inherent form of error in calculators or digital computers because they perform mathematical operations with only a finite number of digits. Whenever such approximation is involved, we would like to known what the extent of approximation is. The difference between the exact value and the approximate value obtained in its place is the error. A typical number representation system in computers is in the normalized decimal form.

$$\pm d_1 d_2 d_3 \ldots d_k \times 10^n$$

$$1 \le d_1 \le 9, \qquad 0 \le d_k \le 9$$

The fractional part $\pm d_1 d_2 d_3 \ldots d_k$ is called the mantissa and the exponential part is called the characteristic. Different types of computers can handle different values of k and n.
Assume the actual number is of the form.

$$y = \pm . d_1 d_2 d_3 \ldots d_k d_{k+1} d_{k+2} \cdots \times 10^n$$

If a given computer can retain the mantissa part up to k decimal places only, it will have to discard the values in decimal places d_{k+1} and above. If the quantities d_{k+1} and above are just dropped off, it is referred to as "chopping". If d_{k+1} is 5 or larger than 5, then it is more meaningful to add a unit value to d_k and then drop off d_{k+1} and above. This latter procedure is called rounding-off. If d_{k+1} is less than 5, then both chopping and rounding procedures will give the same result.

EXAMPLE 1.7

Round off $y = 1/3$ to 6 decimal places in the normalized decimal form.

$$y = 1/3 = 0.3333333\ldots \times 10^0$$

Since $d_7 < 5$ we just drop off d_7 and above. The result is:

$$y = 0.333333 \times 10^0$$

EXAMPLE 1.8

Round off $y = 2/3$ to 6 decimal places in the normalized decimal form.

$$y = 2/3 = 0.6666666\ldots \times 10^0$$

Since $d_7 > 5$, we add 1 to d_6 and the result is:

$$y = 0.666667 \times 10^0$$

1.7 ABSOLUTE AND RELATIVE ERRORS

If p is the exact value required and by following a numerical solution method, we obtain an approximation $p*$ to p, then.

$$\text{Absolute error, } \xi_a = |p - p*|$$

and

$$\text{Relative error, } \xi_r = \frac{|p - p^*|}{|p|}$$

$$\text{Provided } p \neq 0.$$

EXAMPLE 1.9

What are the absolute and relative errors involved if $y = 2/3$ is represented in normalized decimal form with 6 digits.

(i)   by chopping (also called truncation)

(ii) by rounding off

Solution

(i)

$$y = 2/3 = 0.666666 \times 10^0$$

$$\text{Absolute error, } \xi_a = 2/3 - 0.666666 \times 10^0$$

$$= (0.6666666 \cdots - 0.666666) \times 10^0$$

$$= 0.000000666 \cdots \times 10^0$$

$$= 0.666 \ldots \times 10^6$$

$$\text{Relative error, } \xi_r = \frac{(0.6666666666 \cdots - 0.666666) \times 10^0}{0.66666666 \ldots \times 10^0}$$

$$= \frac{0.666666666 \cdots \times 10^{-6}}{0.666666666 \cdots 10^0} = 1 \times 10^{-6}$$

(ii)

$$y = 2/3 = 0.666667 \times 10^0$$

$$\text{Absolute error, } \xi_a = \left| (0.6666666 \cdots - 0.666667) \times 10^0 \right|$$

$$= 0.00000033 \ldots \times 10^0 = 0.33 \times 10^{-7}$$

$$\text{Relative error, } \xi_r = \frac{\left| (0.6666666 \cdots - 0.666667) \times 10^0 \right|}{\left| 0.66666666 \cdots \times 10^0 \right|}$$

$$= \frac{0.3333333 \cdots \times 10^{-6}}{0.6666666 \cdots \times 10^0} = \frac{1/3 \times 10^{-6}}{2/3 \times 10^0}$$

$$= 0.5 \times 10^{-6} = 5 \times 10^{-7}$$

1.8 SIGNIFICANT DIGITS

If p is the exact value and p^* is an approximation to p, then p^* is said to approximate p to t significant digits if t is the largest non negative integer for which

$$\frac{|p-p^*|}{|p|} < 5\times10^{-t}$$

EXAMPLE 1.10

If $p = \pi$ is approximated by $p^* = 3.14159$, what is the significant number of digits to which p^* approximates p?

Solution

$$\frac{|p-p^*|}{|p|} = \frac{\pi - 3.14159}{\pi}$$

$$= 8.45 \times10^{-7} < 5 \times10^{6}$$

Hence p^* approximates p to 6 significant digits.

EXAMPLE 1.11

If $p = 2/3$ is approximated by $p = 0.666666$, what is the significant number of digits to which p^* approximates p?

Solution

$$\frac{|p-p^*|}{|p|} = \frac{0.6666666\cdots-0.666666}{0.6666666\ldots}$$

$$= 10^{-6} < 5\times10^{-6}$$

Hence p^* approximates p to 6 significant digits.

EXAMPLE 1.12

Suppose that x = 1/3, y = 5/7 and that 5 digits chopping is used for arithmetic calculations involving x and y. What are the absolute and relative errors involved?

Solution $\quad$ $x = 0.33333333333\cdots\times10^{0}$; $x^* = 0.33333$

$$y = 0.71428571428\cdots\times10^{0} ; y^* = 0.71428$$

Operation	Result	Actual value	Abs. error	Rel. error
x + y	0.10476×10^{1}	22/21	0.190×10^{-4}	0.182×10^{-4}
y − x	0.38095	8/21	0.238×10^{-5}	0.625×10^{-5}
x × y	0.23809	5/21	0.524×10^{-5}	0.220×10^{-4}
y + x	0.21428×10^{1}	15/7	0.571×10^{-4}	0.267×10^{-4}

EXAMPLE 1.13

Suppose that $x = 1/3, y = 2/3$ and that 5 digit rounding off is used for arithmetic calculations involving x and y. What are the absolute and relative errors involved?

Solution

$$x = 0.33333333333\cdots\times10^0$$
$$x^* = 0.33333$$
$$y = 0.66666666666\cdots\times10^0$$
$$y^* = 0.66667$$

Operation	Result	Actual value	Abs. error	Rel. error
$x + y$	0.10000×10^1	1	0	0
$y - x$	0.33334	1/3	0.67×10^{-5}	0.2×10^{-4}
$x \times y$	0.22222	2/9	0.22×10^{-5}	0.1×10^{-4}
$y + x$	0.20000×10^1	2	0	0

1.9 SEQUENCES

Whenever approximate solutions to real world problems are ---, it is conceivable that the first approximation may--- be the best solution. Invariably this approximate solution must be required using. Most often, they involve iterative type of operations where the current value of the approximate solution is refined to obtain a better approximation. This iterative operation generates a sequence of better and better approximate solutions, if the sequence is convergent.

A sequence $p_1, p_2, p_3, \ldots$, then p is said to be convergent if there is a number *p* with the property

$$|p_n - p| < \xi \quad \text{for all } n > N$$

where ξ is a very small positive real quantity and N is an integer. The quantity p is called the limit of the sequence,

$$\lim_{n \to \infty} p_n = p$$

If the problem being solved by the iterative process generates the sequence $p_1, p_2, p_3 \ldots$, then p is also the exact solution.

CONVERGENCE OF THE SEQUENCE

A sequence may converge to a limit in a linear fashion or in a nonlinear fashion. If $p_1, p_2, p_3 \ldots$ is a sequence that converges to a limit value p, and if $e_n = p_n - p$ for $n \geq 1$, then the order of convergence can be expressed in the form.

$$\frac{|e_{n+1}|}{|e_n|^\alpha} = \lambda$$

where α is the order of convergence and λ is the asymptotic error constant. If $\alpha = 1$, convergence is linear and if $\alpha = 2$, convergence is quadratic.

1.10 ACCELERATING CONVERGENCE (Aitken's Δ^2 Process)

Quadratic convergence, or convergence of order higher than 2 is faster than a linear convergence. A linearly converging sequence can be accelerated to be quadratically convergent by Aitken's Δ^2 process. This technique will give better results in an iterative numerical technique with fewer **number** of iterative operations.

If $\{p_n\}$, $n = 1$ to ∞, is a linearly convergent ($\alpha = 1$) sequence with limit p, and $e_n = p_n - p$, **then**

$$\lim_{n \to \infty} \frac{|e_{n+1}|}{|e_n|^\alpha} = \lambda \quad \text{and} \quad 0 < \lambda < 1$$

In the derivation of Aitken's process, it is supposed that the limiting case above occurs **for all** $n \ge 1$. Then

$$e_{n+1} = \lambda e_n$$

Hence,

$$p_{n+1} = e_{n+2} + p = \lambda e_{n+1} + p$$
$$\text{i.e. } p_{n+2} = \lambda(p_{n+1} - p) + p \text{ for all } n \ge 1$$

Reducing the subscript by 1, we have

$$p_{n+1} = \lambda(p_n - p) + p$$

Eliminating λ between p_{n+2} and p_{n+1} we get

$$p = \frac{p_{n+2} p_n - p_{n+1}^2}{p_{n+2} - 2p_{n+1} + p_n}$$

$$= \frac{p_n^2 + p_n p_{n+2} - 2p_n p_{n+1} + 2p_n p_{n+1} - p_n^2 - p_{n+1}^2}{p_{n+2} - 2p_{n+1} + p_n}$$

$$p = p_n - \frac{(p_{n+1} - p_n)^2}{p_{n+2} - 2p_{n+1} + p_n}$$

In general, the original assumption of $e_{n+1} = \lambda e_n$ will not be true; nevertheless, it is expected that the sequence $\{p_n^*\}$, defined by

$$p_n^* = p_n - \frac{(p_{n+1} - p_n)^2}{p_{n+2} - 2p_{n+1} + p_n}$$

converges more rapidly to p than the original sequence $\{p_n\}$, $n = 1$ to ∞.

EXAMPLE 1.14

The sequence $\{p_n\}$, $n = 1$ to ∞, and $p_n = n \cdot \ln(1 + 1/n)$ converges linearly to $p = 1$. Using Aitken's Δ^2 process, obtain another sequence, which converges faster to $p = 1$

Solution
The sequence which converges in a quadratic fashion is given by

$$p_n^* = p_n - \frac{(p_{n+1} - p_n)^2}{p_{n+2} - 2p_{n+1} + p_n}$$

The following table provides the original sequence and the faster converging sequence:

n	p_n	p_n^*
1	0.693147	0.904408
2	0.810931	0.931174
3	0.863046	0.946129
4	0.892574	0.955719
5	0.911608	0.962401
6	0.924904	
7	0.934720	

It appears from the table above that p_n^* converges more rapidly than p_n to the exact solution of $p = 1$.

PROBLEMS

1.   For a function f (x) it is known that f (1.8) = $-$ 1.1664 and f'(1.8) = 3.8880. Find the approximate value of x when f (x) = 0, using Taylor series expansion.

2.   Given a differential equation $\dfrac{dy}{dx}$ = 2x, 0 < x < 5 and with y (1) = 1. Obtain the approximate value of y (1.2) using Taylor series expansion.

3.   Obtain the Taylor series expansion of the polynomial
$$P(x) = a_0 + a_1x + a_2x^2 + \ldots + a_nx^n$$
Comment on the result.

4.   Given the Taylor series expansion of f(x) = sin(x) about x=0. From the obtain sin(π/4) to an accuracy of 4 digits.

5.   In problem 4, obtain the values of sin(π/4) considering 1,2,3,4,5,6,7,8,9 and 10 terms in the Taylor series expansion. The result must converge to the exact value of $1/\sqrt{2}$. Check whether the convergence can be improved using Aitkin's Δ^2 process. Comment on the outcome.

6.   Obtain $(1/3)^{10}$ using
      (i)   4 digit arithmetic (result after each multiplication rounded off to 4 digits).
      (ii)   6 digit arithmetic
      (iii) 8 digit arithmetic

7.   Approximate sin(x) using a polynomial
$$P_2(x) = a_0 + a_1\,x + a_2\,x^2$$
In the range $0 \le x \le \pi/2$, by minimizing the squared error in the range, as the given by
$$E = \int\limits_0^{\pi/2}[P_2(x) - \sin(x)^2]dx$$

The coefficients a_0, a_1 and a_2 are obtained by solving the equation $\dfrac{\partial E}{\partial a_1}$ = 0, I=-, 1,2.

Compare $P_2(x)$ with the Taylor series expansion of sin(x) about x = 0, considering only three terms.

8.   Show that $f(x) = (x - 1)^{0.5}$ cannot be expanded in Taylor series about $x = 0$ or $x = 1$, but can be expanded about $x = 2$. (Hint: $f(0)$, $f'(0)$, $f''(0)$, etc. need evaluation of $(-1)^{0.5}$ which does not have real values).

9.   Find the Taylor series expansion of e^x about $x = 0$, and find $e^{1.0}$ to an accuracy of 5 digits.

10.   Find the integral

$$I = \int_0^1 e^x dx$$

using Taylor series expansion of I about $x = 0$, to an accuracy of 3 digits.

Chapter 2

NUMERICAL SEARCH FOR ROOTS OF ALGEBRAIC AND TRANSCENDENTAL EQUATIONS

In many practical engineering problems we may be faced with an equation of the type f(x) = 0, or an equation which can be concreted to this form. In such cases, it is required to find the value of x that satisfies the functional relation.

Let $f(x)$ is continuous in $a \leq x \leq b$.

We will discuss methods to solve for the roots of the equations of the form

$$f(x) = 0 \qquad (2.1)$$

In Fig. 2.1, f(x) is plotted against x. The function f(x) is zero at points a and b, where the equation f(x)=0 is satisfied. Different numerical methods to search for the roots are discussed below.

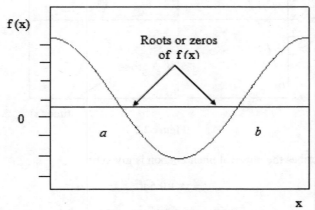

Figure 2.1: Roots of F(x) (f(x) vs x)

2.1 INCREMENTAL SEARCH

When f(x) is plotted against x as in Fig. 2.1, points where the curve crosses the x axis are the roots of the equation f(x) = 0. The incremental search procedure computes the value of f(x) for incremental values of x, starting at x = a, until there is a change in the sign of f(x). A sign change indicates that the curve has crossed the x-axis. The last value of x, before the sign change, is a good approximation for the root. This value can be refined by repeating the incremental search starting at this value using a smaller increment of x.

The incremental search method is widely used to locate the approximate region of the roots; however refinement is usually left to other methods discussed later in this chapter, which converge toward the root much faster.

The incremental search method may be terminated when one or more of the following conditions are met:

1.   When the relative error between successive approximations is below a specified value.
2.   When the increment size is below a specified value.
3.   When the total number of function evaluations is above a maximum specified value. This would be the case when several function evaluations have been made and no root is found. It would then be logical to stop and increase the increment size, before continuing the search.

EXAMPLE 2.1
A ball is thrown vertically up with an initial velocity of 15 m/sec. Determine the time it will take to reach 5 meters above the ground on the way up. Obtain the root with an accuracy of 2 digits.

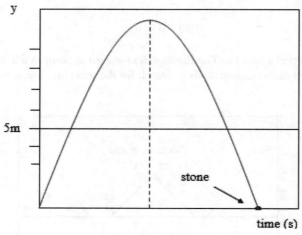

Figure 2.2

<u>Solution</u>:
The function which describes the physical phenomenon is given by

$$y = vt + 0.5at^2$$

(2.2)

where v (initial velocity) = 15 *m/sec*, a (acceleration) = -9.81 *m/sec²*, y (vertical height) = 5*m*, and t (time).

Substituting the appropriate values and putting the equation in the form of Equation (2.1) gives

$$f(t) = 4.90t^2 - 15t + 5 = 0 \tag{2.3}$$

Since the region of the root is unknown, we start at t = 0 and choose relatively large increments of, say 0.1 seconds (Δt = 0.1 sec).

Arranging the time t and the function value f (t) in a table as shown below:

Δt	0.1				
t	0	0.1	0.2	0.3	0.4
f (t)	5	3.55	2.2	0.94	-0.215
root	0.3				

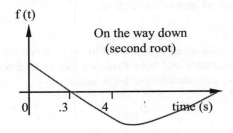

On the way down
(second root)

Figure 2.3: Region on the root (f(t) vs t)

We can see that a sign change between 0.3 and 0.4 indicates the presence of a root in the interval [0.3, 0.4]. To increase the accuracy, the search is continued starting at 0.3 with a smaller time increment. Such a smaller increment can be obtained by dividing the previous increment by a factor of 10.

Δt	0.01								
t	0.31	0.32	0.33	0.34	0.35	0.36	0.37	0.38	0.39
f (t)	0.82	0.70	0.58	0.47	0.35	0.24	0.12	0.008	-0.103
root	0.38								

Note that the relative error between approximations with Δ*t* = 0.1 and Δ*t* = 0.01 is given by:

$$\xi = \frac{0.38 - 0.3}{0.38} = 0.21 \tag{2.4}$$

As can be seen from the table, this will increase the field of the root by one more digit, 0.38. At this stage the accuracy is measured by the relative error which is computed to be 0.21. This does not have sufficient accuracy; therefore the procedure is continued with one more smaller step where the relative error is 0, and the accuracy to 2 digits is satisfied.

Δt	0.001	
t	0.380	0.381
f (t)	0.0082	-0.003
root	0.380	

Relative error between approximations with $\Delta t = 0.01$ and $\Delta t = 0.001$:

$$\xi = \frac{0.380 - 0.38}{0.380} = 0.0 \tag{2.5}$$

It is possible to continue this procedure to obtain higher accuracy in two ways: 1) either continue incremental search method with decreased increments of 0.0001, 0.00001, and so on, or 2) employ one of the refinement methods to be discussed later in this chapter.

CAUTION:
    Notice that the function may have changed sign several times between 0.3 and 0.4. This would be the case if the function were quite oscillatory and more than one root existed between these two points. This depends on the physical situation and often cannot be avoided.

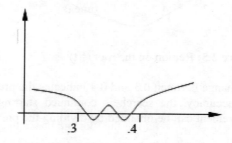

Figure 2.4: Oscillatory Function

However, in the case of Example 2.1 only two roots exist, since the equation is quadratic.

INCREMENTAL SEARCH METHOD

To find all the roots of $f(x) = 0$ in $a < x < b$. Function must be in the form $f(x)=0$

STEP 1.   Input: a, b the range of search
                    Δx incremental step size
                        ε smallest step size

STEP 2. Set $p_1 = a$, $\Delta p = \Delta x$

STEP 3. Set $p_2 = p_1 + \Delta p$

STEP 4. If $f(p_1) \cdot f(p_2) < 0$ then go to 6 (check for sign change)

STEP 5. Set $p_1 = p_2$ and go to 3

STEP 6. If $\Delta p > \varepsilon$ then set $\Delta p = \Delta p /10$, and go to 3

STEP 7. Print p_1 and p_2 (One root is between p_1 and p_2).

STEP 8. Stop.

2.2 BISECTION METHOD

The bisection method is a refinement method over the incremental search method. A sign change in the function value has already been found to occur between a and b by the incremental search method. The first approximation to the root is obtained as the bisection midpoint of a and b as:

$$p_1 = a + \frac{b-a}{2} \tag{2.6}$$

Now $f(p_1)$ is evaluated and its sign is compared with that of $f(a)$. If they are the same, the root lies between p_1 and b. If this is the case, the next approximation p_2 can be obtained by letting $a_{new} = p_1$ and $b_{new} = b$. If $f(p_1)$ and $f(a)$ have opposite signs, then the root lies between a and p_1. In this case let $a_{new} = a$ and $b_{new} = p_1$.

The second approximation to the root is obtained as

$$p_2 = a_{new} + \frac{b_{new} - a_{new}}{2} \tag{2.7}$$

after every p_i is computed, the value of $f(p_i)$ is obtained and we compare its sign with the corresponding $f(a_{new})$ to determine the location of the root and hence the corresponding values b_{new} and a_{new} for the next approximation.

The relative error is obtained as:

$$\xi = \frac{|p_i - p_{i-1}|}{|p_i|} \tag{2.8}$$

The procedure is illustrated in Figure 2.5.

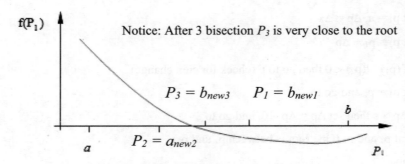

Figure 2.5: Bisection method illustrated

EXAMPLE 2.2

Find a root of the following equation with an accuracy of 4 digits, using the bisection method. The root is found to fall in the interval $[a = 0.3, b = 0.4]$ by the incremental search method.

$$f(t) = 4.905t^2 - 15t + 5 = 0 \tag{2.9}$$

Solution

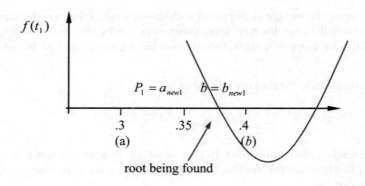

Figure 2.6

The calculations using the bisection method are shown in the following table:

n	a	B	P	f(a)	f(b)	f(p)	ξ
1	.3000	.4000	.3500	.9415	-0.2152	.3509	-
2	.3500	.4000	.3750	.3509	-0.2152	.0648	.0667
3	.3750	.4000	.3875	.0648	-0.2512	-.0760	.0323
4	.3750	.3875	.3813	.0648	-0.0760	-.0064	.0163
5	.3750	.3813	.3782	.0648	-0.0064	.0286	.0082
6	.3782	.3813	.3798	.0286	-0.0064	.0105	.0042

7	.3798	.3813	.3806	.0105	-0.0064	.0015	.0021
8	.3806	.3813	.3810	.0015	-0.0064	-.0030	.0010
9	.3806	.3810	.3808	.0015	-0.0030	-.0007	.0005
10	.3806	.3808	.3807	.0015	-0.0007	.0004	.0003

Explanation of table:

1. p_1 = midpoint of [a, b] = a+(b - a)/2 = 0.3 + 0.05 = 0.35
2. $f(p_1)*f(a) = f(0.35)*f(0.3) = 0.351*0.94 > 0$; p_1 and a are on the same side of the root.
3. $a_{new1} = p_1 = 0.35$; $b_{new1} = b = 0.4$
4. p_2 = midpoint of [a_{new1}, b_{new1}] = a_{new1} + (b_{new1} - a_{new1})/2 = 0.35+0.025 = 0.375
5. $f(p_2)*f(a_{new2}) = f(0.375)*f(0.35) = 0.0648*0.351 > 0$; p_2 and a a_{new1} are on the same side of the root.
6. $a_{new2} = p_2 = 0.375$; $b_{new2} = b_{new1} = 0.4$.
7. p_3 = midpoint of [a_{new2}, b_{new2}] = a_{new2} + (b_{new2} - a_{new2})/2 = 0.375+0.0225 = 0.3875.
8. $f(p_3)*f(a_{new1}) = f(0.3875)*f(0.375) < 0$; p_3 and b_{new2} are on the same side of the root.
9. $b_{new3} = p_3 = 0.3875$; $a_{new3} = a_{new2}$

.
.
.

and so on!
Up to the tenth bisection where the condition of 4 digit accuracy is satisfied.
Hence the root is 0.3807

$$\xi = \frac{0.3807 - 0.3806}{0.3807} = 0.0003 < 5 \times 10^{-4}$$

BISECTION METHOD

To find the root of f(x) = 0 in a < x < b and f(a)·f(b) < 0 (one root is between a and b).

STEP 1.   Input: a, b end points

                          N   maximum number of iterations

                          ξ relative error

STEP 2.   Set i = 1

STEP 3.   Set p_0 = a, p_1 = (a+b)/2

STEP 4.   If | p_1 - p_0 | / | p_1 | < ξ then go to 9 (relative error check)

STEP 5.   Set i = i +1; p_0 = p_1

STEP 6.   If i > N then stop

STEP 7.   If $f(p_1) \cdot f(a) > 0$ then set $a = p_1$ else set $b = p_1$

STEP 8.   Go to 3

STEP 9.   Output: p_1 (root)

2.3 METHOD OF FALSE POSITION (REGULA FALSI)

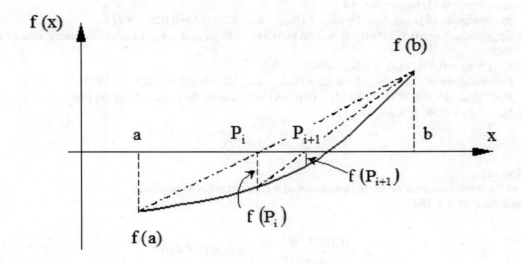

Figure 2.7: Method of false position (f(x) vs x)\

The method of false position requires the incremental search method to locate a range in which a root exists. The method uses linear interpolation to approximate the root. If functions evaluated at points a and b have opposite signs, then a root must exit between them. By joining f(a) and f(b) by a line the point P_i which intersects the x-axis can be evaluated by simple geometry (similar triangles). Once this approximation to the root is evaluated a more accurate approximation to the root can be found by calculating $f(P_i)$. The sign of the function at the new point, $f(P_i)$, is compared with that of f(a) to determine whether the root lies between a and P_i or between P_i and b. If f(a) and $f(P_i)$ have the same sign, the root is between P_i and b. If so, for the next approximation, join $f(P_i)$ and f(b) by a straight line (see fig. 2.7). The next approximation P_{i+1} is where this line intersects the x-axis. In order to evaluate P_1, the similar triangles $\Delta[a, f(a), P_i]$ and $\Delta[P_i, f(b), b]$ with common point P_i are used.

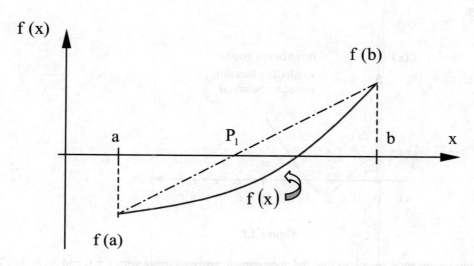

Figure 2.8: Method of false position (f(x) vs x)

By similar triangles,

$$p_i = \frac{af(b) - bf(a)}{f(b) - f(a)}$$

Hence

$$\frac{p_i - a}{-f(a)} = \frac{b - p_i}{f(b)} \qquad (2.10)$$

CAUTION:

For the second approximation, check the sign of $f(P_1)$ with $f(a)$ so that the interval in which the root lies is known. If the signs of $f(P_1)$ and $f(a)$ are the same then $a_{new} = P_1$; if they have different signs, let $b_{new} = P_1$.

EXAMPLE 2.3

Find the first root above $x = 0$ for the following function with an accuracy of 4 digits.

$$f(x) = e^x - 2x^2 = 0 \qquad (2.11)$$

Solution

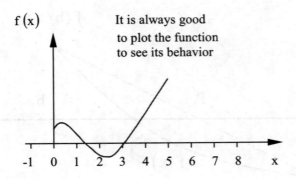

It is always good to plot the function to see its behavior

Figure 2.9

Apply incremental search to find the approximate range, starting with $x = 0$ and $\Delta x = 0.5$. This can be seen from

x	0	0.5	1.0	1.5	2.0
f(x)	1	1.149	0.718	-0.018	-0.61

The root lies between $x = 1.0$ and $x = 1.5$. Now apply the method of false position with $a = 1.0$ and $b = 1.5$.

$$P_i = \frac{af(b) - bf(a)}{f(b) - f(a)} \tag{2.12}$$

Calculations based on the method of false position are shown in the following table.

n	a	b	f(a)	f(b)	P_i	$f(P_i)$	ξ
1	1.0	1.5	0.718	-0.0183	1.4876	0.00055	-
2	1.4876	1.5	0.00055	-0.0183	1.48796	0.00003	0.0002

The relative error after the second step is

$$\xi = \frac{1.48796 - 1.4876}{1.48796} = 0.0002 \tag{2.13}$$

which is $0.0002 < 5 \times 10^{-4}$. Hence 4 digit accuracy is obtained. The root is 1.488.

METHOD OF FALSE POSITION

To find root of $f(x) = 0$ in $a < x < b$ and $f(a) \cdot f(b) < 0$ (one root between a and b).

STEP 1.   Input: a, b   end points

N   maximum number of iterations

ξ   relative error

STEP 2.   Set i = 1

STEP 3.   Set $P_0 = a$, $p_i = \dfrac{af(b) - bf(a)}{f(b) - f(a)}$

STEP 4.   If $\dfrac{|p_1 - p_0|}{|p_1|} < \xi$ then go to 9 (relative error)

STEP 5.   Set i = i + 1

STEP 6.   If i > N then stop

STEP 7.   If $f(P_1) \cdot f(a) > 0$ then set $a = P_1$ else set $b = P_1$

STEP 8.   Go to 3

STEP 9.   Output: P_1 (root)

2.4 NEWTON RAPHSON METHOD

The Newton Raphson method can be used to solve for the roots of equation f(x)=0, if **f(x) is** continuous and has a first derivative. This method uses a Taylor series expansion of the function **about** a point in the vicinity of the root. The function value at the exact root is known to be zero, hence:

$$f(x) = f(x_0) + (x - x_0)f'(x_0) + \frac{(x - x_0)^2}{2!} f''(x_0) + \cdots = 0 \tag{2.14}$$

x = exact root
x_0 = a point in the vicinity of *x* about which f(x) is being expanded. (usually obtained from incremental search method).

Newton Raphson method considers only two terms in the Taylor series expansion. Because of **this,** *x* is no longer the exact root, however it is a better approximation than x_0. The modified **equation** becomes:

$$f(x) = f(x_0) + (x - x_0)f'(x_0) = 0 \tag{2.15}$$

In Eq. (2.15) f'(x_0) is obtained by evaluating f'(x) at x = x_0.
The value of x is obtained by;

$$x = x_0 - \frac{f(x_0)}{f'(x_0)} \tag{2.16}$$

The accuracy of x depends largely upon how close x_0 is to the real root to begin with. This suggests that better accuracy can be obtained if the process is repeated, letting the latest calculated x to be equal to an improved x_0, since it is closer to the actual root than the original x_0 obtained from incremental search.

This type of process is called an iterative process. The iterative scheme is given by

$$x_{i+1} = x_i - \frac{f(x_i)}{f'(x_i)} \qquad (2.17)$$

After each iteration the relative error between the previously calculated root (x_i) and the latest root (x_{i+1}) should be checked to see if the relative error is acceptable. The iteration is stopped if

$$\left| \frac{x_{i+1} - x_i}{x_{i+1}} \right| \le \xi \qquad (2.18)$$

where ξ is the allowable relative error, x_{i+1} is the required root.
Let us progression graphically during the iteration process.

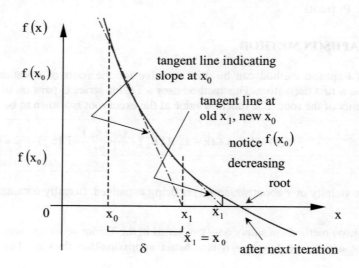

Figure 2.10: Newton Raphson method (f(x) vs x)

From Eq. (2.15)

$$f(x) = f(x_0) + (x - x_0)f'(x_0) = 0$$

Hence,

$$x = x_0 - \frac{f(x_0)}{f'(x_0)}$$

Eventually $f(x_0)$ goes a very small number and $x \cong x_0$.

EXAMPLE 2.4
Find a root of the function

$$f(x) = e^x - 2x^2 \qquad (2.19)$$

between 0 and 100, using Newton Raphson method to an accuracy of 5 digits.

Solution

It was previously found by increment search that a root lies between 1.0 and 2.0.
Start at $x_0 = 1.0$, and form a table as shown below, where,

$$f(x) = e^x - 2x^2 \qquad (2.20)$$
$$f'(x) = e^x - 4x$$

Relative error

$$\xi = \left| \frac{x_{n+1} - x_n}{x_{n+1}} \right|$$

$$x_{n+1} = x_n - \frac{f(x_n)}{f'(x_n)} \qquad (2.21)$$

n	x_n	f (x_n)	f' (x_n)	x_{n+1}	ξ
0	1.0	0.718	-1.2817	1.5602	0.359
1	1.5602	-0.109	-1.481	1.4866	0.0495
2	1.4866	0.00207	-1.5244	1.48796	0.0009
3	1.4880	-0.00006	-1.52377	1.487961	2.6×10^{-5}

In the third iteration the accuracy of 5 digits is reached and the root is 1.4880.

The rate of convergence for this method is of the order α^2 where;

$$\alpha = \frac{f(x_n)}{f'(x_n)} = x_{n+1} - x_n \qquad (2.22)$$

which signifies that after each iteration the distance between the root and the guess is decreased by the square of the distance in the previous iteration.

CAUTION: When the function is oscillatory an initial guess in the vicinity of one root may converge on another root

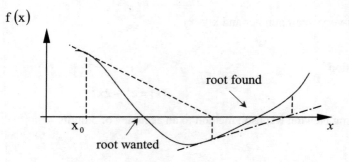

Figure 2.11: Region of no interest

Also if the initial guess is close to a maximum or a minimum the tangent line may shoot off into a region of no interest.

If the initial guess corresponds exactly to a minimum or maximum, then the derivative is zero and the method will not work.

Also care must be taken to stop if the method diverges.

NEWTON – RAPHSON METHOD

To find root of f(x) = 0 in the vicinity of $x = p_0$

STEP 1.   Input: p_0 initial approximation

             N   maximum number of iterations

             ξ   relative error

STEP 2.   Set i = 1

STEP 3.   Set $p_1 = p_0 - \dfrac{f(p_0)}{f'(p_0)}$

STEP 4.   If $\dfrac{|p_1 - p_0|}{|p_1|} < \xi$ then go to 8

STEP 5.   Set i = i + 1

STEP 6.   If i > N then stop

STEP 7.   Set $p_0 = p_1$, and go to 3

STEP 8.   Output: p_1 (root).

2.5 MODIFIED NEWTON RAPHSON METHOD

This method is used when the function has roots of multiplicity greater than 1. Consider the function of the form

$$f(x) = (x - m)^R . g(x) = 0 \qquad (2.23)$$

which has R number of roots at $x = m$. If R is even, $f(x)$ will just touch the x-axis at $x = m$, whereas if R is odd, $f(x)$ will have a horizontal inflection point on the x-axis at $x = m$.

The regular Newton-Raphson method will be very slow to converge on roots of such a function at $x = m$. The reason for this is that as the function value, $f(x_0)$, decreases so does its derivative $f'(x_0)$. Thus by examining the regular Newton Raphson equation, the problem can be seen immediately.

$$x = x_0 - \frac{f(x_0)}{f'(x_0)} \qquad (2.24)$$

Since $f(x_0)$ decreases at almost the same rate as $f'(x_0)$ the ratio remains almost constant.

For this reason, whenever it is suspected that a function contains multiple roots, the modified Newton Raphson method must be used.

This method, like the others, requires an initial starting point which can be obtained from the incremental search method. The function value will get very small at some point and then start increasing if R is even and will change sign only when x is above m.

This method employs the second derivative of the function, and solves for x_{n+1} in a similar fashion as the Newton Raphson method does.

Let the term $f(x)/f'(x)$ from the regular method equal a new function $U(x)$ such that;

$$\frac{f(x)}{f'(x)} = U(x) \qquad (2.25)$$

Now apply the Newton Raphson method to find the roots of $U(x)$:

$$x_{n+1} = x_n - \frac{U(x_n)}{U'(x_n)} \qquad (2.26)$$

where

$$U'(x) = \frac{[f'(x)]^2 - f(x)f''(x)}{[f'(x)]^2} \qquad (2.27)$$

The formula can be written in terms of $f(x)$ as

$$x_{n+1} = x_n - \frac{f(x_n)f'(x_n)}{[f'(x_n)]^2 - f(x_n)f''(x_n)} \qquad (2.28)$$

EXAMPLE 2.5

It is suspected that a root of f(x)=0 lies between 4 and 6 since during the incremental search the function value got very small, then started increasing. It is also suspected the function has multiplicity larger than 1 since no sign change occurred. If

$$f(x) = x^4 - 8.6x^3 - 35.51x^2 + 464.4x - 998.46 = 0 \qquad (2.29)$$

obtain the root to an accuracy of 5 digits.

Solution

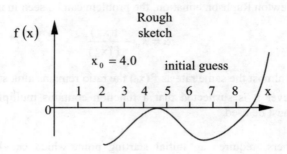

Figure 2.12

Using modified Newton-Raphson method:

$$f(x) = x^4 - 8.6x^3 - 35.51x^2 + 464.4x - 998.46 = 0$$
$$f'(x) = 4x^3 - 25.8x^2 - 71.02x + 464.4$$
$$f''(x) = 12x^2 - 51.6x - 71.02$$

Use a starting value of $x_0 = 4.0$

$$x_{n+1} = x_n - \frac{f(x_n)f'(x_n)}{[f'(x_n)]^2 - f(x_n)f''(x_n)} \qquad (2.30)$$

Form a table:

n	x_n	$f(x_n)$	$f'(x_n)$	$f''(x_n)$	x_{n+1}	ξ
0	4.0	-3.42	23.52	-85.42	4.3081	0.072
1	4.3081	-0.0023	-0.5737	-70.60	4.2999	0.001
2	4.2999	-3.7×10^{-7}	0.0071	-71.025	4.3000	2×10^{-5}

After second iteration the accuracy of 5 digits is obtained. The root is 4.3000.

Note that for the regular method the same root was found after 6 iterations (twice as long).

CAUTION: Even though the modified Newton Raphson method is twice as fast as the Newton Raphson method, if the function is extremely difficult to differentiate the normal Newton Raphson method or the secant method which is discussed next may be preferred.

MODIFIED NEWTON – RAPHSON METHOD

To find the root of $f(x) = 0$ in the vicinity of $x = p_0$, when convergence in Newton-Raphson method becomes slow or erratic. Multiple roots suspected.

STEP 1. Input: p_0 Initial approximation

            N   maximum number of iterations

            ξ relative error

STEP 2. Set $i = 1$

STEP 3. Set $p_1 = p_0 - \dfrac{f(p_0).f'(p_0)}{[f'(p_0)]^2 - f(p_0).f''(p_0)}$

STEP 4. If $\dfrac{|p_1 - p_0|}{|p_1|} < \xi$ then go to 8

STEP 5. Set $i = i + 1$

STEP 6. If $i > N$ then stop

STEP 7. Set $p_0 = p_1$, and then go to 3

STEP 8. Output: p_1 (root)

2.6 SECANT METHOD

The secant method is almost identical to the Newton Raphson method. Except for the derivative of the function being approximated by finite differences instead of being calculated analytically.

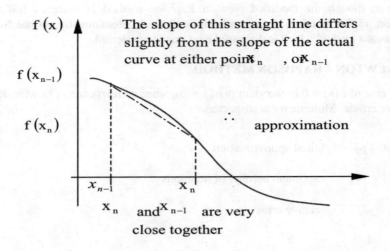

Figure 2.13: Secant method (f(x) vs x)

The procedure does not converge as fast as the Newton Raphson method. However, it is preferred when the derivative of the function is very difficult to obtain. The secant method requires two starting points to get an approximation of the derivative, as follows:

$$f'(x_n) = \frac{f(x_n) - f(x_{n-1})}{x_n - x_{n-1}} = \frac{\text{rise}}{\text{run}} \qquad (2.31)$$

The starting points are obtained by the incremental search method; however instead of just using the first value before the sign change, two previous values will be used. For example if the incremental search reveals a root between 1.5 and 1.6 then $p_n = 1.5$ and $p_{n-1} = 1.4$.

The formula to evaluate x_{n+1} becomes:

$$x_{n+1} = x_n - \frac{f(x_n)}{f'(x_n)} \qquad (2.32)$$

where $f'(x_n)$ is evaluated by Eq. 2.31.

The method becomes quite convenient when handling functions such as:

$$f(x) = x^2 3^x \cos 2x \qquad (2.33)$$

The analytical derivative would be too hard to work with:

$$f'(x) = 2x3^x \cos 2x + x^2 3^x \ln(\cos 2x) - 2x^2 3^x \sin 2x \qquad (2.34)$$

EXAMPLE 2.6
Find a root of the following functions with an accuracy of 4 digits using secant method.

$$(a) \quad f(x) = e^x - 2x^2 = 0 \qquad\qquad (2.35)$$

$$(b) \quad f(x) = x^3 - 8 = 0 \qquad\qquad (2.36)$$

Solution

(a) Incremental search reveals that a root lies between 1.4 and 1.5.
Hence,

$$x_{n-1} = 1.3$$

$$x_n = 1.4$$

$$f'(x_n) = \frac{f(x_n) - f(x_{n-1})}{x_n - x_{n-1}}$$

$$x_{n+1} = x_n - \frac{f(x_n)}{f'(x_n)}$$

Form a table:

n	x_{n-1}	x_n	$f(x_{n-1})$	$f(x_n)$	$f'(x_n)$	x_{n+1}	ξ
0	1.3	1.4	0.2893	0.1352	-1.541	1.4877	-
1	1.4	1.4877	0.1352	0.0004	-1.5371	1.48796	0.0002

The root is 1.488 (4 digit accuracy)

(a) Incremental search reveals that a root lies between 1.9 and 2.1 (The known root of x = 2 is intentionally avoided).

$$x_{n-1} = 1.7$$
$$x_n = 1.9$$

Form a table:

n	x_{n-1}	x_n	$f(x_{n-1})$	$f(x_n)$	$f'(x_n)$	x_{n+1}	ξ
0	1.7	1.9	-3.087	-1.141	9.73	2.0173	-
1	1.9	2.0173	-1.141	0.2094	11.51	1.999	-0.009
2	2.0173	1.999	0.2094	-0.0119	2.098	1.9999	0.0003

The root is 2.000

SECANT METHOD

To find root of f(x) = 0 in the vicinity of x = p_1

STEP 1. Input: p_0, p_1 initial approximation ($p_0 < p_1$)

N   maximum number of iteration

ξ relative error

STEP 2. Set i = 2

STEP 3.   Set $p_2 = p_1 - \dfrac{f(p_1)}{[f(p_1) - f(p_0)]/(p_1 - p_0)}$

STEP 4.   If $\dfrac{|p_2 - p_1|}{|p_2|} < \xi$ then go to 8

STEP 5.   Set $i = i + 1$

STEP 6.   Set $p_0 = p_1$, $p_1 = p_2$ and go to 3 (update p_0 and p_1)

STEP 7.   Output: p_2 (root)

2.7 COMPLEX ROOTS OF EQUATIONS

Complex roots of equations of the type

$$f(z) = 0 \tag{2.37}$$

where z is a complex number of the type x + jy, may be found by using Newton-Raphson method. In this case successive approximations for the complex roots are obtained as:

$$z_n = z_{n-1} - \frac{f(z_{n-1})}{f'(z_{n-1})} \tag{2.38}$$

Using a programming language which permits complex variables and writing the program in the same lines as that for finding real roots, it is possible to obtain complex roots. The variable z must be defined as complex and the initial guess value must be given as a complex value. The error can be estimated from

$$\xi = \left[\{ Re(\zeta Re^2 + \{ Im(\zeta Im^2 \right]^{0.5} \tag{2.39}$$

Where

$$\zeta = \frac{|z_n - z_{n-1}|}{|z_n|} \tag{2.40}$$

and $Re(\zeta)$ and $Im(\zeta)$ are the real and imaginary parts of ζ respectively.

EXAMPLE 2.7
Find a root of the equation, using Newton-Raphson method.

$$f(z) = 4 - z + z^2$$

Solution:
  Assume

$$z_1 = i + j$$

$$f(z) = z^2 - z + 4$$

$$f'(z) = 2z - 1$$

$$z_n = z_{n-1} - \frac{f(z_{n-1})}{f'(z_{n-1})}$$

Substituting the initial guess into the scheme we get the following table:

n	z	f(z)
1	i+j	3+j
2	0+2j	0-2j
3	0.4706+1.8824j	0.2076-0.1107j
4	0.5009+1.9370j	$-2.1 \times 10^{-3} + 3.3 \times 10^{-3}$ j
5	0.5000+1.9365j	$7.2 \times 10^{-7} + 9.5 \times 10^{-7}$ j

The corresponding error is

n	ζ	ξ
1		
2	0.5+0.5j	0.7071
3	$2.5 \times 10^{-5} - 2.5 \times 10^{-1}$ j	0.2500
4	$3.0 \times 10^{-2} - 7.8 \times 10^{-3}$ j	0.0312
5	$-3.5 \times 10^{-4} - 3.7 \times 10^{-4}$ j	0.0005

The root is $-3.5 \times 10^{-4} - 3.7 \times 10^{-4}$ j.

2.8 PRACTICAL APPLICATION

  It is usual practice to involve the incremental search method with one of the refinement methods. Once the region of the root is found by incremental search, a refinement is used in order to converge on the root quickly.

EXAMPLE 2.8
Find the 4 roots of the polynomial

$$P_4(x) = \frac{35}{8} x^4 - \frac{15}{4} x^2 + \frac{3}{8} \tag{2.41}$$

between −1 and 1. Accuracy of 4 digits is required.
(Hint: Use incremental search to find a region, then get 4 digit accuracy using Newton Raphson iterative method)

Solution:

First use incremental search with $\Delta x=0.1$ to find regions in which roots exist.

$\Delta x = 0.1$										
x_1	-1	-.9	-.8	-.7	-.6	-.5	-.4	-.3	-.2	-.1	0
$P(x_1)$	1	.21	-.23	-.41	-.40	-.29	-.11	.07	.23	.34	.37

$\Delta x = 0.1$										
x_1	.1	.2	.3	.4	.5	.6	.7	.8	.9	1
$P(x_1)$	.34	.23	.07	-.11	-.29	-.41	-.42	-.23	.21	1

Four sign changes are found, therefore the regions of all four roots are known. In the case that less than four regions are found, it would be necessary to repeat the procedure with a smaller increment.

Thus the starting points to be employed in the iterative process are

$$x_0 = -0.9, x_0 = -0.4, x_0 = -0.3, x_0 = -0.8$$

Newton-Raphson method:

$$x_{n+1} = x_n - \frac{f(x_n)}{f'(x_n)} \tag{2.42}$$

Recall

$$f(x) = \frac{35}{8} x^4 - \frac{15}{4} x^2 + \frac{3}{8} = 0$$

$$f'(x) = \frac{35}{2} x^3 - \frac{15}{2} x = 0$$

Root 1: $x_0 = -0.9$

i	x_i	$f(x_i)$	$f'(x_i)$	x_{i+1}	ξ
0	-0.9	0.21	-6.0075	-0.8654	0.039
1	-0.8654	0.020	-4.8515	-0.8613	0.005
2	-0.8613	0.0008	-4.722	-0.8611	0.0002

Root 1 = -0.8611

Root 2: $x_0 = -0.4$

i	x_i	$f(x_i)$	$f'(x_i)$	x_{i+1}	ξ
0	-0.4	-0.113	1.88	-0.3399	0.177
1	-0.3399	0.00015	1.86	0.3400	0.0002

Root 2 = -0.3400

Root 3: $x_0 = 0.3$

i	x_i	$f(x_i)$	$f'(x_i)$	x_{i+1}	ξ
0	0.3	0.073	-1.778	0.3411	0.12

| 1 | 0.3411 | -0.002 | -1.864 | 0.3400 | 0.003 |
| 2 | 0.3400 | -0.00008 | -1.862 | 0.3400 | 0.000 |

Root 3 = 0.3400

Root 4: $x_0 = 0.8$

i	x_i	$f(x_i)$	$f'(x_i)$	x_{i+1}	ξ
0	0.8	-0.233	2.96	0.8787	0.089
1	0.8787	0.088	5.28	0.8621	0.019
2	0.8621	0.004	4.75	0.8613	0.0009
3	0.8613	0.0006	4.72	0.8611	0.0001

Root 4 = 0.8611

PROBLEMS

1.   Find the smallest positive root of the equation:
$$1- \frac{x^2}{6} + \frac{x^4}{120} - \frac{x^6}{5040} = 0$$
in the interval (0,8), with an accuracy of 5 digits. Initially locate the root roughly with incremental search technique, and then refine it using Newton-Raphson method.

2.   Compute the positive root of the equation correct to 3 decimal places using Bisection method.
$$t - 0.2 \sin t = \frac{1}{2}$$

3.   Find the points where x-axis cuts the circle,
$$(x-4)^2 + (y-1)^2 = 9,$$
to an accuracy of 4 digits using Bisection method (find both points).

4.   Find the point where the following function has a maximum in the range $2 \leq x \leq 6$ using Bisection method.
$$f(x) = \cos x \cosh x - 1 = 0$$
with an accuracy of 3 digits.

5.   Determine the smallest positive root between (0,2) of the given equation using Bisection method with an error less than 0.02.
$$2x = e^x - 0.9$$

6.   Determine the smallest positive root between (0,2) of the given equation using Bisection method with an error less than 0.02.
$$\cos x = \frac{1}{x} \ln x$$

7.   Solve $4 \cos x = e$

       with accuracy 6 digits using method of False Position in the range of $\pi/4 \le x \le \pi/2$.

8.   Determine the smallest positive root of the following equation

$$2 \sin x + x - 1 = 0$$

       to 3 decimal places using method of False Position.

9.   Compare the results obtained when the Bisection method the method of False Position and the
       Secant method are used to solve the equation for the smallest positive root with an error less
       than 10^{-4}.

$$3 \sin x = x + \frac{1}{x}$$

10   Use the method of False Position to find the positive root of the equation correct to 4 decimal
       places.

$$2x + \cos x - 2 = 0$$

11   Given $x \tan x = 2$
       find the smallest positive root with an error less than 0.02 by.
       (i)   Method of False position
       (ii)   Modified Newton Raphson method.

12.   Evaluate the smallest positive root of the following equation by Newton-Raphson method and
       4 significant digits.

$$3x^2 - \tan x = 0$$

13.   Use Newton-Raphson method to find the real roots of the following equation to 4 decimal
       places.

$$\cos x = x^2$$

14.   Determine the positive root of $3xe = 1$ to 4 decimal places using Newton-Raphson method.

15.   Use Newton-Raphson method to determine the positive root of the equation
$$\sin x - 4x^2 + 1 = 0$$
       correct to 3 decimal places.

16.   Find the root of the equation

$$x^4 - 3.600x^3 + 4.860x^2 - 2.916x + 0.6561 = 0$$

       using Newton-Raphson method. The root is suspected to be of multiplicity more than one.
       Obtain accuracy up to 4 digits.

17.   Find a root of the equation (if one exists)

$$x^2 - 4x + 5 = 0$$

using Newton-Raphson method, starting with a value of 1.5. Carry out 5 iterations. Do they converge towards a root? Discuss the results by plotting the function.

18.   Consider the following equations:

$$3x_1 - 4x_2 = 0$$

$$\cos x_1 + x_2 - 1.4713 = 0$$

Reduce the two equations to a single equation in x_2 and solve for one positive root of x_2 to four significant digits using Newton-Raphson method.

19.   Solve the equation

$$3x^2 - e^x = 0$$

by the Newton-Raphson method. Obtain the initial approximation by an incremental search method. Relative error must be less than 5×10^{-5}.

20.   Freudenstein's equation relating the input angle θ and the output angle ϕ of a four bar mechanism is given by

$$R_1 \cos\theta - R_2\cos\phi + R_3\cos(\theta-\phi) = 0$$

where R_1, R_2 and R_3 are constants depending on dimensions of the mechanism. For the value of $R_1 = 2.0$, $R_2 = 3.0$, $R_3 = 1.2$ find the output angle to an accuracy of 5 digits when input angle $\theta = 30^0$. Use Newton-Raphson method.

21.   Find the root of the equation

$$2x^2 + 1 = e^x$$

correct to 3 decimal places. Use the Newton-Raphson technique and start your solution at x = 1.0. (note: x = 0 is a trivial solution).

22.   Determine the non-zero root of $2x = 2 - 2e^{-2x}$ to 4 decimal places applying Modified Newton Raphson method.

23.   Apply Secant method to find the root of the following equation to 5 decimal places.
$$2x + 1 - e^x = 0$$

24.   Use Secant method to determine the roots of the equation
$$\sin x - 4x + 1 = 0$$

to 4 decimal places.

25.   Find cubic root of 8 using Secant method accurate to 4 digits.
        (Hint: solve $x^3 - 8 = 0$).

26.   Find the root of sin x = 0 in the range $\pi/2 \le x \le 3\pi/2$ using Secant method to 5 decimal places.

27.   Obtain the cubic root of 10 using Secant method. Use initial value of x = 1.0 and x = 2.0.
        Carry out 3 iterations.

28.   Given the equation.

$$f(x) = x^4 - 8.6x^3 - 35.51x^2 + 464.4x - 998.46 = 0.$$

Write a computer program to obtain all the roots of the equation using Modified Newton-Raphson method. Incorporate the Incremental Search procedure to get the initial rough approximations for the roots. Solve for the roots accurate up to 4 significant digits.

Chapter 3

METHODS TO SOLVE LINEAR SIMULTANEOUS EQUATIONS

Solutions of Linear Simultaneous Equations

There are many situations in engineering where a number of unknown quantities are related to each other through 'n' linear simultaneous equations containing these unknown quantities. Numerical methods can be conveniently used to solve such linear simultaneous equations in a logical fashion by expressing the equations in a matrix form and performing matrix operations.

Consider the three linear simultaneous equations as given below:

$$5x + 2y + 4z = 1$$
$$6x + 7y - 3z = 4$$
$$9x + 8y + 6z = 0$$

The can be expressed in matrix form as

$$\begin{bmatrix} 5 & 2 & 4 \\ 6 & 7 & -3 \\ 9 & 8 & 6 \end{bmatrix} \begin{bmatrix} x \\ y \\ z \end{bmatrix} = \begin{bmatrix} 1 \\ 4 \\ 0 \end{bmatrix}$$

3.1 PROPERTIES OF MATRICES

Matrices are a two dimensional array of numbers. Each number has a particular address depending on which row and column intersect at its location. An element A_{ij} is located in the i th row and *j* th column. Matrices are square if the number of rows is the same as number of columns. For example, 'A' is a 5 row by 5 column square matrix. Its dimension is 5x5.

$$A = \begin{bmatrix} 5 & 7 & 9 & 2 & 1 \\ 6 & 2 & 2 & 1 & 4 \\ 5 & 1 & 5 & 8 & 3 \\ 4 & 8 & 4 & 6 & 5 \\ 3 & 6 & 3 & 7 & 4 \end{bmatrix}$$

The number at location row i = 4, column j = 3 is denoted by

$$A_{row,column} = A_{i,j} = A_{4,3} = 4$$

A matrix with a single column is called a column vector, while a matrix with a single row is called a row vector.

$$\text{column vector } C = \begin{bmatrix} 5 \\ 6 \\ 5 \\ 4 \\ 3 \end{bmatrix} \qquad \text{row vector} = R = [5, 7, 9, 2, 1]$$

Square Matrix Configurations:

The central diagonal from left top corner to right bottom corner of a square matrix is called the main diagonal. For all elements on the main diagonal i = j. The elements on the main diagonal of matrix 'A' are $A_{11}, A_{22}, A_{33}, A_{44}, A_{55}$ or 5, 2, 5, 6, 4, in the example.

A square matrix is called symmetric if $A_{ij} = A_{ji}$ for all elements. Consider the matrices A and B given as

$$A = \begin{bmatrix} 5 & 7 & 9 & 2 & 1 \\ 6 & 2 & 2 & 1 & 4 \\ 5 & 1 & 5 & 8 & 3 \\ 4 & 8 & 4 & 6 & 5 \\ 3 & 6 & 3 & 7 & 4 \end{bmatrix} \qquad B = \begin{bmatrix} 1 & 2 & 3 \\ 2 & 6 & 5 \\ 3 & 5 & 4 \end{bmatrix}$$

Matrix 'A' is not symmetric, but matrix 'B' is, since $B_{ij} = B_{ji}$ for all elements.
A square matrix with all the nondiagonal elements zero, is called a diagonal matrix. When all the elements on the main diagonal of a diagonal matrix are 1, it is called an identity matrix I.

$$I = \begin{bmatrix} 1 & 0 & 0 & 0 \\ 0 & 1 & 0 & 0 \\ 0 & 0 & 1 & 0 \\ 0 & 0 & 0 & 1 \end{bmatrix}$$

Product of square matrix A and identity matrix I is A itself, *i.e.* AI = A
Matrix Addition

$$G + H = J$$

is possible if G and H have the same dimensions, since each element at an address in G is added to the element at the same address in H.

Matrix Addition is commutative:

$$H + G = G + H$$

Matrix Multiplication is <u>not</u> commutative:

$$HG \neq GH$$

For GH = J, each element of J has the address which corresponds to the row from G and the column from H whose elements were multiplied and summed to form it. For example, if row 1 of G and column 1 of H are multiplied, then the element formed in J would have the address J_{11}.

These elements are formed by:

$$J_{ij} = \sum_{k=1}^{n} G_{ik} H_{kj} \qquad n \rightarrow \text{ size of matrix}$$

For example, if G and H are both 3x3 matrices

$$J_{11} = G_{11}H_{11} + G_{12}H_{21} + G_{13}H_{31}$$

Notice that in order to carry out this operation the number of columns in G must be equal to the number of rows in H. To illustrate:

$$G*H = \sum_{k=1}^{n} G_{1k} H_{kj} = J_{11}$$

$$\begin{bmatrix} 3 & 2 \\ 4 & 5 \\ 1 & 3 \end{bmatrix}_{3\times 2} * \begin{bmatrix} 5 & 1 \\ 1 & 2 \end{bmatrix}_{2\times 2} = \begin{bmatrix} 17 & 7 \\ 25 & 14 \\ 8 & 7 \end{bmatrix}_{3\times 2}$$

$$J_{11} = G_{11} H_{11} + G_{12} H_{21}$$
$$J_{11} = (3*5) + (2*1) = 17$$

If two square matrices are multiplied and the result is the identity matrix, then the second square matrix is called the inverse of the first one.

$$AA^{-1} = I \qquad A^{-1} = \text{inverse of matrix } A$$

Similarly

$$(A^{-1})^{-1} = A$$

If the address of each element in any matrix, (square or not) is switched so that the row position becomes the column and the column position becomes the row, then the matrix is said to be transposed.

A_{ij} becomes A_{ji}

$$A = \begin{bmatrix} 1 & 2 & 3 \\ 4 & 5 & 6 \end{bmatrix} \quad A^T = \begin{bmatrix} 1 & 4 \\ 2 & 5 \\ 3 & 6 \end{bmatrix}$$

A^T = transpose of matrix A

If a square matrix C has a transpose $C^T = C^{-1}$, then the square matrix C is called orthogonal.

3.2 GAUSSIAN ELIMINATION

Gaussian Elimination is a simple systematic technique used to solve linear simultaneous equations. The procedure is carried out on the system of linear equations, arranged in the augmented matrix form. This is the normal coefficient matrix with the right hand column vector of the system included as shown below.

Consider the system of equations

$$2x_1 - x_2 + x_3 = 1$$
$$3x_1 + 3x_2 + 9x_3 = 0$$
$$3x_1 + 3x_2 + 5x_3 = 4$$

The augmented matrix form is given by

$$\widetilde{A}^1 = \begin{bmatrix} 2 & -1 & 1 & \vdots & 1 \\ 3 & 3 & 9 & \vdots & 0 \\ 3 & 3 & 5 & \vdots & 4 \end{bmatrix}$$

Basic row operations, such as multiplying one row by a number and subtracting it from another row, are performed in order to eliminate the coefficients one by one.

The row operations are:

$$\text{Row } j - \frac{A_{ji}}{A_{ij}} \text{Row } i \rightarrow \text{New Row } j$$

i =1, 2 ... n-1, j =2, ... n, n = number of equations

When i =1 and j goes from 2 to n, the resulting augmented matrix will be of the form;

$$\widetilde{A}^2 = \begin{bmatrix} A_{11} & A_{12} & A_{13} & \cdots & A_{1,n} & \vdots & A_{1,n+1} \\ 0 & A_{22} & A_{23} & \cdots & A_{2,n} & \vdots & A_{2,n+1} \\ 0 & A_{32} & A_{33} & \cdots & A_{3,n} & \vdots & A_{3,n+1} \\ 0 & . & . & & . & \vdots & . \\ 0 & . & . & & . & \vdots & . \\ . & . & . & & . & \vdots & . \\ 0 & A_{n,2} & A_{n,3} & \cdots & A_{n,n} & \vdots & A_{n,n+1} \end{bmatrix}$$

Note that all the elements below $A_{ii} = A_{i1}$ are zero. A_{ii} is known as the pivot element. For each value of i, the value of j goes i+1 to n. when i has gone from 1 to n-1 the matrix has the form;

$$\widetilde{A}^2 = \begin{bmatrix} A_{11} & A_{12} & A_{13} & \cdots & A_{1,n} & \vdots & A_{1,n+1} \\ 0 & A_{22} & A_{23} & \cdots & A_{2,n} & \vdots & A_{2,n+1} \\ 0 & 0 & A_{33} & \cdots & A_{3,n} & \vdots & A_{3,n+1} \\ 0 & 0 & 0 & & & \vdots & . \\ 0 & . & . & & & \vdots & . \\ . & . & . & & & \vdots & . \\ 0 & 0 & 0 & \cdots & A_{n,n} & \vdots & A_{n,n+1} \end{bmatrix}$$

In case of 3x3 matrix, the form will be;

$$\begin{array}{ccc} x_1 & x_2 & x_3 \end{array}$$
$$\widetilde{A}^3 = \begin{bmatrix} A_{11} & A_{12} & A_{13} & \vdots & A_{14} \\ 0 & A_{22} & A_{23} & \vdots & A_{24} \\ 0 & 0 & A_{33} & \vdots & A_{34} \end{bmatrix}$$

This can be put into the equation form and the solutions can be obtained by a simple procedure known as backward substitution.

$$A_{11}x_1 + A_{12}x_2 + A_{13}x_3 = A_{14}$$
$$A_{22}x_2 + A_{23}x_3 = A_{24}$$
$$A_{33}x_3 = A_{34}$$

From these equations;

$$x_3 = \frac{A_{34}}{A_{33}}$$

$$x_2 = \frac{A_{24} - A_{23}x_3}{A_{22}}$$

$$x_1 = \frac{A_{14} - A_{13}x_3 - A_{12}x_2}{A_{11}}$$

Note that x_3 must be obtained before x_2 and x_2 before x_1

This last value, x_3, is always obtained as $x_n = \dfrac{A_{n,n+1}}{A_{n,n}}$ (n is the number of rows). The procedure for obtaining subsequent values (x_{n-1}, x_{n-2}, *etc*), can be put into a simple form, as

$$x_i = \frac{A_{i,n+1} - \sum_{j=i+1}^{n} A_{ij}x_j}{A_{ii}}$$

i = n-1, $\cdots$ 2, 1; j = i+1, i+2 $\cdots$ n; n = number of equations

This procedure is known as backward substitution.

Note that x_n is already known, and therefore i only goes up to n-1.

EXAMPLE 3.1
Solve the following system of equations using Gaussian elimination.

$$2x_1 - x_2 + x_3 = 1$$
$$3x_1 + 3x_2 + 9x_3 = 0$$
$$3x_1 + 3x_2 + 5x_3 = 4$$

Solution
The augmented matrix is given by

$$\widetilde{A}^1 = \begin{bmatrix} 2 & -1 & 1 & \vdots & -1 \\ 3 & 3 & 9 & \vdots & 0 \\ 3 & 3 & 5 & \vdots & 4 \end{bmatrix}$$

where $A_{11} = 2$, is the first pivot element. The objective is to reduce all elements below A_{11} to 0. To get the first element below A_{11} to be 0, multiply row 1 by $\dfrac{A_{21}}{A_{11}}$ and subtract it from row 2. This is the same as following the reduction equation:

$$\text{Row } j - \frac{A_{ji}}{A_{ij}} \text{Row } i \rightarrow \text{New Row } j$$

range of i and j: $i = 1, 2$
$$j = 2, 3$$
$$n = 3 \text{ equations}$$

In this case

for $\begin{matrix} i = 1 \\ j = 2 \end{matrix}$ $\text{Row } 2 - \dfrac{A_{21}}{A_{11}} \text{Row } 1 \rightarrow \text{New Row } 2$

$A_{21} - \dfrac{A_{21}}{A_{11}} A_{11} \rightarrow A_{21} (\text{New})$ $3 - \dfrac{3}{2}(2) \rightarrow 0$ $A_{21} (\text{New})$

$A_{22} - \dfrac{A_{21}}{A_{11}} A_{12} \rightarrow A_{22} (\text{New})$ $3 - \dfrac{3}{2}(-1) \rightarrow 4.5$ $A_{22} (\text{New})$

$A_{23} - \dfrac{A_{21}}{A_{11}} A_{13} \rightarrow A_{23} (\text{New})$ $9 - \dfrac{3}{2}(1) \rightarrow 7.5$ $A_{23} (\text{New})$

$A_{24} - \dfrac{A_{21}}{A_{11}} A_{14} \rightarrow A_{24} (\text{New})$ $0 - \dfrac{3}{2}(-1) \rightarrow 1.5$ $A_{24} (\text{New})$

To get a zero element in row 3 under the pivot element A_{11} the above procedure is continued. A new row 3 is evaluated with a zero under A_{11}. The i value in the reduction equation remains unchanged since values below the same pivot element are being reduced. The value of j corresponds to that of the new row.

for $\quad\begin{array}{l} i = 1 \\ j = 3 \end{array}$ $\qquad$ Row $2 - \dfrac{A_{21}}{A_{11}}$ Row $1 \rightarrow$ New Row 2

$$A_{31} - \frac{A_{31}}{A_{11}} A_{11} \rightarrow A_{31} \text{(New)} \qquad 3 - \frac{3}{2}(2) \rightarrow 0 \qquad A_{31} \text{(New)}$$

$$A_{32} - \frac{A_{31}}{A_{11}} A_{12} \rightarrow A_{32} \text{(New)} \qquad 3 - \frac{3}{2}(-1) \rightarrow 4.5 \qquad A_{32} \text{(New)}$$

$$A_{33} - \frac{A_{31}}{A_{11}} A_{13} \rightarrow A_{33} \text{(New)} \qquad 5 - \frac{3}{2}(1) \rightarrow 3.5 \qquad A_{33} \text{(New)}$$

$$A_{34} - \frac{A_{31}}{A_{11}} A_{14} \rightarrow A_{34} \text{(New)} \qquad 0 - \frac{3}{2}(-1) \rightarrow 1.5 \qquad A_{34} \text{(New)}$$

Once all the elements below the pivot elements are reduced to zero, the matrix formed is referred to as the next augmented matrix.

$$\tilde{A}^2 = \begin{bmatrix} 2 & -1 & 1 & : & -1 \\ 0 & 4.5 & 7.5 & : & 1.5 \\ 0 & 4.5 & 3.5 & : & 5.5 \end{bmatrix}$$

Notice that when working with the first pivot element, the first row remained unchanged and when working with the second pivot element, rows 1 and 2 will be unchanged. Only rows below the pivot element are altered.

Note that even though individual elements A_{ij} have changed in the new augmented matrix $\tilde{A}^2$, they will be referred to again as A_{ij} for convenience. The next pivot element is A_{22}. Again the objective is to reduce all elements below this one to zero. Since there in only one row below this pivot element, only one reduction is carried out.

$$\text{Row3} - \frac{A_{32}}{A_{22}} \text{Row2} \rightarrow \text{New Row3}$$

$i = 2$ (pivot element row)
$j = 3$ (new row being calculated)

$$A_{32} - \frac{A_{32}}{A_{22}} A_{22} \rightarrow A_{32} \text{(New)}$$

$$A_{33} - \frac{A_{32}}{A_{22}} A_{23} \rightarrow A_{33} \text{(New)}$$

$$A_{34} - \frac{A_{32}}{A_{22}} A_{24} \rightarrow A_{34} \text{(New)}$$

Since all values below the pivot element become zero, the next augmented matrix becomes:

$$\widetilde{A}^3 = \begin{bmatrix} 2 & -1 & 1 & \vdots & -1 \\ 0 & 4.5 & 7.5 & \vdots & 1.5 \\ 0 & 0 & -4 & \vdots & 4 \end{bmatrix}$$

Note: The next pivot element A_{33}, has no element below it; therefore the system is completely reduced and ready for back substitution.

Back substitution:

$$-4x_3 = 4$$
$$x_3 = -1$$

Recall that

$$x_1 = \frac{A_{i,n+1} - \sum_{j=i+1}^{n} A_{ij}x_j}{A_{11}}$$

$i = n-1, \ldots 2, 1; j = i+1, i+2 \ldots n; n = 3$ equations

for $i = 2$
        $j = 3$

$$x_2 = \frac{A_{2,4} - \sum_{j=3}^{3} A_{2j}x_j}{A_{22}} = \frac{A_{2,4} - A_{2,3}x_3}{A_{22}}$$

$$x_2 = \frac{1.5 - 7.5(-1)}{4.5} = 2$$

$$x_1 = \frac{A_{1,4} - \sum_{j=2}^{3} A_{ij}x_j}{A_{11}} = \frac{A_{1,4} - A_{12}x_2 - A_{13}x_3}{A_{11}}$$

$$x_1 = \frac{-1-(-1)(2)-1(-1)}{2} = 1$$

Ans: $x_1 = 1$
         $x_2 = 2$
         $x_3 = -1$

The pivot element cannot be zero. If the pivot element is zero, the row containing the zero pivot element can be interchanged with any row below it. If this procedure still fails to produce a nonzero element in the pivot position, then this method will not work.

There must be the same number of equations as there are unknown variables.

GAUSSIAN ELIMINATION METHOD

To solve a matrix equation $Ax = b$ where A is nxn, b is nx1 and x is an nx1 vector of unknowns.

STEP 1.   Input: $\widetilde{A}$ Augmented matrix with a_{ij}, $i = 1, 2, \ldots n$,

j = 1, 2, ... n+1, where $a_{i,n+1} = b_i$.

Row i of $\widetilde{A}$ is denoted as E_i.

STEP 2.   For $i = 1, 2, \ldots, n-1$, do steps until 5

STEP 3.   Find the first available nonzero a_{pi} in $i \leq p \leq n$. (nonezero pivot element).

STEP 4.   If $p \neq i$ then interchange $E_i \Leftrightarrow E_p$

STEP 5.   Perform $E_j = (E_j - \dfrac{a_{ji}}{a_{ii}} E_i)$ for $j = i+1, i+2, \ldots, n$

STEP 6.   Set $x_n = a_{n,n+1} / a_{nn}$

and $x_i = \dfrac{a_{i,n+1} - \sum\limits_{j=i+1}^{n} a_{ij} x_j}{a_{ii}}$ $i = n-1, n-2, \ldots, 1$

STEP 7.   Output: $x_1, x_2, \ldots x_n$

3.3 PIVOTING TECHNIQUES

In the previous section, two rows are interchanged if the pivot element is zero. In this section it will be shown that also when the pivot element is relatively small compared to those elements beneath it row interchanges should be performed. This is to avoid the substantial round off errors that will occur when calculations are performed using finite digit arithmetic in a computer.

EXAMPLE 3.2
Solve the following system using 3digit arithmetic with rounding. Use Gaussian elimination.
(a)   without interchanging rows
(b)   interchanging rows
System of equations
$$0.0001x_1 + 1.00x_2 = 1.00$$
$$1.00x_1 + 1.00x_2 = 2.00$$

Solution
(a) The augmented matrix is given by

$$\widetilde{A}^1 = \begin{bmatrix} 0.0001 & 1.00 & \vdots & 1.00 \\ 1.00 & 1.00 & \vdots & 2.00 \end{bmatrix}$$

The pivot element of 0.0001 is very small compared to the element below it. Hence the factor $\dfrac{A_{21}}{A_{11}}$

becomes very large and critical due to the small denominator.

$$\text{Row2 -} \frac{A_{21}}{A_{11}} \text{Row1} \rightarrow \text{New Row 2} \quad \frac{A_{21}}{A_{11}} = \frac{1.00}{0.0001} = 10{,}000$$

$$A_{21} - \frac{A_{21}}{A_{11}} A_{11} \rightarrow A_{21} \text{ (New)} \qquad 1.00\text{-}10{,}000(0.0001) = 0$$

$$A_{22} - \frac{A_{21}}{A_{11}} A_{12} \rightarrow A_{22} \text{ (New)} \qquad 1.00\text{-}10{,}000(1.00)=\text{-}9999=\text{-}10{,}000$$

$$A_{23} - \frac{A_{21}}{A_{11}} A_{13} \rightarrow A_{23} \text{ (New)} \qquad 2.00\text{-}10{,}000(1.00)=\text{-}9998=\text{-}10{,}000$$

In the above computations, values are rounded to three significant digits.
The augmented matrix in the 2nd stage is given by

$$\widetilde{A}^2 = \begin{bmatrix} 0.0001 & 1.00 & \vdots & 1.00 \\ 0.00 & -10000 & \vdots & -10000 \end{bmatrix}$$

These results are quite inaccurate due to the amplification of round off errors by the large term, $\dfrac{A_{ji}}{A_{ii}}$

Carrying out back substitution results in

$$-10{,}000x_2 = -10{,}000$$
$$x_2 = +1.00$$
$$0.0001x_1 = 1.00 - 1.00(1.00)$$
$$x_1 = 0.00$$

(b) The round off error encountered in (a) can be avoided by interchanging the two rows. As a result, a relatively large number is placed in the pivot position.
The resulting matrix after the rows are interchanged is given by

$$\widetilde{A}^1 = \begin{bmatrix} 1.00 & 1.00 & \vdots & 2.00 \\ 0.0001 & 1.00 & \vdots & 1.00 \end{bmatrix}$$

where the point element is 1.00. Carrying out rows operations as follows, we get

$$\text{Row2-} \frac{A_{21}}{A_{11}} \text{Row1} \rightarrow \text{New Row2}$$

$$A_{21} - \frac{A_{21}}{A_{11}} A_{11} \rightarrow A_{21} \text{ (New)} \quad 0.0001\text{-}0.0001(1.00) = 0$$

$$A_{22} - \frac{A_{21}}{A_{11}} A_{12} \rightarrow A_{22} \text{ (New)} \quad 1.00-0.0001(1.00)=0.9999=1.00$$

$$A_{23} - \frac{A_{21}}{A_{11}} A_{13} \rightarrow A_{23} \text{ (New)} \quad 1.00-0.0001(2.00)=0.9998=1.00$$

The computations are carried out with 3 digit accuracy after rounding, and the augmented matrix at the next stage is

$$\widetilde{A}^2 = \begin{bmatrix} 1.00 & 1.00 & \vdots & 2.00 \\ 0.0 & 1.00 & \vdots & 1.00 \end{bmatrix}$$

Back substitution results in a much closer approximation to the correct answer.

$$1.00x_2 = 1.00$$
$$x_2 = 1.00$$
$$x_1 = \frac{2.00-1.00}{1.00} = 1.00$$

The true solution is

$$x_1 = 1.0001$$
$$x_2 = 0.9999$$

The strategy followed in the above example is called maximum column pivoting (also known as partial pivoting). In this procedure, the elements in the column directly below the current pivot element are examined in order to search for an element whose absolute value is larger than that of the current pivot element. If such an element is found, then the row containing that element is interchanged with the row containing the current pivot element. If the absolute value of the current pivot element is larger than that of any value below it, then no row interchange is necessary. This strategy must be performed on each augmented matrix. (Note: this strategy is carried out on each pivot element only once).

EXAMPLE 3.3
Use maximum column pivoting with Gaussian elimination to solve the following system.

$$x_1 + 2x_3 + 3x_4 = 1$$
$$-x_1 + 2x_2 + 2x_3 - 3x_4 = -1$$
$$x_2 + x_3 + 4x_4 = 2$$
$$6x_1 + 2x_2 + 2x_3 + 4x_4 = 1$$

Solution
The augmented matrix is given by

$$\widetilde{A}^1 = \begin{bmatrix} 1 & 0 & 2 & 3 & \vdots & 1 \\ -1 & 2 & 2 & -3 & \vdots & -1 \\ 0 & 1 & 1 & 4 & \vdots & 2 \\ 6 & 2 & 2 & 4 & \vdots & 1 \end{bmatrix}$$

The first pivot element $a_{11} = 1$, is obviously not the largest value in the first column. Hence row4 is interchanged with row1, which results in

$$\tilde{A}^1 = \begin{bmatrix} 6 & 2 & 2 & 4 & \vdots & 1 \\ -1 & 2 & 2 & -3 & \vdots & -1 \\ 0 & 1 & 1 & 4 & \vdots & 2 \\ 1 & 0 & 2 & 3 & \vdots & 1 \end{bmatrix}$$

At this stage, the basic row operations are used to get 0's below the first pivot element. Once these operations are performed, the matrix is referred to as the second augmented matrix.

$$\tilde{A}^2 = \begin{bmatrix} 6 & 2 & 2 & 4 & \vdots & 1 \\ 0 & 7/3 & 7/3 & -7/3 & \vdots & -5/6 \\ 0 & 1 & 1 & 4 & \vdots & 2 \\ 0 & -1/3 & 5/3 & 7/3 & \vdots & 5/6 \end{bmatrix}$$

Now the second pivot element, $7/3$, is examined. There is no element larger than $7/3$ below the pivot element; therefore, no row change is required. The third augmented matrix is formed by getting zeros below the second pivot element.

$$\tilde{A}^3 = \begin{bmatrix} 6 & 2 & 2 & 4 & \vdots & 1 \\ 0 & 7/3 & 7/3 & -7/3 & \vdots & -5/6 \\ 0 & 0 & 0 & 5 & \vdots & 33/14 \\ 0 & 0 & 2 & 2 & \vdots & 5/7 \end{bmatrix}$$

The pivot element in row3 is 0 and hence the row operations cannot be contined in this form. Hence, it is necessary to interchange row3 with row4 such that the pivot element becomes 2 as shown below:

$$\tilde{A}^3 = \begin{bmatrix} 6 & 2 & 2 & 4 & \vdots & 1 \\ 0 & 7/3 & 7/3 & -7/3 & \vdots & -5/6 \\ 0 & 0 & 2 & 2 & \vdots & 5/7 \\ 0 & 0 & 0 & 5 & \vdots & 33/14 \end{bmatrix}$$

Now the system can be solved directly by back substitution Using 4 digit accuracy,

$$5x_4 = 33/14$$
$$x_4 = 0.4714$$

$$x_3 = \frac{5/7 - 0.4714(2)}{2} = -0.1143$$

$$x_2 = \frac{-5/6 + 7/3(0.4714) - 7/3(-0.1143)}{7/3} = 0.2286$$

$$x_1 = \frac{1 - 4(0.4714) - 2(-0.1143) - 2(.2286)}{6} = -0.1857$$

GAUSSIAN ELIMINATION WITH MAXIMAL COLUMN PIVOTING

To solve a matrix equation $AX = b$, where A is nxn, b is nx1 and X is an nx1 vector of unknowns.

STEP 1.   Input: $\widetilde{A}$ Augmented matrix, a_{ij}, $i = 1, 2, \dots n$,

$j = 1, 2, \dots n+1$ where $a_{i,n+1} = b_i$

Row i of $\widetilde{A}$ is designated as E_i

STEP 2.   For $i = 1,2,..n-1$ do steps until 5.

STEP 3.   Find the first available nonzero a_{pi} in $i \le p \le n$, such that

$$|a_{pi}| = \max_{i \le j \le n} |a_{ji}|$$

STEP 4.   If $p \ne i$ then interchange $E_i \Leftrightarrow E_p$

STEP 5.   Perform $E_j = (E_j - \dfrac{a_{ji}}{a_{ii}} E_i)$ for $j = i+1, i+2,\dots,n$.

STEP 6.   Set $x_n = a_{n,n+1}/a_{nn}$

and $x_i = \dfrac{a_{i,n+1} - \sum\limits_{j=i+1}^{n} a_{ij}x_j}{a_{ii}}$ $i = n-1, n-2,...1$

STEP 7.   Output: $x_1, x_2, \dots x_n$.

3.3.1 Scaled Column Pivoting

Another pivoting strategy which ensures accuracy when partial pivoting fails is called scaled column pivoting. In this strategy the pivot element and the elements directly below it are scaled with respect to the largest magnitude elements in their respective rows. Once this scaling process is completed, the rest of the procedure becomes similar to the partial pivoting strategy.

SCALING PROCEDURE

Scaling is performed by dividing the absolute value of the element in the pivot position by the maximum absolute value in that row, excluding the last value in each row, $((n+1)^{th}$ value, n = number of equations). The scaled value in each row resulting from the above operations is then stored in an array. The row contributing the largest scaled value in the array is then switched with the row containing the present pivot element;

This strategy must be performed on each augmented matrix.

EXAMPLE 3.4
Use Gaussian elimination with scaled column pivoting to solve following system of equations.

$$2x_1 - x_2 + x_3 = -1$$
$$3x_1 + 3x_2 + 9x_3 = 0$$
$$3x_1 + 3x_2 + 5x_3 = 4$$

Solution
The augmented matrix is given by

$$\tilde{A}^1 = \begin{bmatrix} 2 & -1 & 1 & \vdots & -1 \\ 3 & 3 & 9 & \vdots & 0 \\ 3 & 3 & 5 & \vdots & 4 \end{bmatrix}$$

This first pivot element is 2. It must be scaled with respect to elements in its row. The elements directly below 2 must also be scaled with respect to elements in their row. The scaling procedure does not involve the value in the last column in the augmented matrix.

Scaling of first pivot element A_{11}.

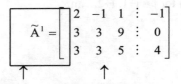

$$\tilde{A}^1 = \begin{bmatrix} 2 & -1 & 1 & \vdots & -1 \\ 3 & 3 & 9 & \vdots & 0 \\ 3 & 3 & 5 & \vdots & 4 \end{bmatrix}$$

↑ box encloses elements involved in scaling procedure

↑ n+1st column excluded from scaling procedure

element in pivot position	maximum absolute in that row	scaled element in pivot position
2 (pivot element)	2	2/2 = 1
3	9	3/9 = 1/3
3	5	3/5 = 3/5

The largest scaled value is 1. Since this is the value obtained from the row containing the present pivot element, no row interchanging takes place.
Proceed with Gaussian elimination to obtain the next augmented matrix.

$$\tilde{A}^2 = \begin{bmatrix} 2 & -1 & 1 & \vdots & -1 \\ 0 & 4.5 & 7.5 & \vdots & 1.5 \\ 0 & 4.5 & 3.5 & \vdots & 5.5 \end{bmatrix}$$

↑ n+1st column excluded in scaling procedure

Again obtaining the scaled pivots, we have

element in pivot position	maximum absolute in that row	scaled element in pivot position
4.5 (pivot element)	7.5	$4.5/7.5 = 0.6$
4.5	4.5	$4.5/4.5 = 1.00$

The scaled pivot element in the last row is larger than that in the row containing the present pivot element. Thus these two rows are interchanged.

The second augmented matrix becomes.

$$\widetilde{A}^2 = \begin{bmatrix} 2 & -1 & 1 & \vdots & -1 \\ 0 & 4.5 & 3.5 & \vdots & 5.5 \\ 0 & 4.5 & 7.5 & \vdots & 1.5 \end{bmatrix}$$

Proceed with Gaussian elimination to obtain the next augmented matrix.

$$\widetilde{A}^3 = \begin{bmatrix} 2 & -1 & 1 & \vdots & -1 \\ 0 & 4.5 & 3.5 & \vdots & 5.5 \\ 0 & 0 & 4 & \vdots & -4 \end{bmatrix}$$

Back substitution yields

$$4x_3 = -4$$
$$x_3 = -1$$
$$x_2 = \frac{5.5 - 3.5(-1)}{4.5} = 2$$
$$x_1 = \frac{-1 - 1(-1) - (-1)2}{2} = 1$$

Ans. $x_1 = 1$
$x_2 = 2$
$x_3 = -1$

Check original equation.

$2x_1 - x_2 + x_3 = -1$	$2(1) - 2 + (-1) = -1$	TRUE
$3x_1 + 3x_2 + 9x_3 = 0$	$3(1) + 3(2) + 9(-1) = 0$	TRUE
$3x_1 + 3x_2 + 5x_3 = 4$	$3(1) + 3(2) + 5(-1) = 4$	TRUE

GAUSSIAN ELIMINATION WITH SCALED COLUMN PIVOTING

To solve a matrix equation $AX = b$, where A is $n \times n$, b is $n \times 1$ and X is $n \times 1$ vector of unknowns. Largest scaled pivot is chosen as the pivot element.

STEP 1.   Input: $\widetilde{A}$ Augmented matrix, a_{ij}, $i = 1, 2, \ldots n$,

$j = 1, 2, \ldots n+1$ where $a_{i,n+1} = b_i$.

Row i of $\widetilde{A}$ is designated as E_i

STEP 2.   For i = 1,2,..., n-1 do steps until 6.

STEP 3.   Find $s_k = \overset{max}{i \leq j \leq n} |a_{kj}|$ for k = i, i+1, ... n

STEP 4.   Find the first available nonzero a_{pi} in i $\leq$ p $\leq$ n, such that

$$\frac{|a_{pi}|}{s_p} = \overset{max}{i \leq j \leq n} \frac{|a_{ji}|}{s_j}$$

(This gives the largest scaled pivot element)

STEP 5.   If p $\neq$ i then interchange $E_i \Leftrightarrow E_p$

STEP 6.   Perform $E_j = (E_j - \dfrac{a_{ji}}{a_{ii}} E_i)$ for j = i+1, i+2, ... n.

STEP 7.   Set $x_n = a_{n,n+1}/a_{n,n}$

and $x_i = \dfrac{a_{i,n+1} - \sum\limits_{j=i+1}^{n} a_{ij}x_j}{a_{ii}}$ i = n-1, n-2, ...,1

STEP 8   Output: $x_1, x_2, ..., x_n$.

3.4 GAUSS-JORDAN METHOD

Gauss Jordan method is similar to Gaussian elimination, except that back substitution is not required in this method. This is because the row reduction operations are performed on the rows above the pivot element as well as on the rows below it. This yields a matrix of the following form.

$$\begin{bmatrix} A_{11} & 0 & 0 & : & A_{14} \\ 0 & A_{22} & 0 & : & A_{24} \\ 0 & 0 & A_{33} & : & A_{34} \end{bmatrix}$$

Row operations are identical to Gaussian elimination:

$$\text{Row j} - \frac{A_{ji}}{A_{ii}} \text{Row i} \rightarrow \text{New Row j} \qquad \begin{matrix} i = 1, 2, ... n \\ j = 1, 2, ... i-1, i+1,... n \end{matrix}$$

EXAMPLE 3.5
Solve the following system of equations using the Gauss-Jordan method.

$$2x_1 - x_2 + x_3 = -1$$
$$3x_1 = + 3x_2 + 9x_3 = 0$$
$$3x_1 + 3x_2 + 5x_3 = 4$$

Solution

The augmented matrix is given by

$$\tilde{A}^1 = \begin{bmatrix} 2 & -1 & 1 & \vdots & -1 \\ 3 & 3 & 9 & \vdots & 0 \\ 3 & 3 & 5 & \vdots & 4 \end{bmatrix}$$

There are no rows above the first pivot element. Hence, the procedure is identical to **Gaussian** elimination.

The second augmented matrix becomes.

$$\tilde{A}^2 = \begin{bmatrix} 2 & -1 & 1 & \vdots & -1 \\ 0 & 4.5 & 7.5 & \vdots & 1.5 \\ 0 & 4.5 & 3.5 & \vdots & 5.5 \end{bmatrix}$$

Now a row exists above the pivot element. Gauss-Jordan method must eliminate values above **and** below the pivot element.

Row below pivot element

$$\text{Row } j - \frac{A_{ji}}{A_{ii}} \text{ Row } i \rightarrow \text{New Row } j$$

$$i = 2 \text{ (pivot element row)}$$
$$j = 3 \text{ (new row being calculated)}$$

$$A_{31} - \frac{A_{32}}{A_{22}} A_{21} \rightarrow A_{31} \text{ (New)} \qquad 0 - \frac{4.5}{4.5}(0) \rightarrow 0$$

$$A_{32} - \frac{A_{32}}{A_{22}} A_{22} \rightarrow A_{32} \text{ (New)} \qquad 4.5 - \frac{4.5}{4.5}(4.5) \rightarrow 0$$

$$A_{33} - \frac{A_{32}}{A_{22}} A_{23} \rightarrow A_{33} \text{ (New)} \qquad 3.5 - \frac{4.5}{4.5}(7.5) \rightarrow -4$$

$$A_{34} - \frac{A_{32}}{A_{22}} A_{24} \rightarrow A_{34} \text{ (New)} \qquad 5.5 - \frac{4.5}{4.5}(1.5) \rightarrow 4$$

$$\tilde{A}^2 = \begin{bmatrix} 2 & -1 & 1 & \vdots & -1 \\ 0 & 4.5 & 7.5 & \vdots & 1.5 \\ 0 & 0 & -4 & \vdots & 4 \end{bmatrix}$$

Row above pivot element.

$$i = 2 \text{ (pivot element row)}$$
$$j = 1 \text{ (new row being calculated)}$$

$$A_{11} - \frac{A_{12}}{A_{22}} A_{21} \rightarrow A_{11} \text{ (New)} \qquad 0 - \frac{-1}{4.5}(0) \rightarrow 2$$

$$A_{12} - \frac{A_{12}}{A_{22}} A_{22} \rightarrow A_{12} \text{ (New)} \qquad -1 - \frac{-1}{4.5}(4.5) \rightarrow 0$$

$$A_{13} - \frac{A_{12}}{A_{22}} A_{23} \rightarrow A_{13} \text{ (New)} \qquad 1 - \frac{-1}{4.5}(7.5) \rightarrow 2.667$$

$$A_{14} - \frac{A_{12}}{A_{22}} A_{24} \rightarrow A_{14} \text{ (New)} \qquad -1 - \frac{-1}{4.5}(1.5) \rightarrow .667$$

$$\widetilde{A}^3 = \begin{bmatrix} 2 & 0 & 2.667 & \vdots & -0.667 \\ 0 & 4.5 & 7.5 & \vdots & 1.5 \\ 0 & 0 & -4 & \vdots & 4 \end{bmatrix}$$

The new pivot element becomes −4. All elements above this one must be eliminated.

$$i = 3 \text{ (pivot element row)}$$
$$j = 2 \text{ (row being reduced)}$$

$$A_{21} - \frac{A_{23}}{A_{33}} A_{31} \rightarrow A_{21} \text{ (New)} \qquad -\frac{7.5}{-4}(0) \rightarrow 0$$

$$A_{22} - \frac{A_{23}}{A_{33}} A_{32} \rightarrow A_{22} \text{ (New)} \qquad 4.5 - \frac{7.5}{-4}(0) \rightarrow 4.5$$

$$A_{23} - \frac{A_{23}}{A_{33}} A_{33} \rightarrow A_{23} \text{ (New)} \qquad 7.5 - \frac{7.5}{-4}(7.5) \rightarrow 0$$

$$A_{24} - \frac{A_{23}}{A_{33}} A_{34} \rightarrow A_{24} \text{ (New)} \qquad 1.5 - \frac{7.5}{-4}(4) \rightarrow 9.0$$

$$\widetilde{A}^3 = \begin{bmatrix} 2 & 0 & 2.667 & \vdots & -0.667 \\ 0 & 4.5 & 0 & \vdots & 9.0 \\ 0 & 0 & -4 & \vdots & 4 \end{bmatrix}$$

Notice that the first two calculations are obvious and can be written down right away.

$$i = 3 \text{ (pivot element row)}$$
$$j = 1 \text{ (row being reduced)}$$

$$A_{11} - \frac{A_{13}}{A_{33}} A_{31} \rightarrow A_{11} \text{ (New)} \qquad 2 - \frac{2.667}{-4}(0) \rightarrow 2$$

$$A_{12} - \frac{A_{13}}{A_{33}} A_{32} \rightarrow A_{12} \text{ (New)} \qquad 0 - \frac{2.667}{-4}(0) \rightarrow 0$$

$$A_{13} - \frac{A_{13}}{A_{33}} A_{33} \rightarrow A_{13} \text{ (New)} \qquad 2.667 - \frac{2.667}{-4}(-4) \rightarrow 0$$

$$A_{14} - \frac{A_{13}}{A_{33}} A_{34} \rightarrow A_{14} \text{ (New)} \qquad -.667 - \frac{2.667}{-4}(4) \rightarrow 2$$

$$\widetilde{A}^4 = \begin{bmatrix} 2 & 0 & 0 & \vdots & 2 \\ 0 & 4.5 & 0 & \vdots & 9.0 \\ 0 & 0 & -4 & \vdots & 4 \end{bmatrix}$$

The answer becomes immediately obvious at this point without the need for back substitution routine.

$$
\begin{array}{ll}
2x_1 = 2 & x_1 = 1 \\
4.5x_2 = 9 & x_2 = 2 \\
-4x_3 = 4 & x_3 = -1
\end{array}
$$

The Gauss-Jordan method involves more mathematical operations than the Gaussian elimination method. However, this method is useful in obtaining the inverse of a large matrix, as will be explained in a later section.

In this method, the pivot element cannot be zero. If the pivot element is zero, the row containing the pivot element can be exchanged with any row below it. If this fails to produce a non-zero element in the pivot position, then this method will not work.

GAUSS – JORDAN METHOD

To solve a matrix equation $AX = b$, where A is $n \times n$, b is $n \times 1$ and X is $n \times 1$ vector of unknowns.

STEP 1   Input: $\widetilde{A}$ Augmented matrix with a_{ij}, $i = 1, 2, \ldots n$,

$j = 1, 2, \ldots n+1$, where $a_{i,n+1} = b_i$.

Row i of $\widetilde{A}$ is denoted as E_i.

STEP 2   For $i = 1, 2, \ldots n$, do steps until 5

STEP 3   Find the first available nonzero a_{pi} in $i \le p \le n$. (nonzero pivot element)

STEP 4   If $p \ne i$ then interchange $E_i \Longleftrightarrow E_p$

STEP 5   Perform $E_j = (E_j - \dfrac{a_{ji}}{a_{ii}} E_i)$ for $j = 1, 2, \ldots n, j \ne i$

STEP 6   Set $x_i = a_{i,n+1}/a_{ii}$ for $i = 1, 2, \ldots n$.

STEP 7   Output: $x_1, x_2, \ldots x_n$.

3.5 MATRIX FACTORIZATION TECHNIQUES

A square matrix with all elements below the main diagonal equal to zero, is called an upper triangular matrix denoted by U (see fig. 3.5.1). Similarly if all the elements above the main diagonal are zero, then the square matrix is called a lower triangular matrix denoted by L (see fig. 3.5.2.).

$$U = \begin{bmatrix} U_{11} & U_{12} & \cdots & U_{1n} \\ 0 & U_{22} & \cdots & U_{2n} \\ \vdots & \vdots & & \vdots \\ 0 & 0 & \cdots & U_{nn} \end{bmatrix} \qquad L = \begin{bmatrix} L_{11} & 0 & \cdots & 0 \\ L_{21} & L_{22} & \cdots & 0 \\ \vdots & \vdots & & \vdots \\ L_{n1} & L_{n2} & \cdots & L_{nn} \end{bmatrix}$$

Figure 3.5.1   Figure 3.5.2

A system of equations can be expressed in the form:

$$Ax = b$$

where the matrix 'A' can be factorized into an upper and lower triangular matrix:

$$A = L * U$$

$$\begin{bmatrix} A_{11} & A_{12} & \cdots & A_{1n} \\ A_{21} & A_{22} & \cdots & A_{2n} \\ \vdots & \vdots & & \vdots \\ A_{n1} & A_{n2} & \cdots & A_{nn} \end{bmatrix} = \begin{bmatrix} L_{11} & 0 & \cdots & 0 \\ L_{21} & L_{22} & \cdots & 0 \\ \vdots & \vdots & & \vdots \\ L_{n1} & L_{n2} & \cdots & L_{nn} \end{bmatrix} * \begin{bmatrix} U_{11} & U_{12} & \cdots & U_{1n} \\ 0 & U_{22} & \cdots & U_{2n} \\ \vdots & \vdots & & \vdots \\ 0 & 0 & \cdots & U_{nn} \end{bmatrix}$$

Matrix 'A' has nxn number of known elements, whereas matrices L and U have $\frac{(n \times n)}{2} + \frac{n}{2}$ unknown elements each. Hence the right hand side of the system has (nxn + n) unknown quantities which cannot be determined using only (nxn) known quantities on the left hand side. This problem is overcome by arbitrarily assigning values to n quantities on the right hand side. There are three methods in which these n values are assigned to the matrices on the right hand side:

(1) Doolittle's method assigns values of unity to the diagonal elements of the lower triangular matrix.

i.e. $L_{ii}=1$, $i = 1, 2, 3, \ldots n$

(2) Crout's method assigns values of unity to the diagonal elements of the upper triangular matrix.

i.e. $U_{ii} = 1$, $i = 1, 2, 3 \ldots n$

(3) Choleski's method assigns values on the main diagonal of the upper triangular matrix to be the same as those on the main diagonal of the lower triangular matrix.

i.e. $L_{ii} = U_{ii}$, $i = 1, 2, \ldots n$

3.5.1 Doolittle Method (L*U Decomposition)

Theorem:

Any square matrix A, in which Gaussian elimination can be performed without row interchanges, can be factorized into an upper and a lower triangular matrix.

$$A = L * U$$

$$\begin{bmatrix} A_{11} & A_{12} & A_{13} \\ A_{21} & A_{22} & A_{23} \\ A_{31} & A_{32} & A_{33} \end{bmatrix} = \begin{bmatrix} L_{11} & 0 & 0 \\ L_{21} & L_{22} & 0 \\ L_{31} & L_{32} & L_{33} \end{bmatrix} * \begin{bmatrix} U_{11} & U_{12} & U_{13} \\ 0 & U_{22} & U_{23} \\ 0 & 0 & U_{33} \end{bmatrix}$$

In the Doolittle method the main diagonal elements of the lower triangular matrix are assigned to be 1. This enables the computation of all other elements in the lower and upper triangular matrix such that $L * U = A$.

$$L_{ii} = 1$$

In order to solve the remaining unknowns in L and U, each column of U is multiplied by a row in L. The results of these operations are equated to the value which should correspond in matrix A. This procedure can be represented in the following form:

$$\begin{array}{ll} \text{Row i * Column j} = A_{ij} & i = 1, 2, 3 \ldots n \\ \text{(of L)} \qquad \text{(of U)} & j = 1, 2, 3 \ldots n \end{array}$$

for each value of i, j going form 1 to n

Once the system is factorized into an upper and lower triangular matrix, the equation can be solved in the following way:

$$Ax = b$$
$$L * Ux = b$$

Let $Ux = Z$

$$L * Z = b$$

Solve Z

Substitute Z back

This requires less effort due to the simple calculations involved when dealing with triangular matrices.

EXAMPLE 3.6
Solve the following system using the Doolittle method.

$$2x_1 - x_2 + x_3 = -1$$
$$3x_1 + 3x_2 + 9x_3 = 0$$
$$3x_1 + 3x_2 + 5x_3 = 4$$

$$A = \begin{bmatrix} 2 & -1 & 1 \\ 3 & 3 & 9 \\ 3 & 3 & 5 \end{bmatrix} \qquad b = \begin{bmatrix} -1 \\ 0 \\ 4 \end{bmatrix}$$

Solution
Factorize A into L * U

$$\begin{bmatrix} 2 & -1 & 1 \\ 3 & 3 & 9 \\ 3 & 3 & 5 \end{bmatrix} = \begin{bmatrix} 1 & 0 & 0 \\ L_{21} & 1 & 0 \\ L_{31} & L_{32} & 1 \end{bmatrix} * \begin{bmatrix} U_{11} & U_{12} & U_{13} \\ 0 & U_{22} & U_{23} \\ 0 & 0 & U_{33} \end{bmatrix}$$

Notice: $\qquad L_{11} = L_{22} = L_{33} = L_{ii} = 1$

Now apply procedure:

$$\text{Row}i * \text{column}j = A_{ij} \qquad i = 1, 2, 3$$
$$\text{(of L)} \quad \text{(of U)} \qquad j = 1, 2, 3$$

i = 1	Row1 * Column1 = A_{11}	
j = 1	(of L) (of U)	
	$(1 * U_{11} + 0 * 0 + 0 * 0) = 2$	
	$U_{11} = 2$	
i = 1	Row 1 * Column2 = A_{12}	i = 1
j = 2	(of L) (of U)	j = 1, 2, 3
	$(1 * U_{12} + 0 * U_{22} + 0 * 0) = -1$	gives
	$U_{12} = -1$	U_{11}, U_{12}, U_{13}
i = 1	Row1 * Column3 = A_{13}	
j = 3	(of L) (of U)	
	$(1 * U_{13} + 0*U_{23} + 0 * U_{33}) = 1$	
	$U_{13} = 1$	

i = 2	Row2 * Column1 = A_{21}	
j = 1	(of L) (of U)	
	$(L_{21} * U_{11} + 1 * 0 + 0 * 0) = 3$	
	$2L_{21} = 3; L_{21} = 1.5$	
i = 2	Row2 * Column2 = A_{22}	i = 2
j = 2	(of L) (of U)	j = 1, 2, 3
	$(L_{21} * U_{12} + 1 * U_{22} + 0 * 0) = 3$	gives
	$(1.5 * -1) + U_{22} = 3; U_{22} = 4.5$	L_{21}, U_{22}, U_{23}
i = 2	Row2 * Column3 = A_{23}	
j = 3	(of L) (of U)	
	$(L_{21} * U_{13} + 1 * U_{23} + 0 *U_{33}) = 9$	
	$(1.5 * 1) + U_{23} = 9; U_{23} = 7.5$	

i = 3	Row3 * Column1 = A_{31}	
j = 1	(of L) (of U)	
	$(L_{31} * U_{11} + L_{32} * 0 + 1 * 0) = 3$	
	$L_{31} * 2 = 3; L_{31} = 1.5$	
i = 3	Row3 * Column2 = A_{32}	i = 3
j = 2	(of L) (of U)	j = 1, 2, 3
	$(L_{31} * U_{12} + L_{32} * U_{22} + 1 * 0) = 3$	gives
	$(1.5 * -1) + L_{32} * 4.5 = 3; L_{32} = 1$	L_{31}, L_{32}, U_{33}
i = 3	Row3 * Column3 = A_{33}	
j = 3	(of L) (of U)	
	$(L_{31} * U_{13} + L_{32} * U_{23} + 1 *U_{33}) = 5$	
	$(1.5 * 1) + 1(7.5) + U_{33} = 5; U_{33} = -4$	

The system is now factorized

$$\begin{bmatrix} 2 & -1 & 1 \\ 3 & 3 & 9 \\ 3 & 3 & 5 \end{bmatrix} = \begin{bmatrix} 1 & 0 & 0 \\ 1.5 & 1 & 0 \\ 1.5 & 1 & 1 \end{bmatrix} * \begin{bmatrix} 2 & -1 & 1 \\ 0 & 4.5 & 7.5 \\ 0 & 0 & -4 \end{bmatrix}$$

$$A = L * U$$

Recall
$$b = \begin{bmatrix} -1 \\ 0 \\ 4 \end{bmatrix}$$

$$Ax = b$$

Hence
$$L * Ux = b$$
$$\text{Let } Ux = Z$$
$$L * Z = b$$

Solve Z

$$\begin{bmatrix} 1 & 0 & 0 \\ 1.5 & 1 & 0 \\ 1.5 & 1 & 1 \end{bmatrix} \begin{bmatrix} Z_1 \\ Z_2 \\ Z_3 \end{bmatrix} = \begin{bmatrix} -1 \\ 0 \\ 4 \end{bmatrix}$$

$$L Z = b$$

This can now be solved immediately by forward substitution:

$$Z_1 = -1$$

$$Z_2 = \frac{0 - 1.5(-1)}{1} = 1.5 \qquad Z = \begin{bmatrix} -1 \\ 1.5 \\ 4 \end{bmatrix}$$

$$Z_3 = \frac{4 - 1.5(-1) - 1(1.5)}{1} = 4$$

Now it is possible to solve x immediately by back substitution

$$Ux = Z$$

$$\begin{bmatrix} 2 & -1 & 1 \\ 0 & 4.5 & 7.5 \\ 0 & 0 & -4 \end{bmatrix} \begin{bmatrix} x_1 \\ x_2 \\ x_3 \end{bmatrix} = \begin{bmatrix} -1 \\ 1.5 \\ 4 \end{bmatrix}$$

Backward substitution gives;

$$x_3 = -1$$

$$x_2 = \frac{1.5 - 7.5(-1)}{4.5} = 2$$

$$x_1 = \frac{-1-(-1)(2)-1(-1)}{2} = 1$$

$$\text{Ans } x = \begin{bmatrix} 1 \\ 2 \\ -1 \end{bmatrix}$$

In order to carry out the above factorization, matrix 'A' must be square. Matrix 'A' must be able to undergo Gaussian elimination without row interchanges. (No 0 can appear in the pivot element position in any augmented matrix).

3.5.2 Crout's Method

This method is similar to the Doolittle method. In the Crout's method the main diagonal elements of the upper triangular matrix are assigned to be 1 (see below).

$$A = L * U$$

$$\begin{bmatrix} A_{11} & A_{12} & A_{13} \\ A_{21} & A_{22} & A_{23} \\ A_{31} & A_{32} & A_{33} \end{bmatrix} = \begin{bmatrix} L_{11} & 0 & 0 \\ L_{21} & L_{22} & 0 \\ L_{31} & L_{32} & L_{33} \end{bmatrix} * \begin{bmatrix} 1 & U_{12} & U_{13} \\ 0 & 1 & U_{23} \\ 0 & 0 & 1 \end{bmatrix}$$

This enables the computation of all other elements in the lower and upper triangular matrix such that $L * U = A$.

$$U_{ii} = 1$$

In order to solve the remaining unknowns in L and U, each column of U is multiplied by a row in L. The results of these operations are equated to the value which should correspond in A. This procedure can be represented in the following form:

$$\text{Row}_i * \text{Column}_j = A_{ij} \quad i = 1, 2, 3 \ldots n$$
$$\text{(of L)} \quad \text{(of U)} \qquad\qquad j = 1, 2, 3 \ldots n$$

Once these operations yields all the unknowns in the upper and lower triangular matrix, the system can be solved in the following way.

$$Ax = b$$
$$L * Ux = b$$

Let $ux = Z$

$$L * Z = b$$

Solve Z

Substitute Z back into

$Ux = Z$ to solve x

EXAMPLE 3.7
Solve the following system using the Crout's method.

$$3x_1 + x_2 = -1$$

$$2x_1 + 4x_2 + x_3 = 7$$
$$2x_2 + 5x_3 = 9$$

$$A = \begin{bmatrix} 3 & 1 & 0 \\ 2 & 4 & 1 \\ 0 & 2 & 5 \end{bmatrix} \qquad b = \begin{bmatrix} -1 \\ 7 \\ 9 \end{bmatrix}$$

Solution

Factorize A into L * U by letting $U_{ii} = 1$.

$$\begin{bmatrix} 3 & 1 & 0 \\ 2 & 4 & 1 \\ 0 & 2 & 5 \end{bmatrix} = \begin{bmatrix} L_{11} & 0 & 0 \\ L_{21} & L_{22} & 0 \\ L_{31} & L_{32} & L_{33} \end{bmatrix} * \begin{bmatrix} 1 & U_{12} & U_{13} \\ 0 & 1 & U_{23} \\ 0 & 0 & 1 \end{bmatrix}$$

Now apply the procedure:

$$\text{Row}i * \text{Column}j = A_{ij} \qquad i = 1, 2, 3$$
$$\text{(of L)} \quad \text{(of U)} \qquad\qquad j = 1, 2, 3$$
for each value of i, j goes form 1 to 3.

i = 1	Row1 * Column1 = A_{11}
j = 1	(of L) (of U)
	$(L_{11} * 1 + 0 * 0 + 0 * 0) = 3$
	$L_1 = 3$
i = 1	Row1 * Column2 = A_{12}
j = 2	(of L) (of U)
	$(L_{11} * U_{12} + 0 * 1 + 0 * 0) = 1$
	$U_{12} = 1/3 = 0.33333$
i = 1	Row1 * Column3 = A_{13}
j = 3	(of L) (of U)
	$(L_{11} * U_{13} + 0 * U_{23} + 0 * 1) = 0$
	$U_{13} = 3$

i = 1
j = 1, 2, 3
gives
L_{11}, U_{12}, U_{13}

i = 2	Row2 * Column1 = A_{11}
j = 1	(of L) (of U)
	$(L_{21} * 1 + L_{22} * 0 + 0 * 0) = 2$
	$L_{21} = 2$
i = 2	Row2 * Column2 = A_{22}
j = 2	(of L) (of U)
	$(L_{21} * U_{12} + L_{22} * 1 + 0 * 0) = 4$
	$2/3 + L_{22} = 4; L_{22} = 10/3 = 3.333$
i = 2	Row2 * Column3 = A_{23}
j = 3	(of L) (of U)
	$(L_{21} * U_{13} + L_{22} * U_{23} + 0 * 1) = 1$
	$2(0) + 10/3(U_{23}) = 1; U_{23} = 3/10 = 0.3$

i = 2
j = 1, 2, 3
gives
L_{21}, L_{22}, U_{23}

$i = 3$ Row3 * Column1 = A_{31}

$j = 1$ (of L)   (of U)

(L_{31} * 1 + L_{32} * 0 + L_{33} * 0) = 0

$L_{31} = 0$

$i = 3$ Row3 * Column2 = A_{32}

$j = 2$ (of L)   (of U)

(L_{31} * U_{12} + L_{32} * 1 + L_{33} * 0) = 2

$L_{32} = 2$

$i = 3$ Row3 * Column3 = A_{23}

$j = 3$ (of L)   (of U)

(L_{31} * U_{13} + L_{32} * U_{23} + L_{33} * 1) = 5

2(0.3)+ L_{33} = 5; L_{33} = 4.4

$i = 3$
$j = 1, 2, 3$
gives
L_{31}, L_{32}, L_{33}

The system is now factorized

$$\begin{bmatrix} 3 & 1 & 0 \\ 2 & 4 & 1 \\ 0 & 2 & 5 \end{bmatrix} = \begin{bmatrix} 3 & 0 & 0 \\ 2 & 3.33 & 0 \\ 0 & 2 & 4.4 \end{bmatrix} * \begin{bmatrix} 1 & 0.3333 & 0 \\ 0 & 1 & 0.3 \\ 0 & 0 & 1 \end{bmatrix}$$

$$A = L * U$$

Recall

$$b = \begin{bmatrix} -1 \\ 7 \\ 9 \end{bmatrix}$$

$$Ax = b$$
$$L * Ux = b$$
$$\text{Let } Ux = Z$$
$$L * Z = b$$

Solve Z

$$\begin{bmatrix} 3 & 0 & 0 \\ 2 & 3.33 & 0 \\ 0 & 2 & 4.4 \end{bmatrix}\begin{bmatrix} Z_1 \\ Z_2 \\ Z_3 \end{bmatrix} = \begin{bmatrix} -1 \\ 7 \\ 9 \end{bmatrix}$$

$$LZ = b$$

This can now be solved immediately by forward substitution.
Forward substitution gives

$$3Z_1 = -1$$
$$Z_1 = -0.33333$$
$$Z_2 = \frac{7 - 2(-0.3333)}{3.333} = 2.3$$

$$Z_3 = \frac{9 - 2(2.3)}{4.4} = 1$$

Now it is possible to solve x immediately by back substitution.

$$Ux = Z$$

$$\begin{bmatrix} 1 & 0.3333 & 0 \\ 0 & 1 & 0.3 \\ 0 & 0 & 1 \end{bmatrix} \begin{bmatrix} x_1 \\ x_2 \\ x_3 \end{bmatrix} = \begin{bmatrix} -0.3333 \\ 2.3 \\ 1 \end{bmatrix}$$

$$x_3 = 1$$

$$x_2 = \frac{2.3 - 0.3(1)}{1} = 2$$

$$x_1 = \frac{-0.3333 - 0.3333(2)}{1} = -1$$

$$\text{Ans. } x = \begin{bmatrix} -2 \\ 2 \\ 1 \end{bmatrix}$$

DIRECT FACTORIZATION: DOOLITTLE'S METHOD AND CROUT'S METHOD

To solve a matrix equation $Ax = b$, by factoring $A = LU$, where L is a lower triangular matrix and U is an upper triangular matrix. Denote Row i of A as E_i.

STEP 1   Input: a_{ij}, i = 1, 2, ... n, j = 1, 2, ... n, element of A

                 $l_{ii} = 1$, for i = 1, 2, ... n if DOOLITTLE METHOD

                 $u_{ii} = 1$, for i =1, 2, ... n if CROUT'S METHOD

STEP 2   Find the first available nonzero a_{p1} in $1 \leq p \leq n$, such that $\left| a_{p1} \right| = 1 \leq \overset{max}{j} \leq n \left| a_{j1} \right|$

STEP 3   If $p \neq 1$ then $E_p \Leftrightarrow E_1$

STEP 4   Set $I_{11}..u_{11} = a_{11}$.

STEP 5   For j = 2, 3, ... n,   set $u_{ij} = a_{ij} /l_{11}$, and $l_{ji} = a_{j1}/u_{11}$

STE[6   For i =2, 3, ... n-1 do steps until 10

STEP 7   Find the first available a_{pi} in $i \leq p \leq n$, such that

$$\left| a_{pi} - \sum_{k=1}^{i-1} \ell_{pk} u_{ki} \right| = 1 \leq \overset{max}{j} \leq n \left| a_{ji} - \sum_{k=1}^{i-1} \ell_{jk} u_{ki} \right|$$

STEP 8   If $p \neq i$ then interchange $E_p \Leftrightarrow E_i$. Also interchange p th and i th rows in L.

STEP 9   Set ℓ_{ii} and u_{ii} such that $\ell_{ii} = a_{ii} - \sum_{k=1}^{i-1} I_{ik} u_{ki}$

STEP 10   For $j = i+1, i+2, \dots n.$

                set $u_{ij} = \dfrac{1}{I_{ii}} [a_{ij} - \sum_{k=1}^{i-1} 1_{ik} u_{kj}]$

                and $\ell_{ji} = \dfrac{1}{u_{ii}} [a_{ji} - \sum_{k=1}^{i-1} \ell_{jk} u_{kj}]$

STEP 11   Set ℓ_{nn} and u_{nn} such that

                $\ell_{nn} u_{nn} = a_{nn} - \sum_{k=1}^{n-1} \ell_{nk} u_{kn}$

STEP 12   Set $z_1 = a_{1,n+1} / \ell_{11}$

STEP 13   For $i = 2, 3, \dots n$ solution of

                Set $z_i = \dfrac{1}{\ell_{ii}} [a_{i,n+1} - \sum_{j=1}^{i-1} \ell_{ij} z_j]$ $L z = b$

STEP 14   Set $x_n = z_n / u_{nn}$

STEP 15   For $i = 1, 2, \dots n-1$ solution of

                set $x_i = \dfrac{1}{u_{ii}} [z_i - \sum_{j=i+1}^{n} u_{ij} x_j]$ $U x = z$

STEP 16   Output: $x_1, x_2, \dots x_n.$

Note: Steps 1, 2, 6 and 7 incorporate maximal column pivoting in factorization. They may be
         deleted if factorization is done as is.

3.6 CHOLESKY METHOD (valid only for matrices which are positive definite)

Theorem:
    If 'A' is a <u>positive definite</u>, symmetric matrix, then 'A' can be factorized in the form:

$$A = L * L^T$$

$L^T \rightarrow$ is the transpose of L where L is lower triangular matrix. The system can then be solved in a
way similar to the Doolittle method, called the Cholesky method.

Definition:
A symmetric matrix 'A' is called positive definite if $x^T A x > 0$ for every n dimensional column vector x
with at least some non-zero elements.

A matrix is determined to be positive definite by applying the definition $x^TAx > 0$, which is long and tedious, or by using the Sylvestor co-factor technique.

EXAPLE 3.8

Determine whether the following system is suitable to undergo Cholesky decomposition.

$$25x_1 + 12x_2 + 5x_3 = 165$$
$$12x_1 + 7x_2 - 5x_3 = 119$$
$$5x_1 - 5x_2 + 50x_3 = -56$$

Solution

Matrix A must be symmetrical and positive definite;

$$A = \begin{bmatrix} 25 & 12 & 5 \\ 12 & 7 & -5 \\ 5 & -5 & 50 \end{bmatrix}$$

Form inspection it is obvious that 'A' meets the first condition of symmetry.
To check for positive definiteness:
By definition: $x^TAx > 0$

$$\begin{bmatrix} x_1 & x_2 & x_3 \end{bmatrix} \begin{bmatrix} 25 & 12 & 5 \\ 12 & 7 & -5 \\ 5 & -5 & 50 \end{bmatrix} \begin{bmatrix} x_1 \\ x_2 \\ x_3 \end{bmatrix} > 0$$

$$25x_1^2 + 12x_1x_2 + 5x_1x_3 + 12x_1x_2 + 7x_2^2 - 5x_2x_3 +$$

$$5x_1x_3 - 5x_2x_3 + 50x_3^2 > 0$$

$$6(2x_1 + x_2)^2 + (x_1 + 5x_3)^2 + (x_2 - 5x_3)^2 > 0$$

Since all terms within the brackets are squared, the resulting value must be > 0. Hence matrix A is positive definite.

The above method is difficult to apply. A better method is the <u>Sylvestor co-factor method.</u>

In this method, if the determinants of the co-factor matrices in 'A' are greater than zero, the matrix 'A' is positive definite.

$$A = \begin{bmatrix} \begin{bmatrix} 25 \end{bmatrix} & 12 & 5 \\ 12 & 7 & -5 \\ 5 & -5 & 50 \end{bmatrix}$$

1st cofactor matrix is $\begin{bmatrix} 25 \end{bmatrix}$

2nd cofactor matrix is $\begin{bmatrix} 25 & 12 \\ 12 & 7 \end{bmatrix}$

3rd cofactor matrix is
$$\begin{bmatrix} 25 & 12 & 5 \\ 12 & 7 & -5 \\ 5 & -5 & 50 \end{bmatrix}$$

$$\text{Det} \quad |25| > 0$$

$$\text{Det} \begin{vmatrix} 25 & 12 \\ 12 & 7 \end{vmatrix} = 31 > 0$$

$$\text{Det} \begin{vmatrix} 25 & 12 & 5 \\ 12 & 7 & -5 \\ 5 & -5 & 50 \end{vmatrix} = 150 > 0$$

The determinants of all cofactor matrices are greater than zero. Hence 'A' is positive definite.

Since 'A' is symmetrical and positive definite, the Cholesky method can be used to solve the system.

After it is certain that 'A' is symmetrical and positive definite, the matrix 'A' is factored in the form:

$$A = L * L'$$

$$\begin{bmatrix} A_{11} & A_{12} & A_{13} \\ A_{21} & A_{22} & A_{23} \\ A_{31} & A_{32} & A_{33} \end{bmatrix} = \begin{bmatrix} L_{11} & 0 & 0 \\ L_{21} & L_{22} & 0 \\ L_{31} & L_{32} & L_{33} \end{bmatrix} * \begin{bmatrix} L_{11} & L_{12} & L_{13} \\ 0 & L_{22} & L_{23} \\ 0 & 0 & L_{33} \end{bmatrix}$$

$$A = L * L^T$$

In this method it is not necessary to make $L_{ii} = 1$, since the main diagonal in both of the factored matrices is identical. The factorization procedure is identical to the Doolittle method except that U is replaced by L^T.

$$\text{Row i * Column j} = A_{ij} \qquad i=1,2,3\ldots n$$
$$\text{(of L) (of } L^T) \qquad\qquad j=1,2,3\ldots n$$

Once the system is factorized into the correct form of $L*L^T$, then the system can be solved in the following way. The original form can be broken down so that forward and backward substitution can be used to yield the answer.

$$Ax = b$$
$$L * L^T x = b$$

Let $\quad L^T x = Z$

$$L * Z = b$$

Solve Z

Substitute Z back

into $L^T x = Z$ to solve x

EXAMPLE 3.9

Solve the following system by Cholesky method.

$$2x_1 - x_2 + 0x_3 = 3$$
$$-x_1 + 2x_2 - x_3 = -3$$
$$0x_1 - x_2 + x_3 = 2$$

Solution

The first step is to determine whether 'A' symmetrical, and positive definite.

$$A = \begin{bmatrix} 2 & -1 & 0 \\ -1 & 2 & -1 \\ 0 & 1 & 1 \end{bmatrix} \quad b = \begin{bmatrix} 3 \\ -3 \\ 2 \end{bmatrix}$$

By inspection it is obvious that 'A' is symmetrical. To check positive definiteness by Sylvestor's method:

$$\text{Det } |2| > 0$$

$$\text{Det } \begin{vmatrix} 2 & -1 \\ -1 & 2 \end{vmatrix} = 3 > 0$$

$$\text{Det } \begin{vmatrix} 2 & -1 & 0 \\ -1 & 2 & -1 \\ 0 & -1 & 1 \end{vmatrix} = 2 \begin{vmatrix} 2 & -1 \\ -1 & 1 \end{vmatrix} + 1 \begin{vmatrix} -1 & -1 \\ 0 & 1 \end{vmatrix} > 0$$

Since the determinants of all the cofactor matrices are greater than zero, the matrix 'A' is positive definite.

Now to factorize 'A'

$$\begin{bmatrix} 2 & -1 & 0 \\ -1 & 2 & -1 \\ 0 & -1 & 1 \end{bmatrix} = \begin{bmatrix} L_{11} & 0 & 0 \\ L_{21} & L_{22} & 0 \\ L_{31} & L_{32} & L_{33} \end{bmatrix} * \begin{bmatrix} L_{11} & L_{12} & L_{13} \\ 0 & L_{22} & L_{23} \\ 0 & 0 & L_{33} \end{bmatrix}$$

Now apply the procedure:

$$\text{Row } i * \text{Column } j = A_{ij} \qquad i = 1,2,3$$
$$(\text{of } L) \quad (\text{of } L^T) \qquad j = 1,2,3$$

$i = 1$ Row1 * Column1 = A_{11}
$j = 1$ (of L)   (of L^T)
         $(L_{11} * L_{11} + 0 * 0 + 0 * 0) = 2$

         $L_{11} = \sqrt{2} = 1.414$

$i = 1$ Row1 * Column2 = A_{12}
$j = 2$ (of L)   (of L^T)
         $(L_{11} * L_{21} + 0 * L_{22} + 0 * 0) = -1$ $i = 1$

       $\sqrt{2}\ L_{21} = -1; L_{21} = \dfrac{-1}{\sqrt{2}}$ $j = 1,2,3$

                                              gives L_{11}, L_{21}, L_{31}

$i = 1$ Row1 * Column3 = A_{13} Due to symmetry
$j = 3$ (of L)   (of L^T) $L_{12} = L_{21}$
         $(L_{11} * L_{31} + 0 * L_{32} + 0 * L_{33}) = 0$ $L_{13} = L_{31}$
         $L_{11} * L_{31} = 0; L_{31} = 0$

Notice that since 'A' is symmetrical less operation will be required.
Because 'A' is symmetrical start at $j = 2$.

$i = 2$ Row2 * Column2 = A_{23}
$j = 2$ (of L)   (of L^T) $i = 2$
         $(L_{21} * L_{21} + L_{22} * L_{22} + 0 * 0) = 2$ $j = 2,3$
         $(L_{21})^2 + (L_{22})^2 = 2; \quad (L_{22})^2 = 1.5$ gives
         $L_{22} = 1.225$ L_{22}, L_{32}
$i = 2$ Row 2 * Column3 = A_{23} and due to
$j = 3$ (of L)   (of L^T)   symmetry
         $(L_{21} * L_{31} + L_{22} * L_{32} + 0 * L_{33}) = -1$ L_{23}

      $\dfrac{-1}{\sqrt{2}} * 0 + 1.225 L_{32} = -1; L_{32} = -.8165$

$i = 3$ Row3 * Column3 = A_{33} $i = 3$
$j = 3$ (of L)   (of L^T) $j = 3$
         $(L_{31} * L_{31} + L_{32} * L_{32} + L_{33} * L_{33}) = 1$ gives
         $0 + (-0.8165)^2 + L_{33}{}^2 = 1; L_{33}{}^2 = 0.333$ L_{33}
         $L_{33} = 0.577$

The system is now factorized.

$$\begin{bmatrix} 2 & -1 & 0 \\ -1 & 2 & -1 \\ 0 & -1 & 1 \end{bmatrix} = \begin{bmatrix} 1.414 & 0 & 0 \\ -.707 & 1.225 & 0 \\ 0 & -.8165 & .577 \end{bmatrix} * \begin{bmatrix} 1.414 & -.707 & 0 \\ 0 & 1.225 & -.8165 \\ 0 & 0 & .577 \end{bmatrix}$$

$$A = L * L^T$$

Recall
$$b = \begin{bmatrix} 3 \\ -3 \\ 2 \end{bmatrix}$$

$$Ax = b$$
$$L * L^Tx = b$$
$$\text{Let } L^Tx = Z$$
$$L * Z = b$$

Solve Z

$$\begin{bmatrix} 1.414 & 0 & 0 \\ -.707 & 1.225 & 0 \\ 0 & -.8165 & .577 \end{bmatrix} \begin{bmatrix} Z_1 \\ Z_2 \\ Z_3 \end{bmatrix} = \begin{bmatrix} 3 \\ -3 \\ 2 \end{bmatrix}$$

$$L\,Z = b$$

Forward substitution:

$$1.414\,Z_1 = 3$$
$$Z_1 = 2.122$$
$$Z_2 = \frac{-3 + 0.707(2.122)}{1.225} = -1.224$$
$$Z_3 = \frac{2 + 0.8165(-1.224)}{0.577} = -1.734$$

Now it is possible to solve x immediately by back substitution in $L^Tx = Z$.

$$\begin{bmatrix} 1.414 & -0.707 & 0 \\ 0 & 1.225 & -0.8165 \\ 0 & 0 & .577 \end{bmatrix} \begin{bmatrix} x_1 \\ x_2 \\ x_3 \end{bmatrix} = \begin{bmatrix} 2.122 \\ -1.195 \\ 1.734 \end{bmatrix}$$
$$L^T x = Z$$

Backward substitution:

$$0.577x_3 = 1.734$$
$$x_3 = 3.0$$
$$x_2 = \frac{-1.195 + 0.8165(3)}{1.225} = 1.0$$
$$x_1 = \frac{2.12 + 0.700(1.0)}{1.414} = 2.0$$

Answer:
$$x = \begin{bmatrix} 2.0 \\ 1.0 \\ 3.0 \end{bmatrix}$$

Cholesky method can only be used if the matrix 'A' is symmetric and positive definite. Also, Gaussian elimination must be possible on 'A' without row interchanges.

CHOLESKI FACTORIZATION

To solve a matrix equation $Ax = b$, by Choleski decomposition of $A = LL^T$. L is lower triangular. Choleski decomposition is possible only when A is positive definite.

STEP 1   Input: a_{ij}, b_i, i = 1, 2, ... n, j = 1, 2, ... n,

elements of A and b.

STEP 2   Set $L_{11} = \sqrt{a_{11}}$

STEP 3   Set $L_{j1} = a_{ji}/L_{11}$

STEP 4   For i = 2, 3, ... n-1 do steps until 6.

STEP 5   Set $L_{ii} = \left[a_{ii} - \sum_{k=1}^{n-1} L_{ik}^2 \right]^{0.5}$

STEP 6   Set $L_{ji} = \dfrac{1}{L_{ii}} \left[a_{ji} - \sum_{k=1}^{i-1} L_{jk} L_{ik} \right]$

STEP7   Set $L_{nn} = \left[a_{nn} - \sum_{k=1}^{n-1} L_{nk}^2 \right]^{0.5}$

STEP 8   Set $Z_1 = a_{1,n+1}/L_{11}$

STEP 9   For i = 2, 3, ... n, $\left.\begin{array}{c} \\ \\ \\ \\ \end{array}\right\}$ Solution of

Set $z_i = \dfrac{1}{L_{ii}} \left[a_{i,n+1} - \sum_{j=1}^{i-1} L_{ij} z_j \right]$ $Lz = b$

STEP 10   Set $x_n = z_n/L_{nn}$

STEP 11   For i = n-1, n-2, ..., n-1   solution of   Set

$x_1 = \dfrac{1}{L_{ii}} \left[z_i - \sum_{j=i+1}^{n} L_{ji} x_j \right]$ $L^T x = z$

STEP 12   Output: $x_1, x_2, ... x_n$.

<u>Iterative Methods</u>

Systems of equations can be solved by iterative techniques. When using iterative techniques the system of equations is put into the form below.

Original System   iterative form

$$A_{11}x_1 + A_{12}x_2 + A_{13}x_3 ... = C_1 \qquad\qquad x_1 = \frac{C_1 - A_{12}x_2 - A_{13}x_3 ...}{A_{11}}$$

$$A_{21}x_1 + A_{22}x_2 + A_{23}x_3 \ldots = C_2 \qquad\qquad x_2 = \frac{C_2 - A_{21}x_1 - A_{23}x_3 \ldots}{A_{22}}$$

$$A_{13}x_1 + A_{32}x_2 + A_{33}x_3 \ldots = C_3 \qquad\qquad x_3 = \frac{C_3 - A_{31}x_1 - A_{32}x_2 \ldots}{A_{33}}$$

$$\vdots \qquad\qquad\qquad \vdots \qquad\qquad\qquad \vdots$$

$$A_{n1} \qquad\qquad\qquad C_n \qquad\qquad\qquad x_n$$

Once in this form an initial guess must be made for x_1, x_2, $x_3 \ldots x_n$ in order to begin the procedure. The guess values are substituted into the right hand side of the equations in order to provide new values of x_1, x_2, $x_3 \ldots x_n$. This procedure is continued until the relative error between the vector norms of x_i and x_{i+1} is sufficiently small.

Convergence of the system depends solely upon the characteristics of the equations in the system (If two equations are dependent on one another the system will diverge). The rate of convergence depends on the initial guess made for the vector x.

The two relative methods to be discussed are called Jacobi method and Gauss Seidel method. These iterative methods involve vectors whose convergence towards a solution can be checked only in terms of their 'norms', explained below.

3.7 NORMS OF VECTORS AND MATRICES

In chapter 1 the relative error existing between two values in an iterative technique was expressed as

$$\xi = \left| \frac{P_{n+1} - P_n}{P_{n+1}} \right|$$

This is sufficient when single values are involved; however, in this chapter iterative techniques are applied on systems of equations rather than a single equation. In this case, convergence of a vector must be checked rather than that of a single value. Vector norms are used for this purpose.

Several types of norms for vectors are shown below.

$$L_p = \|x\|$$

This notation represents the p norm of vector x

The ℓ_1 norm of a vector is evaluated by summing the absolute values in the vector x, as

$$\ell_1 = \|x\|_1 = \sum_{i=1}^{n} |x_i|$$

Consider a physical example. Let each value in vector x represent the length of piece of pipe. In such a case the ℓ_1 norm might represent the maximum length if all the pipes are joined.

The ℓ_2 norm of a vector is evaluated by summing the squares of each value in vector x and then taking square root.

$$\ell_2 = \|x\|_2 = \left\{\sum_{i=1}^{n} x_i^2\right\}^{1/2}$$

This is the magnitude or the length of the vector x.

The ℓ_∞ norm of a vector x is evaluated by finding the maximum absolute value in the vector x.

$$\ell_\infty = \|x\|_\infty = \overset{max}{1 \le i \le n} |x_i|$$

If each value in vector x represents the pressure at different points in a vessel then this norm would show where maximum pressure exists.

Error representation between vectors

$$\text{Absolute Error} = \|x_{n+1} - x_n\|_p$$

$$\text{Relative Error} = \frac{\|x_{n+1} - x\|_p}{\|x_{n+1}\|}$$

The choice of p depends on the physical nature of the problem. If we want to limit the total length of a vector we choose p=2. If the maximum absolute in a vector is of interest we choose p=∞, etc.

EXAMPLE 3.10
Evaluate the 1,2, and ∞ norms of x_{n+1}. Also evaluate the absolute error norms 1,2, and ∞ between x_n and x_{n+1}. Evaluate the relative error in norm 1 between the two vectors.

$$x_{n+1} = \begin{bmatrix} 0.397 \\ 0.214 \\ 0.309 \end{bmatrix} \qquad x_n = \begin{bmatrix} 0.504 \\ 0.186 \\ 0.342 \end{bmatrix}$$

Solution

$$\ell_1 \text{ of } x_{n+1} = \|x_{n+1}\|_1 = 0.397 + 0.214 + 0.309 = 0.920$$

$$\ell_2 \text{ of } x_{n+1} = \|x_{n+1}\|_2 = \{(0.397)^2 + (0.214)^2 + (0.309)^2\}^{1/2} = 0.547 \quad \ell_\infty \text{ of } x_{n+1} = \|x_{n+1}\|_\infty = 0.397$$

$$x_{n+1}-x_n = \begin{bmatrix} -0.107 \\ 0.028 \\ -0.33 \end{bmatrix} \text{ absolute error vector}$$

$$\ell_1 \text{ of } \|x_{n+1} - x_n\|_1 = 0.107 + 0.028 + 0.333 = 0.168$$

$$\ell_2 \text{ of } \|x_{n+1} - x_n\|_2 = \{(0.107)^2 + (0.028)^2 + (0.333)^2\}^{1/2} = 0.115 \quad \ell_\infty \text{ of } \|x_{n+1} - x_n\|_\infty = 0.107$$

Relative error in norm 1, ℓ_1;

$$\ell_1 = \frac{\|x_{n+1} - x_n\|_1}{\|x_{n+1}\|_1} = \frac{0.168}{0.920} = 0.183$$

Norms also exist for matrices. Sometimes it is necessary to compare two matrices.

$$\ell_p = \|A\|_p$$

This notation represents the p norm of a matrix A. The $\|A\|_1$ norm of an nxm matrix A is given by

$$\ell_1 = \|A\|_1 = 1 \leq \overset{max}{j} \leq m \ \sum_{i=1}^{n} |a_{ij}|$$

After the absolute values of the elements in each column are summed, the largest resulting value is the 1 norm.

The $\|A\|_\infty$ norm of an nxm matrix a is given by

$$\ell_\infty = \|A\|_\infty = 1 \leq \overset{max}{i} \leq n \ \sum_{j=1}^{m} |a_{ij}|$$

After the absolute values of the elements in each row are summed, the largest resulting value is the ∞ norm.

The $\|A\|_E$ norm of an nxm matrix A is given by

$$\ell_E = \|A\|_E = \left\{ \sum_{i=1}^{n} \sum_{j=1}^{m} a_{ij}^2 \right\}^{1/2}$$

Every value in the matrix is squared and summed. Then the square root is taken.

EXAMPLE 3.11

Evaluate the $\ell_1, \ell_\infty,$ and ℓ_E norm of the following matrix.

$$A = \begin{bmatrix} 1 & 3 & 1 \\ -2 & 1 & -6 \\ 3 & 2 & 1 \end{bmatrix}$$

Solution

$\ell_1 = \|A\|_1 = $ largest absolute column sum

$$\begin{array}{ccc} |1| & |3| & |1| \\ (+) \quad |-2| & |1| & |-6| \\ \underline{|3|} & \underline{|2|} & \underline{|1|} \\ 6 & 6 & 8 \end{array} \leftarrow \text{maximum value resulting}$$

$$\ell_1 = \|A\|_1 = 8$$

$\ell_\infty = \|A\|_\infty = $ largest absolute row sum

$$|1| \quad + \quad |3| \quad + \quad |1| \quad = \quad 5$$
$$|-2| \quad + \quad |1| \quad + \quad |-6| \quad = \quad 9 \leftarrow \quad \text{maximum value}$$
$$|3| \quad + \quad |2| \quad + \quad |1| \quad = \quad 6$$

$$\ell_\infty = \|A\|_\infty = 9$$

$$\ell_E = \|A\|_E = \{1^2 + 3^2 + 1^2 + (-2)^2 + 1^2 + (-6)^2 + 3^2 + 2^2 + 1^2\}^{1/2}$$

$$\ell_E = \|A\|_E = \{66\}^{1/2} = 8.12$$

MATRIX CONDITION

The above matrix norms can be used to determine the condition of a matrix. The condition of a matrix specifies whether a matrix is well behaved or ill conditioned. An ill conditioned matrix is one which yields incorrect answers when an operation such as Gaussian elimination is performed on it. These errors are produced by the entries in the matrix. Some entries will amplify errors more than others.

To determine the condition of a matrix, the number resulting from the following operation is examined.

$$\text{Condition Number} = \|A\|_p * \|A^{-1}\|_p$$

p denotes that any of the norm criteria can be used.

If the condition number is greater than 100 the system of equations forming the matrix is considered ill-conditioned. Normally when a matrix 'A' is multiplied by its inverse A^{-1}, the resulting matrix should be the identity matrix. Similarly when the norm of matrix 'A' is multiplied by the norm of A^{-1}, the result should be unity. This is not always the case, depending on the size of the matrix entries and the precision of the arithmetic used to evaluate the inverse. When the matrix is inverted, it may contain significant round off errors, hence a high condition number. This condition number can be lowered by increasing the precision used when inverting the matrix. Therefore the precision used to achieve a satisfactory condition number, should also be used when performing other operations on the matrix so that accurate answers can be obtained.

EXAMPLE 3.12

Evaluate the condition number of the following system of equations with single digit accuracy.

$$x - 2z = 3$$
$$3x + y + 2z = 4$$
$$x - y = 5$$

Solution

$$A = \begin{bmatrix} 1 & 0 & -2 \\ 3 & 1 & 2 \\ 1 & -1 & 0 \end{bmatrix}$$

The first step is to find A^{-1}. This is done by using Gauss Jordan elimination on the matrix A and performing identical operations on an identity matrix placed along side. At the end of a series of such operations, when the matrix A is converted into an identity matrix, the identity matrix placed along side will be converted into the inverse of the matrix A, A^{-1}.

$$\widetilde{A}^1 = \begin{bmatrix} 1 & 0 & -2 & \vdots & 1 & 0 & 0 \\ 3 & 1 & 2 & \vdots & 0 & 1 & 0 \\ 1 & -1 & 0 & \vdots & 0 & 0 & 1 \end{bmatrix}$$

$$\text{Row } j - \frac{A_{ij}}{A_{ii}} \text{Row } i \rightarrow \text{New Row } j$$

$$\widetilde{A}^2 = \begin{bmatrix} 1 & 0 & -2 & \vdots & 1 & 0 & 0 \\ 0 & 1 & 8 & \vdots & -3 & 1 & 0 \\ 0 & -1 & 2 & \vdots & -1 & 0 & 1 \end{bmatrix}$$

$i = 1;$
$j = 2; j = 3$
Gives new row 2
Gives new row 3

$$\widetilde{A}^3 = \begin{bmatrix} 1 & 0 & -2 & \vdots & 1 & 0 & 0 \\ 0 & 1 & 8 & \vdots & -3 & 1 & 0 \\ 0 & 0 & 10 & \vdots & -4 & 1 & 1 \end{bmatrix}$$

$i = 2$
$j = 3$
Gives new row 3

$$\widetilde{A}^4 = \begin{bmatrix} 1 & 0 & 0 & \vdots & 0.2 & 0.2 & 0.2 \\ 0 & 1 & 0 & \vdots & 0.2 & 0.2 & -0.8 \\ 0 & 0 & 10 & \vdots & -4 & 1 & 1 \end{bmatrix}$$

$i = 3$
$j = 2; j = 1$
gives new row 2
gives new row 1

Divide row 3 by 10 to get identity matrix.

$$\widetilde{A}^4 = \begin{bmatrix} 1 & 0 & 0 & \vdots & 0.2 & 0.2 & 0.2 \\ 0 & 1 & 0 & \vdots & 0.2 & 0.2 & -0.8 \\ 0 & 0 & 1 & \vdots & -0.4 & 0.1 & 0.1 \end{bmatrix}$$

Hence
$$\widetilde{A}^1 = \begin{bmatrix} 0.2 & 0.2 & 0.2 \\ 0.2 & 0.2 & -0.8 \\ -0.4 & 0.1 & 0.1 \end{bmatrix}$$

Now evaluate $\|A\|_\infty$ and $\|A \sim^1\|$

For $\|A\|_\infty$

$$A = \begin{bmatrix} 1 & 0 & -2 \\ 3 & 1 & 2 \\ 1 & -1 & 0 \end{bmatrix} \begin{matrix} \rightarrow & |1| & + & |0| & + & |-2| & = 3 \\ \rightarrow & |3| & + & |1| & + & |2| & = 6 \\ \rightarrow & |1| & + & |-1| & + & |0| & = 2 \end{matrix}$$

Maximum absolute value $\{3,6,2\} = \|A\|_\infty = 6$

$$\widetilde{A}^1 = \begin{bmatrix} 0.2 & 0.2 & 0.2 \\ 0.2 & 0.2 & -0.8 \\ -0.4 & 0.1 & 0.1 \end{bmatrix} \begin{array}{l} \rightarrow \\ \rightarrow \\ \rightarrow \end{array} \begin{array}{l} |0.2| + |0.2| + |0.2| = 0.6 \\ |0.2| + |0.2| + |-0.8| = 1.2 \\ |-0.4| + |0.1| + |0.1| = 0.6 \end{array}$$

Maximum absolute value $\{0.6, 1.2, 0.6\} = \left\|A^{-1}\right\|_\infty = 1.2$

Now the condition number is evaluated;

$$\left\|A\right\|_\infty * \left\|A^{-1}\right\|_\infty = 6*1.2 = 7.2$$

7.2 < 100. Hence the matrix is well conditioned and will give accurate results when performing operations on it with single digit accuracy.

Notice that the norms used to find the matrix condition were p=∞; this was arbitrarily chosen, and p=1 or p=E could just as well have been used.

INVERSE OF MATRIX USING GAUSS – JORDAN OPERATIONS

To obtain inverse of nxn matrix A using Gauss – Jordan operations.

STEP 1.   Input: a_{ij}, i = 1, 2, ... n, j = 1, 2, ... n, elements of A. Row i of A is E_i

                d_{ii} = i, i = 1, 2, ... n, nxn identity

                Matrix D. Row i of D is denoted as F_i

STEP 2.   For j = 1, 2, ... n, $j \neq i$

                Perform $E_j = (E_j - \dfrac{a_{ji}}{a_{ii}} E_i)$

                $F_j = F_j - \dfrac{a_{ji}}{a_{ii}} F_i$

STEP 3.   For j = 1, 2, ... n, i = 1, 2, ... n, perform $d_{ij} = d_{ij}/a_{ii}$

STEP 4.   Output: d_{ij}, i = 1, 2, ... n, j = 1, 2, ... n, the elements of matrix $D = A^{-1}$

3.8 JACOBI METHOD

In this method the values of the guess vector are repeatedly used in the equations on the right hand side until a completely new solution vector is formed. The values of the solution vector for the next iteration are ready only after all the equations on the right hand side have been evaluated with the old values in the guess vector.

The procedure is continued until the relative error between two consecutive vector norms is satisfactorily small.

EXAMPLE 3.13

Solve the following system of equations using the Jacobi iterative method.

$$10x_1 - x_2 + 2x_3 = 6$$
$$-x_1 + 11x_2 - x_3 = 22$$
$$2x_1 - x_2 + 10x_3 = -10$$

Solution

Iterative form

$$x_1 = \frac{6 + x_2 - 2x_3}{10}$$
(new)

$$x_2 = \frac{22 + x_1 + x_3}{11}$$
(new)

$$x_3 = \frac{-10 - 2x_1 + x_2}{10}$$
(new)

Let the initial guess be $\qquad x^{(0)} = \begin{bmatrix} 1 \\ 1 \\ 1 \end{bmatrix}$

$$x_1 = \frac{6 + 1 - 2}{10} = 0.5$$
(new)

$$x_2 = \frac{22 + 1 + 1}{11} = 2.18$$
(new)

$$x_3 = \frac{-10 - 2 + 1}{10} = -1.1$$
(new)

Notice: Even though a better value was found in the first equation for x_1, the old value is used to approximate x_2 (new). Similarly the original x_1 and x_2 are used to evaluate x_3 (new) instead of the updated versions x_1 (new) and x_2 (new).

The new guess vector for the next iteration is

$$x^{(1)} = \begin{bmatrix} 0.5 \\ 2.18 \\ -1.1 \end{bmatrix}$$

The best way to proceed is to form a table:

n	x_1	x_2	x_3	ξ
0	1	1	1	
				← 0.987
1	0.5	2.18	-1.1	
				← 0.238
2	1.038	1.945	-0.882	
				← 0.68
3	0.9709	2.014	-1.013	
				← 0.17
4	1.004	1.996	-0.993	
				← 0.0045
5	0.9982	2.001	-1.0012	

where

$$\xi = \frac{\left\| x^{(n+1)} - x^{(n)} \right\|_2}{\left\| x^{(n+1)} \right\|_2}$$

It becomes obvious that the values are converging towards the exact solution below.

Answer: $x_1 = 1.0$
                   $x_2 = 2.0$
                   $x_3 = -1.0$

3.9 GAUSS-SEIDEL METHOD

This method converges on the roots of the system faster than the Jacobi method. The reason for this is that the new value of x_i immediately replaces the old value of x_I in the $x^{(n)}$ vector. As soon as a new value of x_i is calculated, it is used immediately in the next equation to give a closer approximation to the next value, x_{i+1}. Convergence is assured in the Gauss-Seidel method if matrix A is diagonally dominant and positive definite. If it is not in a diagonally dominant form, it should be converted to a diagonally dominant form by row interchanges, before starting the Gauss-Seidel iterations.

EXAPLE 3.14
Solve the following system by Gauss Seidel iteration method.

$$10x_1 - x_2 + 2x_3 = 6$$

$$-x_1 + 11x_2 - x_3 = 22$$

$$2x_1 - x_2 + 10x_3 = -10$$

Solution
    Iterative form

$$x_1 = \frac{6 + x_2 - 2x_3}{10}$$
(new)

$$x_2 = \frac{22 + x_1 + x_3}{11}$$
(new)

$$x_3 = \frac{-10 - 2x_1 + x_2}{10}$$
(new)

Let the initial guess be
$$x^{(0)} = \begin{bmatrix} 1 \\ 1 \\ 1 \end{bmatrix}$$

$$x_1 = \frac{6 + 1 - 2}{10} = 0.5$$
(new)

$$x_2 = \frac{22 + (0.5) + (1)}{11} = 2.14$$
(new)

$$x_3 = \frac{-10 - 2(1/2) + (2.14)}{10} = -0.886$$
(new)

Note: Notice that the new value of x_1 was used immediately to find the new value of x_2. Also the new values of x_1 and x_2 were used to find the new value of x_3.

Again a table should be formed and relative errors between the norms of $x^{(n)}$ and $x^{(n+1)}$ should be compared.

n	x_1	x_2	x_3	ξ
0	1	1	1	
				← 0.95
1	0.5	2.14	-0.886	
				← 0.212
2	0.9912	2.009	-0.997	
				← 0.0039
3	1.0003	2.0003	-1.00	

Notice that this method converged much faster than the Jacobi method.

JACOBI AND GAUSS – SEIDEL ITERATIVE METHODS

To solve the matrix equation $AX = b$. Used normally when A is sparse, so as to reduce computational effort. Needs an initial guess solution vector $X^{(0)}$.

STEP 1.   Input $\tilde{A}$, augmented matrix, (a_{ij}, i = 1, 2, ... n, j =1, 2, ... n), where $a_{i,n+1} = b_i$, Row i of $\tilde{A}$ is E_i.

N is maximum number of iterations

$x_i^{(0)}$ i = 1, 2, ... n, initial guess vector.

ξ is relative error.

(Arrange A in a diagonally dominant form for better convergence in steps 2 to 4).

STEP 2   For i = 1, 2, ... n do steps until 4.

STEP 3   Find a_{pi} in i $\leq$ p $\leq$ n, such that $\left| a_{pi} \right| = 1 \leq \overset{max}{j} \leq n \left| a_{ji} \right|$

STEP 4   If p $\neq$ i then interchange $E_p \Leftrightarrow E_i$

STEP 5   Set k = 1.

STEP 6   For i =1, 2, ... n perform

$$x_i^{(1)} = \frac{a_{i,n+1} - \sum_{j=1, j \neq i}^{n} a_{ij} x_j^{(0)}}{a_{ii}}$$

(FOR JACOBI METHOD)

$$x_i^{(1)} = \frac{a_{i,n+1} - \sum_{j=1}^{i-1} a_{ij} x_j^{(1)} - \sum_{j=i+1}^{n} a_{ij} x_j^{(0)}}{a_{ii}}$$

(FOR GAUSS-SEIDEL METHOD)

STEP 7   if $\left\| X^{(1)} - X^{(0)} \right\|_p / \left\| X^{(1)} \right\|_p < \xi$ then go to 11

STEP 8   Set k = k+1

STEP 9   If k > N go to 11

STEP 10   Set $x_i^{(0)} = x_i^{(1)}$ for i = 1, 2, ... n and go to 6

STEP 11   Output: $x_1, x_2, ... x_n$.

PROBLEMS

1.   A set of simultaneous equations are given below in the matrix form AX = b

$$\begin{bmatrix} 4.1 & 3.0 & 1.1 \\ 3.2 & 4.9 & 2.3 \\ 2.6 & 1.7 & 6.7 \end{bmatrix} \begin{Bmatrix} x_1 \\ x_2 \\ x_3 \end{Bmatrix} = \begin{bmatrix} 14.13 \\ 18.74 \\ 19.36 \end{bmatrix}$$

a)   Solve for x_1, x_2 and x_3 by Gaussian Elimination method.

b)   Solve for x_1, x_2 and x_3 by Gauss-Seidel method. Start with a trial vector of [1,1,1] and perform one iteration.

c)   Factorize A into a lower triangular and an upper triangular matrix using Doolittle's method.

2.   Consider the following set of simultaneous equation

A $\bar{x}$ = b where

$$A = \begin{bmatrix} 1 & -1 & 0 \\ -1 & 4 & -2 \\ 0 & -2 & 2 \end{bmatrix} \text{ and b = } [-1,1,2]^T$$

a)   Check whether the matrix A is positive definite.

b)   Obtain the l_2 norm of vector b

c)   Solve equation A $\bar{x}$ = b by Gaussian Elimination method.

3.   Solve the following equations using Gaussian Elimination with partial pivoting.

$$\begin{bmatrix} -2 & 4 & 3 \\ -1 & 9 & 4 \\ 3 & -2 & 7 \end{bmatrix} \begin{Bmatrix} x_1 \\ x_2 \\ x_3 \end{Bmatrix} = \begin{Bmatrix} 12 \\ 27 \\ 15 \end{Bmatrix}$$

4.   Solve the following linear system using 5 digit rounding by

a)   Gaussian Elimination with Backward Substitutions.

b)   Gaussian Elimination with Backward Substitution and using maximal column pivoting.

c)   Gaussian Elimination with Backward Substitution using scaled column pivoting.

$3.3330\, x_1 + 15920\, x_2 - 10.333\, x_3 = 15913$

$2.2220\, x_1 + 16.710\, x_2 + 9.6120\, x_3 = 28.544$

$1.5611\, x_1 + 5.1791\, x_2 + 1.6852\, x_3 = 8.4254$

5.   Solve the following linear system using Gaussian Elimination method, and determine whether row interchanges are necessary.

$$x_1 - x_2 + 3x_3 = 2$$
$$3x_1 - 3x_2 + x_3 = -1$$
$$x_1 + x_2 = 3$$

6.   Using modal analysis, we can obtain the following equations
$$5v_1 - 2v_2 = 3$$
$$-2v_1 + 6v_2 - 2v_3 = 4$$
$$-2v_2 + 7v_3 = 17$$

   a)   Use Gaussian Elimination procedure to solve for the unknowns of the above system.

   b)   Solve the above set of equations using Crout's method.

7.   Find x_1, x_2, x_3 and x_4 of the following linear system using the Gauss-Jordan method.
$$x_1 - x_2 + 2x_3 - x_4 = -8$$
$$2x_1 - 2x_2 + 3x_3 - 3x_4 = -20$$
$$x_1 + x_2 + x_3 = -2$$
$$x_1 - x_2 + 4x_3 + 3x_4 = 4$$

8.   Show that the following matrix is positive definite
$$A = \begin{bmatrix} 6 & 2 & 1 & -1 \\ 2 & 4 & 1 & 0 \\ 1 & 1 & 4 & -1 \\ -1 & 0 & -1 & 3 \end{bmatrix}$$

9.   Convert the following matrix into a product of lower triangular and upper triangular matrices using the Cholesky decomposition.
$$A = \begin{bmatrix} 5 & 4 & 2 & 3 \\ 4 & 8 & 3 & 2 \\ 2 & 3 & 10 & 1 \\ 3 & 2 & 1 & 13 \end{bmatrix}$$

10.   Find the Crout decomposition of the matrix
$$A = \begin{bmatrix} 1 & 1 & 0 \\ -1 & 4 & -2 \\ 0 & 2 & 2 \end{bmatrix}$$

11.   Consider the following set of simultaneous equations of the form $Ax = b$
$$4x_1 + x_2 - x_3 = 7$$
$$x_1 + 3x_2 - x_3 = 8$$
$$-x_1 - x_2 + 5x_3 + 2x_4 = -4$$
$$2x_3 + 4x_4 = 6$$

   a)   Show that the matrix A is positive definite

   b)   Factorize the matrix A using Cholesky method and using the factorization solve the system of equations.

   c)   Find the l_1, l_2, and l_∞ norms of vector x.

12. For the A matrix of problem 11
    a) Find $\|A\|_1, \|A\|_E$ and $\|A\|_\infty$
    b) Find the condition number of matrix A.

13. Find $\|A\|_1, \|A\|_E$ and $\|A\|_\infty$

$$[A] = \begin{bmatrix} 2 & 1 & 1 \\ 1 & 3 & 2 \\ 1 & 1 & 2 \end{bmatrix}$$

14. Check the following matrix for positive definiteness.

$$\begin{bmatrix} 2 & 1 & 1 \\ 1 & 3 & 1 \\ 1 & 1 & 9 \end{bmatrix}$$

15. Invert the matrix using the Gauss-Jordan method.

$$\begin{bmatrix} 1 & -1 & 2 \\ 3 & 0 & 1 \\ 1 & 0 & 2 \end{bmatrix}$$

16. If $[A] = [2,-4,1,3]^T$ find

$$\|A\|_1, \|A\|_E, \|A\|_\infty$$

17. Check the condition of matrix in problem 2 using the infinity norm of matrices.

18. Find the $\|A\|_1, \|A\|_\infty \|A\|_E$ norms of the matrix

$$A = \begin{bmatrix} 1.012 & 2.132 & 3.104 \\ -2.132 & 4.096 & -7.013 \\ 3.104 & 7.013 & 0.014 \end{bmatrix}$$

19. Solve the linear system
$$2x_1 - x_2 = 3$$
$$-x_1 + 2x_2 - x_3 = -3$$
$$-x_2 + 2x_3 = 1$$
by Jacobi's iteration method starting with the approximation $x^{(1)} = [1,0,0]^T$. Carry out 3 iterations and compare the result with that obtained by Gauss-Seidel method.

20. The linear system
$$4x_1 + 3x_2 = 24$$
$$3x_1 + 4x_2 - x_3 = 30$$

$$-x_2 + 4x_3 = -24$$

has a solution $x_1 = 3$, $x_2 = 4$, and $x_3 = -5$. Using Jacobi iterative method starting with $x^{(0)} = (1,1,1)^T$, carry out 4.

21. In problem 20, find the relative error of the solution vector after 4th iteration, using the 1_∞ norm, in both Jacobi method and Gauss-Seidel method.

22. The linear system $[A]\bar{x} = \bar{b}$ given by

$$A = \begin{bmatrix} 4 & 3 & 0 \\ 3 & 4 & -1 \\ 0 & 1 & 4 \end{bmatrix}; \ \bar{x} = \begin{bmatrix} x_1 \\ x_2 \\ x_3 \end{bmatrix}; \ \bar{b} = \begin{bmatrix} 24 \\ 30 \\ -24 \end{bmatrix}$$

has a solution $(3.75, 3, -6.75)^T$

Using Jacobi iterative method and Gauss-Seidel method, carry out 4 iterative steps and compare the results.

23. Solve the following set of simultaneous equations using Gauss-Seidel method. Use 5 iterations.

$$3x_1 + 4x_2 - 2x_3 + x_4 = 6$$
$$2x_1 - 2x_2 + 3x_3 - x_4 = 4$$
$$-x_1 + x_2 - x_3 + x_4 = 0$$
$$2x_1 - 2x_2 + 3x_3 + 3x_4 = 6$$

24. Solve problem 20 by Gauss-Seidel iterations, start with $x^{(0)} = (1,1,1)^T$ and perform 2 iterations.

25. Evaluate the roots of the following system of equation by Gauss Seidel method, use 6 iterations.

a) $3.5x_1 + 2.8x_2 + 6.3x_3 = 9.5$
       $2.8x_1 + 8x_2 + 4x_3 = -6.1$
       $-4x_1 - 3.5x_2 - 3x_3 = 5.5$

b) $2x_1 - 4.5x_2 - 2x_3 = 19$
       $3x_1 + 3x_2 + 4x_3 = 3$
       $-6x_1 + 4x_2 + 3x_3 = -18$

26. The following simultaneous equations are obtained for an electrical circuit

$$7i_1 - 2i_2 - 4i_3 = 10$$
$$-2i_1 + 13i_2 - 6i_3 = -15$$
$$-4i_1 - 6i_2 + 15i_3 = 15$$

Solve the above set of equations
a)   by Gauss-Jordan method
b)   by Gauss Seidel iteration method using 9 iterations

27. Write a computer program to solve the following set of equations by Gauss-Seidel method.

$$\begin{bmatrix} 1 & -10 & 2 & 4 \\ 3 & 1 & 4 & 12 \\ 9 & 2 & 3 & 4 \\ -1 & 2 & 7 & 3 \end{bmatrix} \begin{Bmatrix} x_1 \\ x_2 \\ x_3 \\ x_4 \end{Bmatrix} = \begin{Bmatrix} 2 \\ 12 \\ 21 \\ 37 \end{Bmatrix}$$

28. Write a general computer program implementing Gaussian Elimination technique using partial pivoting to solve a system of simultaneous equations.
As a sample problem, re-solve problem 5.

29. Construct a general purpose computer program implementing Cholesky's method of solving linear system of equations. Use it to solve the system of equations below.

$$2x_1 - x_2 = 3$$
$$-x_1 + 2x_2 - x_3 = -3$$
$$-x_2 + x_3 = -5$$

Chapter 4

FUNCTION APPROXIMATION OR INTERPOLATION

This chapter develops methods to approximate a function given the function value at certain discrete points. These values may be the measured outputs from a physical system responding to certain excitation inputs during an experiment (Fig.4.1)

Figure 4.1: Function approximation based on input and output

The function approximation, based on these points permits an approximation to the system's response to intermediate input values not obtained during the actual experiment. The accuracy of this function approximation to the actual system response depends on the amount of experimental data available and the method used to get the approximation.

The following methods are usually used to obtain a function in the form of a polynomial. The polynomial approximation is made when experimental data are the only information. If the actual form of the system response is already known, then some of these methods can be used to solve the coefficients of a function of this form.

4.1 DISCRETE LEAST SQUARES APPROXIMATION

The method of least squares fits the best curve through the given experimental data. This is done by developing a polynomial that will agree closely with data points obtained during an experiment, or solving the coefficients of an already known function form. In either case the objective of the method is to minimize the square of the error between the approximated function values and the experimental output values over the entire range of input values. The square of the error can be represented in the form:

$$E = \sum_{i=0}^{n} \left[P(x_i) - f(x_i) \right]^2$$

where $f(x_i)$ is the output value obtained during the experiment for each input value of x_i, $i = 0, 1, \ldots n$. $P(x_i)$ is the output value of the approximating function $P(x)$ at $x = x_i$.

When the form of the function that the data are following is completely unknown, the form is arbitrarily chosen to have the form of a continuous polynomial of degree m, such that $m < n$.

$$P(x_i) = a_0 x_i^0 + a_1 x_i^1 + a_2 x_i^2 \ldots a_m x_i^m$$

Now the squared error expression takes the form;

$$E = \sum_{1=0}^{n} \left[a_0 x_i^0 + a_1 x_i^1 + a_2 x_i^2 \ldots a_m x_i^m - f(x_i) \right]^2$$

Since there are m+1 coefficients to solve, there must be m+1 normal equations formed. Each equation is formed by minimizing E with respect to each coefficient as $\dfrac{\partial E}{\partial a_k} = 0$, $k = 0, 1, 2, \ldots$ m. The first equation is formed by minimizing E with respect to the first coefficient a_0 as shown below.

Equation 1

$$\frac{\partial E}{\partial a_0} = \frac{\partial}{\partial a_0} \sum_{i=0}^{n} \left[a_0 x_i^0 + a_1 x_i^1 + a_2 x_i^2 \ldots a_m x_i^m - f(x_i) \right]^2$$

$$\frac{\partial E}{\partial a_0} = \sum_{i=0}^{n} \left[a_0 x_i^0 + a_1 x_i^1 + a_2 x_i^2 \ldots a_m x_i^m - f(x_i) \right] = 0$$

The first equation becomes

$$\sum_{i=0}^{n} a_0 x_i^0 + \sum_{i=0}^{n} a_1 x_i^1 + \sum_{i=0}^{n} a_2 x_i^2 + \ldots \sum_{i=0}^{n} a_m x_i^m = \sum_{i=0}^{n} f(x_i)$$

To form (m+1) equations, the procedure $\dfrac{\partial E}{\partial a_0} = 0$, $\dfrac{\partial E}{\partial a_1} = 0$, $\ldots$ $\dfrac{\partial E}{\partial a_m} = 0$ is carried out. These equations can then be put into a matrix form and the coefficients a_0, a_1, a_2, $\ldots$ a_m can be solved by Gaussian elimination.

The matrix takes the form:

$$
\begin{bmatrix}
n+1 & \sum_{i=0}^{n} x_i & \sum_{i=0}^{n} x_i^2 \cdots & \sum_{i=0}^{n} x_i^m \\
\sum_{i=0}^{n} x_i & \sum_{i=0}^{n} x_i^2 & \sum_{i=0}^{n} x_i^3 \cdots & \sum_{i=0}^{n} x_i^{m+1} \\
\sum_{i=0}^{n} x_i^2 & \sum_{i=0}^{n} x_i^3 & \sum_{i=0}^{n} x_i^4 \cdots & \sum_{i=0}^{n} x_i^{m+2} \\
\vdots & & & \\
\sum_{i=0}^{n} x_i^m & \sum_{i=0}^{n} x_i^{m+1} & \sum_{i=0}^{n} x_i^{m+2} & \sum_{i=0}^{n} x_i^{2m}
\end{bmatrix}
\begin{bmatrix}
a_0 \\ a_1 \\ a_2 \\ \vdots \\ a_m
\end{bmatrix}
=
\begin{bmatrix}
\sum_{i=0}^{n} f(x_i) \\
\sum_{i=0}^{n} x_i f(x_i) \\
\sum_{i=0}^{n} x_i^2 f(x_i) \\
\vdots \\
\sum_{i=0}^{n} x_i^{\,m} f(x_i)
\end{bmatrix}
$$

$$[X] * \{a\} = \{F\}$$

When $m = n$, we need not use the method of least square errors, since $n+1$ coefficients in the polynomial can be solved using the $n+1$ data points exactly. We use least square error method only when $m<n$. Consequently, the largest approximated polynomial is of degree $(n-1)$. This does not mean that it is the most accurate approximation. Lower degree polynomials can be constructed by solving a smaller matrix, which will yield fewer, but completely different coefficients. For $n+1$ sets of data, polynomials of degree $m =1, 2, \ldots\ n-1$ can be formed. The results of each of these polynomials should be substituted back into the error expression, and the one exhibiting the least squared error should be used.

It is always good practice to plot the data if possible, to see what type of curve might best suit it.

EXAMPLE 4.1
An experiment was carried out in which the force deflection relation for a non-linear spring was found. The data obtained are shown below.

x_i	Force, F (newtons)	1.0	2.5	3.5	4.0
$f(x_i)$	Deflection,δ (meters)	3.8	15.0	26.0	33.0

Find the polynomial which best describes the nonlinear spring.

Solution
There are 4 sets of data, $n=3$, and $n-1$ polynomials can be formed, $m=2$ or $m=1$. The largest polynomial is of degree 2, using the least squares method.

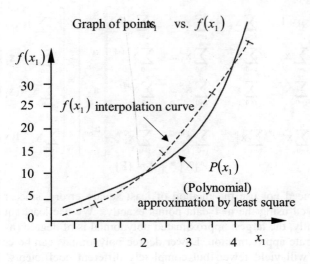

Figure 4.2: Polynomial approximation by least squares

Start by approximating a straight line (degree = 1).

$$P(x) = a_0 + a_1 x$$

To solve the coefficients we just solve the matrix developed earlier:
We have

$$E = \sum_{i=0}^{n} \left[P(x_i) - f(x_i) \right]^2$$

$$E = \sum_{i=0}^{n} \left[a_0 x_i^0 + a_1 x_i^1 - f(x_i) \right]^2$$

$$\frac{\partial E}{\partial a_0} = \sum_{i=0}^{n} 2 \left[a_0 x_i^0 + a_1 x_i^1 - f(x_i) \right](1) = 0$$

$$\sum_{i=0}^{n} a_0 x_i^0 + \sum_{i=0}^{n} a_1 x_i^1 = \sum_{i=0}^{n} f(x_i)$$

Similarly;

$$\frac{\partial E}{\partial a_0} = \sum_{i=0}^{n} 2 \left[a_0 x_i^0 + a_1 x_i^1 - f(x_i) \right] x_i^1 = 0$$

$$\sum_{i=0}^{n} a_0 x_i^1 + \sum_{i=0}^{n} a_1 x_i^2 = \sum_{i=0}^{n} f(x_i) x_i$$

Then;

$$\begin{bmatrix} \sum\limits_{i=0}^{3} x_i^0 & \sum\limits_{i=0}^{3} x_i^1 \\ \sum\limits_{i=0}^{3} x_i^1 & \sum\limits_{i=0}^{3} x_i^2 \end{bmatrix} \begin{bmatrix} a_0 \\ a_1 \end{bmatrix} = \begin{bmatrix} \sum\limits_{i=0}^{3} f(x_i) \\ \sum\limits_{i=0}^{3} f(x_i) x_i \end{bmatrix}$$

substituting the data gives

$$\begin{bmatrix} 4 & 11 \\ 11 & 35.5 \end{bmatrix} \begin{bmatrix} a_0 \\ a_1 \end{bmatrix} = \begin{bmatrix} 77.8 \\ 264.3 \end{bmatrix}$$

This can be put into augmented form and solved by Gaussian elimination.

$$i = 1, j = 2$$

$$\text{Row}_j - \frac{A_{ij}}{A_{ii}} \text{Row}_i \rightarrow \text{New Row}_j$$

$$\text{Row}_2 - \frac{A_{21}}{A_{11}} \text{Row}_1 \rightarrow \text{New Row}_2$$

$$11 - \frac{11}{4} \, 4 = 0 \rightarrow \text{New A}_{21}$$

$$35.5 - \frac{11}{4} \, 11 = 5.25 \rightarrow \text{New A}_{22}$$

$$264.3 - \frac{11}{4} \, 77.8 = 50.35 \rightarrow \text{New A}_{23}$$

$$\tilde{A}^1 = \begin{bmatrix} 4 & 11 & \vdots & 77.8 \\ 11 & 35.5 & \vdots & 264.3 \end{bmatrix}$$

$$\tilde{A}^2 = \begin{bmatrix} 4 & 11 & \vdots & 77.8 \\ 0 & 5.25 & \vdots & 50.35 \end{bmatrix}$$

Back substitution gives

$$a_1 = \frac{50.35}{5.25} = 9.5905$$

$$a_0 = 6.9238$$

Hence $P(x) = 9.5905x - 6.9238$

Now the error must be checked

$$E = \sum_{i=0}^{n} |P(x_i) - f(x_i)|^2$$

$$E = \sum_{1=0}^{n} [9.5905x_i - 6.9238 - f(x_i)]^2$$

x	1.0	2.5	3.5	4.0
P(x)	2.667	17.05	26.64	31.44
f(x)	3.8	15.0	26.0	33.0

$$E = (2.667 - 3.8)^2 + (17.05 - 15.0)^2 + (26.64 - 26.0)^2$$
$$+ (31.44 - 33.0)^2 = 8.329 \text{ (quite high)}$$

Next approximation is a quadratic (degree 2).

$$P(x) = a_0 + a_1 x^1 + a_2 x^2$$

The matrix has the form:

$$\begin{bmatrix} 4 & 11 & 35.5 \\ 11 & 35.5 & 123.5 \\ 35.5 & 123.5 & 446 \end{bmatrix} \begin{bmatrix} a_0 \\ a_1 \\ a_2 \end{bmatrix} = \begin{bmatrix} 77.8 \\ 264.3 \\ 944 \end{bmatrix}$$

Notice some values are already known in the matrix.
This again can be put into augmented form and solved
Applying

$$Row_j - \frac{A_{ij}}{A_{ii}} Row_i \rightarrow New Row_j$$

$$\tilde{A}^1 = \begin{bmatrix} 4 & 11 & 35.5 & \vdots & 77.8 \\ 11 & 35.5 & 123.5 & \vdots & 264.3 \\ 35.5 & 123.5 & 446 & \vdots & 944 \end{bmatrix}$$

$$\tilde{A}^2 = \begin{bmatrix} 4 & 11 & 35.5 & \vdots & 77.8 \\ 0 & 5.25 & 25.875 & \vdots & 50.35 \\ 0 & 25.875 & 130.94 & \vdots & 253.53 \end{bmatrix}$$

$$\tilde{A}^3 = \begin{bmatrix} 4 & 11 & 35.5 & \vdots & 77.8 \\ 0 & 5.25 & 25.875 & \vdots & 50.35 \\ 0 & 0 & 3.4107 & \vdots & 5.3714 \end{bmatrix}$$

Back substitution gives
$$a_2 = 1.5749$$
$$a_1 = 1.8286$$
$$a_0 = 0.4441$$
Hence $\qquad P(x) = 1.5749x^2 + 1.8286x + 0.4441$

To check the error form a table;

$$E = \sum_{i=0}^{n} \left[P(x_i) - f(x_i) \right]^2$$

x	1.0	2.5	3.5	4.0
P(x)	3.8476	14.8587	26.1367	32.9569
f(x)	3.8	15.0	26.0	33.0

$$E = (3.8476\text{-}3.8)^2 + (15.0\text{-}14.8587)^2 + (26.1367\text{-}26.0)^2$$
$$+ (33.0\text{-}32.9569)^2 = 0.0428$$

much better than the straight line.
Hence the best least squared approximation is

$$P(x) = 0.4441 + 1.8286\, x + 1.5749\, x^2$$

EXAMPLE 4.2
The following data were collected to examine the rate at which water empties from an experimental fire extinguisher, versus time.

Q flow rate, L/s	4.1	3.4	1.8	0.8
T time, s	1	2	5	9

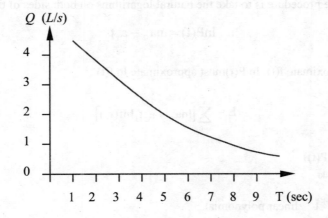

Figure 4.3

From the physical situation and the plot of the data, it seems that the best function would be one describing an exponential decay of the form.

$$P(t) = a_0 e^{-a_1 t}$$

Obtain the constants a_0 and a_1 using a least squares approximation.

Solution

$$\text{Error} = \sum_{i=0}^{n} \left[P(t_i) - f(t_i) \right]^2$$

$$E = \sum_{i=0}^{n} \left[a_0 e^{-a_1 t_i} - f(t_i) \right]^2$$

$$\frac{\partial E}{\partial a_0} = 2\sum_{i=0}^{n} \left[a_0 e^{-a_1 t_i} - f(t_i) \right] e^{-a_1 t_i} = 0$$

$$\frac{\partial E}{\partial a_1} = 2\sum_{i=0}^{n} \left[a_0 e^{-a_1 t_i} - f(t_i) \right] \left(-a_0 t_i e^{-a_1 t_i} \right) = 0$$

No exact solution exists for this nonlinear system of equations. However there is a procedure to follow in order to get the approximate a_0, a_1.
The equation

$$P(t) = a_0 e^{-a_1 t}$$

can be expressed in the form of a polynomial to get the approximation, and then transformed back to the desired form. The procedure is to take the natural logarithms on both sides of the equation.

$$\ln P(t) = \ln a_0 - a_1 t$$

Since P(t) must approximate f(t), In P(t) must approximate ln f(t).

$$E = \sum_{i=0}^{n} \left[\ln a_0 - a_1 t_i \ln f(t_i) \right]^2$$

Let $y = \ln(P(t))$
Let $a_0' = \ln a_0$
Let $a_1' = -a_1$
$y = a_0' + a_1' t$ linear polynomial

Now just continue as if approximating a polynomial of degree (1), keeping in mind that a_0' in In a_0.

$$a_1' = -a_1 \quad \text{and} \quad f(t) = \ln f(t_i)$$

For the given 4 sets of data n=3,

$$\begin{bmatrix} \sum\limits_{i=0}^{n} t_i^0 & \sum\limits_{i=0}^{n} t_i^1 \\ \sum\limits_{i=0}^{n} t_i^1 & \sum\limits_{i=0}^{n} t_i^2 \end{bmatrix} \begin{bmatrix} a_0' \\ a_1' \end{bmatrix} = \begin{bmatrix} \sum\limits_{i=0}^{n} \ln f(t_i) \\ \sum\limits_{i=0}^{n} \ln f(t_i) \end{bmatrix}$$

i	t_i	$f(t_i)$	$\ln f(t_i)$	(t_i^2)	$t_i \ln f(t_i)$
0	1	4.1	1.411	1	1.411
1	2	3.4	1.224	4	2.448
2	5	1.8	0.5878	25	2.939
3	9	0.8	-0.223	81	-2.007
$\sum\limits_{i=0}^{3} =$	17		2.9998	111	4.791

$$\tilde{A}^1 = \begin{bmatrix} 4 & 17 & \vdots & 2.9998 \\ 17 & 111 & \vdots & 4.791 \end{bmatrix}$$

Solve this by Gaussian elimination to get a_0' and a_1'.

$$\tilde{A}^2 = \begin{bmatrix} 4 & 17 & \vdots & 2.9998 \\ 0 & 38.75 & \vdots & -7.9582 \end{bmatrix}$$

$$38.75 a_1' = -7.9582$$
$$a_1' = -.2054$$
$$a_0' = \frac{2.9998 - 17(0.2054)}{4} = 1.6229$$

Recall that $a_0' = \ln a_0$ hence $a_0 = 5.067$ and $a_1 = -0.2054$.
Substituting into the original form gives:

$$P(t) = 5.067 e^{-0.2054t}$$

To check the error
$$E = \sum_{i=0}^{n} \left[P(t_i) - f(t_i) \right]^2$$

t_i	1	2	5	9
$P(t_i)$	4.13	3.36	1.81	0.79
$f(t_i)$	4.1	3.4	1.8	0.8

$$E = (4.13-4.1)^2 + (3.36-3.4)^2 + (`1.81-1.8)^2 + (.8-.79)^2$$
$$E = 0.0027$$

DISCRETE LEAST SQUARES APPROXIMATION

To fit a least squares approximation polynomial to a set of data given by (x_i, y_i), $i = 0, 1, 2, \ldots n$. Degree of polynomial is $m < n$: $P_m(x) = c_0 + c_1 x + c_2 x^2 + \ldots + c_m x^m$

STEP1   Input: (x_i, y_i), $i = 0, 1, 2, \ldots n$.

m degree of polynomial to be fit.

(Form augmented matrix of least squares equations in steps 2 to 3)

STEP 2   For $k = 0, 1, 2, \ldots m$, $j = 0, 1, 2, \ldots m$,

$$\text{Set } a_{jk} = \sum_{i=0}^{n} x_i^{j+k}$$

$$\text{Set } a_{j,m+1} = \sum_{i=0}^{n} y_i x_i^{j}$$

STEP 3   For $i = 0, 1, 2, \ldots m-1$ do steps 4.

STEP 4   For $j = i+1, i+2, \ldots m$ perform

$$E_j = \left(E_j - \frac{a_{ji}}{a_{ii}} E_i \right)$$

STEP 5   Set $c_n = a_{m,m+1}/a_{mm}$

$$\text{Set } c_1 = \frac{a_{i,m+1} - \sum_{j=i+1}^{m} a_{ij} x_j}{a_{ii}} \text{ for } i = 0, 1, 2, \ldots m.$$

STEP 6   Output: $P_m(x) = c_0 + c_1 x + c_2 x^2 + \ldots + c_m x^m$

4.2 LEAST SQUARES FUNCTION APPROXIMATION

Sometimes it is desired to approximate a complicated function by a simple polynomial of degree n, in an interval [a,b]. This approximation can be performed by the least squares method discussed in section 4.1 however the function to be approximated must be continuous in the interval [a, b]. The coefficients of the polynomial are calculated by minimizing the square of the error between the approximation and the original function.

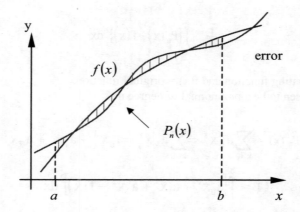

Figure 4.4: Approximation by least squares method

Error expression $E = \int_a^b \left[f(x) - P_n(x) \right]^2 dx$, where $P_n(x)$ can be represented by $P_n(x) = \sum_{i=0}^{n} a_i x^i$, giving

$$E = \text{Error} = \int_a^b (f(x) - \sum_{i=0}^{n} a_i x^i)^2 \, dx$$

To minimize the error, the coefficients a_i, of the polynomial must be chosen such that; $\dfrac{\partial E}{\partial a_i} = 0$, $i = 0, 1, 2, \ldots$ n. Consequently,

$$\frac{\partial E}{\partial a_i} = \int_a^b 2 \left[f(x) - \sum_{j=0}^{n} a_j x^j \right] x^i \, dx = 0 \quad i = 0, 1, 2, \ldots n$$

$$\sum_{j=0}^{n} a_j \int_a^b x^{j+i} dx = \int_a^b x^i f(x) dx \qquad i = 0, 1, 2, \ldots n$$

There are n+1 coefficients, and therefore there are n+1 equations formed, one equation for every i value.

EXAMPLE 4.3
Use the least squares function approximation to find a polynomial of degree 2 that will closely describe the function below;

$$F(x) = \sin \pi x \quad \text{in the interval from } 0 \rightarrow 1$$

Solution

$$E = \int_a^b [P_n(x) - f(x)]^2 \, dx$$

where $P_n(x)$ = approximating function and $f(x)$ = original function.
In this case $P_n(x)$ is chosen to be a polynomial of degree $n=2$.

$$P_n(x) = \sum_{k=0}^{n} a_k x^k = \sum_{k=0}^{2} a_k x^k = a_0 x^0 + a_1 x^1 + a_2 x^2$$

$$E = \int_a^b \left[a_0 x^0 + a_1 x^1 + a_2 x^2 - f(x) \right]^2 dx$$

To minimize error with respect to each coefficient

$$\frac{\partial E}{\partial a_0} = \int_a^b 2 \left[a_0 x^0 + a_1 x^1 + a_2 x^2 - f(x) \right] x^0 dx = 0$$

$$\int_a^b \left[a_0 x^0 + a_1 x^1 + a_2 x^2 - f(x) \right] x^0 dx = 0$$

$$\int_a^b \left[a_0 x^0 + a_1 x^1 + a_2 x^2 \right] x^0 dx = \int_a^b f(x) x^0 dx$$

$$a_0 \int_a^b x^0 x^0 dx + a_1 \int_a^b x^1 x^0 dx + a_2 \int_a^b x^2 x^0 dx = \int f(x) x^0 dx$$

$$a_0 \int_a^b x^0 dx + a_1 \int_a^b x^1 dx + a_2 \int_a^b x^2 dx = \int_a^b f(x) dx$$

Similarly: $\dfrac{\partial E}{\partial a_1} = 0$ and $\dfrac{\partial E}{\partial a_2} = 0$, provide

$$a_0 \int_a^b x^1 dx + a_1 \int_a^b x^2 dx + a_2 \int_a^b x^3 dx = \int f(x) x^1 dx$$

$$a_0 \int_a^b x^2 dx + a_1 \int_a^b x^3 dx + a_2 \int_a^b x^4 dx = \int_a^b f(x) x^2 dx$$

The three equations formed are necessary to solve the three coefficients a_0, a_1, and a_2 that will describe $f(x)$ with the least amount of error.

These equations can be put into matrix form:

$$\begin{bmatrix} \int\limits_0^1 x^0 dx & \int\limits_0^1 x^1 dx & \int\limits_0^1 x^2 dx \\ \int\limits_0^1 x^1 dx & \int\limits_0^1 x^2 dx & \int\limits_0^1 x^3 dx \\ \int\limits_0^1 x^2 dx & \int\limits_0^1 x^3 dx & \int\limits_0^1 x^4 dx \end{bmatrix} \begin{bmatrix} a_0 \\ a_1 \\ a_2 \end{bmatrix} = \begin{bmatrix} \int\limits_0^1 x^0 \sin\pi in\pi \\ \int\limits_0^1 x^1 \sin\pi in\pi \\ \int\limits_0^1 x^2 \sin\pi in\pi \end{bmatrix}$$

The left-hand-side involves simple integration, while the right-hand-side must be integrated 'by parts'.

$$\begin{bmatrix} 1 & 1/2 & 1/3 \\ 1/2 & 1/3 & 1/4 \\ 1/3 & 1/4 & 1/5 \end{bmatrix} \begin{bmatrix} a_0 \\ a_1 \\ a_2 \end{bmatrix} = \begin{bmatrix} 2/\pi \\ 1/\pi \\ (\pi^2 - 4)/\pi^3 \end{bmatrix}$$

Put this into the augmented matrix form and solve by Gaussian elimination.

$$\tilde{A}^1 = \begin{bmatrix} 1 & .50 & .3333 & \vdots & .636619772 \\ .50 & .3333 & .25 & \vdots & .318309886 \\ .3333 & .25 & .20 & \vdots & .189303748 \end{bmatrix}$$

$$\tilde{A}^2 = \begin{bmatrix} 1 & .50 & .3333 & \vdots & .636619772 \\ 0 & .08333 & .08333 & \vdots & 0 \\ 0 & .08333 & .08888 & \vdots & -.022902842 \end{bmatrix}$$

$$\tilde{A}^3 = \begin{bmatrix} 1 & .50 & .3333 & \vdots & .636619772 \\ 0 & .08333 & .08333 & \vdots & 0 \\ 0 & 0 & .005555 & \vdots & -.022902842 \end{bmatrix}$$

Solution: $a_2 = -4.1225$, $a_1 = 4.1225$ and $a_0 = -.050602$

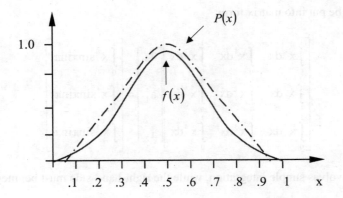

Figure 4.5

$$P(x) = -4.1225 x^2 + 4.1225 x - 0.050602$$

This polynomial is valid to approximate $f(x) = \sin\pi x$ between 0 and 1.

$$\text{Error} = \int_0^1 \left[\left(-4.1225x^2 + 4.1225x - 0.050602\right) - \sin\pi i\,\right]^2 dx$$

$$E = \int_0^1 \left[16.995x^4 + 16.995x^2 + .00256 + \sin^2\pi x - 33.99x^2\right.$$

$$+ 8.245x^2\sin\pi i - 33.99x - 8.245x\sin\pi x + .1012\sin\pi 1]dx$$

$$= \left|3.399x^5 - 5.665x^3 + .00128x^2 + .5x - .0796\sin 2\pi x\right.$$

$$-11.33x^3 + .8354x\sin\pi x - 2.624\left(x^2 - .2026\right)\cos\pi o$$

$$-16.995x^2 - .8354\sin\pi 8 + 2.624x\cos\pi x - 0.3221$$

$$\cos\pi o\Big|_0^1 = .0003$$

CAUTION

In the above case, it was obvious that a polynomial of degree 2 would be a good approximation, since the curve of $\sin\pi x$ is known to have a more or less parabolic shape between 0 and 1.

Most of the time it will be necessary to plot $f(x)$ so that the best degree polynomial can be chosen for the approximation. It should also be realized that the approximation does not have to be a polynomial; however, the whole purpose of the procedure is to obtain a simple approximation that can easily be adapted to a computer.

4.3 INTERPOLATION WITH DIVIDED DIFFERENCES

Interpolation deals with finding value of an unknown function f (x), between discrete values of the function, obtained during an experiment. These discrete values are assumed to be equally spaced and follow reasonably well behaved continuous functions. With these assumptions, it is possible to evaluate the function at any intermediate value x_p using a Taylor series expansion about a known point designated as $x = x_0$. (see fig. 4.6).

Taylor series expansion of f (x_p) about x_0 is given by

$$f(x_p) = f(x_0) + f(x_p - x_0)f'(x_0) + \frac{(x_p - x_0)^2}{2!}f''(x_0) + \cdots$$

To simplify the following derivations let $(x_p - x_0) = S$, giving

$$f(x_p) = f(x_0) + Sf'(x_0) + \frac{S^2}{2}f''(x_0) + \cdots$$

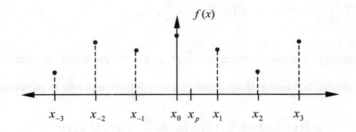

Figure 4.6: Function at any intermediate value x_p

Since the actual function is unknown, are evaluated the derivatives of the above expression by finite methods (divided differences). To evaluate the nth derivative, n+1 points must be known. As n increases, the accuracy of the derivative increases.

In order to evaluate the first derivative f'(x), two function values must be known and the point being

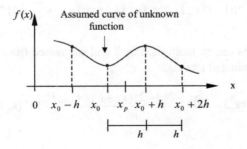

Figure 4.7: First derivative from two function values

approximated should lie between them. Expanding these known points by a Taylor series expansion, an expression for the derivative can be found.

$$f(x_0 + h) = f(x_0) + hf'(x_0) + \frac{h^2}{2!}f''(x_0) + \cdots$$

Ignoring all the terms to the right of the first derivative as a single error term in the first approximation.

$$f(x_0 + h) = f(x_0) + hf'(x_0) + E(f)$$

where
$$E(f) = \left[\frac{h^2}{2!}f''(x_0) + \ldots\right]$$

$$h\,f'(x_0) + E(f) = f(x_0+h) - f(x_0)$$

$$f'(x_0) = \frac{f(x_0 + h) - f(x_0)}{h} - E(f)$$

$$f'(x_0) = k_1 - \frac{h}{2}f''(x_0)$$

where k_1 is the first derivative approximation and $\frac{h}{2}f''(x_0)$ is the predominant term in $E(f)$.

To approximate the second derivative, three points must be known and two expansions are formed.

$$f(x_0 + h) = f(x_0) + hf'(x_0) + \frac{h^2}{2}f''(x_0) + E(f)$$

$$f(x_0 + 2h) = f(x_0) + 2hf'(x_0) + 2h^2f''(x_0) + E(f)$$

$E(f)$ is replaced by its most significant term.

$$f(x_0 + h) = f(x_0) + hf'(x_0) + \frac{h^2}{2}f''(x_0) + \frac{h^3}{6}f'''(x_0)$$

$$f(x_0 + 2h) = f(x_0) + 2hf'(x_0) + 2h^2f''(x_0) + \frac{4^2}{3}h^3f'''(x_0)$$

The first of these equations can be multiplied by 2 and subtracted from the second equation so that the first derivative term is eliminated leaving:

$$f''(x_0) = \frac{f(x_0 + 2h) - 2f(x_0 + h) + f(x_0)}{h^2} - hf'''(x_0)$$

$$f''(x_0) = \left[\frac{\dfrac{f(x_0+2h)-f(x_0+h)}{h} - \dfrac{f(x_0+h)-f(x_0)}{h}}{h} \right] - hf'''(x_0)$$

$$f''(x_0) = k_2 - hf'''(x_0)$$

where k_2 is second derivative approximation.

To approximate a third derivative, four points must be known and three expansions formed.

$$f'''(x_0) = k_3 - hf^{(iv)}(x_0)$$

$k_3 \rightarrow$ third derivative approximation

Replacing the approximated first derivative in equation will give;

$$f(x_p) = f(x_0) + \left[k_1 - \frac{h}{2} f''(x_0) \right] S + \frac{S^2}{2!} f''(x_0) + \dots$$

$$f(x_p) = f(x_0) + k_1 S + \frac{S(S-h)}{2!} f''(x_0) + \dots$$

Replacing the approximated second derivative in this latest expression gives;

$$f(x_p) = f(x_0) + k_1 S + \frac{S(S-h)}{2!} [k_2 - hf'''(x_0)] + \frac{S^3}{3!} f'''(x_0) + \dots$$

$$f(x_p) = f(x_0) + k_1 S + k_2 \frac{S(S-h)}{2!} + k_3 \frac{S(S-h)(S-2h)}{3!} + \dots$$

The expression being formed has the form of a polynomial. If we let $f(x_0) = k_0$;

$$f(x_P) = k_0 + k_1 S + k_2 \frac{S(S-h)}{2!} + k_3 \frac{S(S-h)(S-2h)}{3!} + \dots$$

Notice that the above expression was found by approximating a value $f(x_p)$ to the right of x_0. In this case the derivatives were evaluated in a forward direction by expanding about the points $f(x_0+h)$, $f(x_0+2h)$, $f(x_0+3h)$…etc, and hence the equation above is referred to as "forward divided difference formula". This formula is used when the majority of the known points are to the right of the point being approximated.   As the number of points used increases, so does the accuracy to the approximation.

If the point being approximated has a majority of known points to the left of it, then the derivatives would have been approximated by expanding the Taylor series about such points as $f(x_0-h)$, $f(x_0-2h)$, f

(x_0-3h), ... *etc.* If this is the case the resulting polynomial formed is referred to as the 'backward divided difference formula'. This polynomial has the form:

$$f(x_p) = k_0 + k_1 S + k_2 \frac{S(S+h)}{2!} + k_3 \frac{S(S+h)(S+2h)}{3!} + ...$$

If the point being approximated is located immediately on either side of the known center value of the data, then the derivatives are approximated by expanding about points such as $f(x_0+h)$, $f(x_0-h)$, $f(x_0+2h)$, $f(x_0-2h)$...etc. This will result in the 'central divided difference formula' also referred to as Stirling's formula.

$$f(x_p) = k_0 + S\bar{k}_1 + \frac{S^2}{2!} k_2 + \frac{S(S+h)(S-h)}{3!} \bar{k}_3 +$$

$$\frac{S^2(S+h)(S-h)}{4!} k_4 + \frac{S(S+h)(S-h)(S+2h)(S-2h)}{4!} \bar{k}_5$$

$$+ \frac{S^2(S+h)(S-h)(S+2h)(S-2h)}{6!} k_6 +$$

Notice in the above formula that average values $\bar{k}$ of the odd derivative approximations in the forward and backward directions must be taken. The three equations derived so far can be expressed using divided difference values a_0, a_1, a_2, a_3...a_n in place of the derivative values and their factorials k_0, $\frac{k_1}{1!}$, $\frac{k_2}{2!}$, $\frac{k_3}{3!}$,....$\frac{k_n}{n!}$. The general form of the three equations becomes:

Forward Difference Formula

$$f(x_p) = a_0 + a_1 S + a_2 S(S-h) + a_3 S(S-h)(S-2h) + ...$$

Backward Difference Formula

$$f(x_p) = a_0 + a_1 S + a_2 S(S+h) + a_3 S(S+h)(S+2h) + ...$$

Central Difference Formula

$$f(x_p) = a_0 + \bar{a}_1 S + a_2 S^2 + \bar{a}_3 S(S+h)(S-h) +$$

$$a_4 S^2(S+h)(S-H) + \bar{a}_5 S(S+h)(S-h)(S+2h)(S-2h)$$

$$+ a_6 S^2(S+h)(S-h)(S+2h)(S-2h) + ...$$

where $\bar{a}$ indicates average value. The concept of divided differences is introduced here so that the tedious task of evaluating derivative expressions can be avoided. These difference values differ from the original derivative approximations only by a factorial term. The advantage of these values is that they can be evaluated by a simple procedure in a table form. The following table illustrates how these difference values are evaluated for four data sets.

Divided Difference Table

	a_0	a_1	a_2	a_3
i x_i	$f[x_i]$	$f[x_i,x_{i+1}]$	$f[x_i,x_{i+1},x_{i+2}]$	$f[x_i,x_{i+1},x_{i+2},x_{i+3}]$

$$1\,x_i \qquad f[x_i]$$

$$=\frac{f(x_2)-f(x_1)}{x_2-x_1}$$

$$2\,x_2 \qquad f(x_2) \qquad \qquad =\frac{f[x_2,x_3]-f[x_1,x_2]}{x_3-x_1}$$

$$=\frac{f(x_3)-f(x_2)}{x_3-x_2} \qquad \qquad \qquad =\frac{f[x_2,x_3,x_4]-f[x_1,x_2,x_3]}{x_4-x_1}$$

$$3\,x_3 \qquad f(x_3) \qquad \qquad =\frac{f[x_3,x_4]-f[x_2,x_3]}{x_4-x_2}$$

$$=\frac{f(x_4)-f(x_3)}{x_4-x_3}$$

$$4\,x_4 \qquad f(x_4)$$

↑	↑	↑	↑
known data sets	a₁ represents the first divided difference	a₂ represents the second divided difference	a₃ represents the third divided difference

Notice that with n+1 data sets, only n divided differences can be formed. The column headed by $f[x_i]$ consists of the known function values of the data sets. The values in the column headed $f[x_i,x_{i+1}]$ are determined by the formula.

$$a_1 = f[x_i,x_{i+1}]= \frac{f[x_{i+1}]-f[x_i]}{x_{i+1}-x_i}$$

where $f[x_i] = f(x_i)$ $i = 1, 2, \dots n-1$

These values are referred to as the first divided differences. The next column represents the second divided differences. These values are determined by the formula.

$$a_2 = f[x_i,x_{i+1},x_{i+2}]= \frac{f[x_{i+1},x_{i+2}]-f[x_i,x_{i+1}]}{x_{i+2},x_i}$$

The third divided difference values are in the next column, and so on. Note that if any of these values are multiplied by the appropriate factorial term, (1! For a_1 values, 2! for a_2 values, etc) the result will be the approximation to the appropriate derivative term at that point. The best way to illustrate the usefulness of divided differences formulae will be to consider an example.

EXAMPLE 4.4

Given the following equally spaced data, approximate the output value for an input value of;

$$\text{(a)} \quad x_p = 1.1$$

$$\text{(b)} \quad x_p = 2.1$$

$$\text{(c)} \quad x_p = 1.5$$

Data sets

i	1	2	3	4	5
x_i	1.0	1.3	1.6	1.9	2.2
$f(x_i)$	0.7652	0.6201	0.4554	0.2818	0.1104

Solution

Form a table exhibiting all the divided differences for the data.

--See table on next page --

interval size   These approximations used in
$h = x_{i+1} - x_i$ forward divided difference formula
$h = 0.3$ ( )   These approximations used in
                             backward divided difference formula
                             These approximations used in

Central divided difference formula
 (a) $x_p = 1.1$

This value has a majority of known points on the right hand side; hence the forward divided difference formula is used. The closest known value to the left of $x_p = 1.1$ is designated as x_0.

In this case $x_0 = 1.0$, therefore $S = x_p - x_0 = 0.1$. The approximations to the derivatives are read off the divided difference table. These values correspond to those with an arrow pointing to them. Substituting these values in to to the forward divided difference formula gives the following approximation.

$$f(1.1) = 0.76520 - .4837(0.1) - 0.1087(0.1)(0.1 - 0.3) + 0.0659(0.1)(0.1 - .3)(0.1 - 2(0.3)) +$$
$$0.0021(0.1)(0.1 - 0.3)(0.1 - 2(0.3))(0.1 - 3(0.3)) = 0.7196$$

Divided Difference Table

i	x_i	$f[x_i] = a_0$	$f[x_i,x_{i+1}] = a_1$	$f[x_i,x_{i+1},x_{i+2}] = a_2$	$f[x_i,x_{i+1},x_{i+2},x_{i+3}] = a_3$	a_4
1	1.0	.7652				
			$\dfrac{.6201-.7652}{1.3-1.0} = -.4837$			
2	1.3	.6201		$\dfrac{-.5490-(-.4837)}{1.6-1.0} = -.1087$		
			$\dfrac{.4554-.6201}{1.6-1.3} = -.5490$		$\dfrac{-.0494-(-.1087)}{1.9-1.0} = .0659$	
3	1.6	.4554		$\dfrac{-.5787-(0.5490)}{1.9-1.3} = -.0494$		$\dfrac{.0659-.0681}{1.0-2.2} = 0.0021$
			$\dfrac{.2818-.4554}{1.9-1.6} = -.5787$		$\dfrac{.1180-(-.0494)}{2.2-1.3} = 0.0685$	
4	1.9	.2818		$\dfrac{-.5713-(-.5787)}{2.2-1.6} = 0.0021$		
			$\dfrac{.1104-.2818}{2.2-1.9} = -.5713$			
5	22	(.1104)				

(b) $x_p = 2.1$

This value has a majority of known points on the left hand side; hence the backward divided difference formula is used. The closest known value to the right of $x_p = 2.1$ is designated as x_0.

In this case $x_0 = 2.2$, therefore $s = x_p - x_0 = -0.1$.

The approximation to the derivative are read from the divided difference table. These values are indicated by parentheses ( ). Substituting these values into the backward divided difference formula gives the following approximation.

$$f(2.1) = 0.1104 - 0.5713(-0.1) + 0.0122(-0.1)(-0.1 + 3\ )0.0685(-0.1)(-0.1 + 0.3)(-0.1 + 2(0.3)) +$$
$$0.002(-0.1)(-0.1 + 0.3)(\ -0.1 + 2(0.3))(\ -0.1 + 3(0.3)) = 0.6808$$

(c) $x_p = 1.5$

This value is on the left hand side of the center value; hence it is close enough to the center so that the central divided difference formula can be used. The center value is designated as x_0, if the point being approximated is on either side of it.

In this case $x_0 = 1.6$, therefore $S = x_p - x_0 = -0.1$.

The approximations to the derivatives are read from the divided difference table, and for the odd numbered derivatives the average must be taken. The derivative values for the 'central divided difference' formulae are underlined. Substituting these values into the central divided difference formula gives the following approximation.

$$f(1.5) = 0.4554 + (-.1)\left[\frac{-0.5490 - 0.5787}{2}\right] + (-.1)^2(-.0494) +$$
$$(-0.1)(-.1 + 0.3)(-.1 - .3)\left[\frac{0.0659 + 0.0685}{2}\right] +$$
$$(-.1)^2(-.1 + .3)(-.1 - .3)(0.002\ 1) = 0.5118$$

CAUTION:

If $x_p = 1.5$, and the backward divided difference formula is used; then only two approximations will be available from the table. These values would be $a_0 = 0.4554$ and $a_1 = -0.5490$ and $a_2 = -0.1087$. If the forward divided formula is used, then four approximations can be used from the table. These would be $a_0 = 0.6201$ $a_1 = -0.5490$ and $a_2 = -0.0494$ and $a_3 = 0.0681$. Since the forward formula has more terms, it will give a better approximation than the backward formula; however, the best approximation comes from the central formula with five approximations as illustrated in example 4.3.1.

DIVIDED DIFFERENCE TABLES

To obtain the value of an unknown function $f(x_p)$ through interpolation between equally spaced discrete values obtained during an experiment.

STEP 1   Input x_i, $f(x_i)$, $i = 1, 2, \ldots n$

STEP 2   Set $h = x_{i+1} - x_i$

STEP 3   Form a divided difference table with n-1 difference terms, where;

$a_0 = f[x_i]$ terms are the function values from the data

$a_1 = f[x_i, x_{i+1}]$ terms are the first divided differences

$a_2 = f[x_i, x_{i+1}, x_{i+2}]$ terms are the second divided differences

etc.

STEP 4   Decide which of the three formulae will give best approximation. This will be based on the location of x_p among the known data. $S = x_p - x_0$, where x_0 is the known data point nearest x_p, to the left of x_p in the case of the forward difference, to the right in the case of the backward difference and to either side in the case of the central difference.

Forward Difference Formula:

$$f(x_p) = a_0 + a_1S + a_2S(S-h) + a_3S(S-h)(S-2h) + ...$$

Backward Difference Formula:

$$f(x_p) = a_0 + a_1S + a_2S(S+h) + a_3S(S-h)(S-2h) + ...$$

Central Difference Formula:

$$f(x_p) = a_0 + \bar{a_1}S + a_2S^2 + \bar{a_3}S(S+h)(S-h) +$$
$$a_4S^2(S+h)(S-h) + \bar{a_5}S(S+h)(S-h)(S+2h)(S-2h) +$$
$$a_6S^2(S+h)(S-h)(S+2h)(S-2h) + ...$$

where $\bar{a}$ indicates average value.

The formula which contains the most terms will give the best approximation.

STEP 5   Plug the appropriate divided differences, read from the table formed in step 3, into the chosen formula.

STEP 6   Calculate $f(x_p)$

4.4   LAGRANGE POLYNOMIALS

Lagrange polynomial interpolation is used when the available data are not equally spaced. Lagrange polynomial satisfies all the known data point exactly, and can be used to approximate intermediate values within the range of the data. If there are n+1 sets of data, the corresponding largrange polynomial will be of degree n. The Lagrange polynomial has the following form:

$$P(x) = \sum_{k=0}^{n} f(x_k) L_{n,k} \qquad \text{n+1} \rightarrow \text{pieces of data}$$

where

$$L_{n,k} = \prod_{\substack{i=0 \\ i \neq k}}^{n} \frac{(x - x_i)}{(x_k - x_i)} \qquad k = 0, 1, 2, \ldots n$$

Here, $L_{n,k}$ are the Lagrange coefficients and the symbol $\prod$ indicates that the product of terms for all i must be evaluated for each value of k.

Consider the simplest case of two data points. In this case the polynomial will be of degree one (linear).

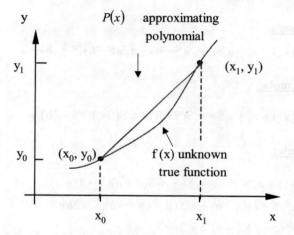

Figure 4.8: Approximating using Lagrange polynomial

The polynomial is formed using the above equation.

N+1 = pieces of data
k = 0, 1

Data		
x	x_0	x_1
y	y_0	y_1

Hence

$$P(x) = \sum_{k=0}^{1} f(x_k) L_{n,k} = f(x_0) L_{2,0} + f(x_1) L_{2,1}$$

k = 0

$$L_{2,0} = \prod_{\substack{i=0 \\ i\neq k}}^{1} \frac{(x-x_i)}{(x_k-x_i)} = \frac{(x-x_i)}{(x_k-x_i)}$$

Note: $i \neq k$ otherwise $L_{n,k}$ is undefined.

$k = 1$

$$L_{2,1} = \prod_{\substack{i=0 \\ i\neq k}}^{1} \frac{(x-x_i)}{(x_k-x_i)} = \frac{(x-x_i)}{(x_1-x_0)}$$

Plugging this into eqn. 4.4.1; the Lagrange polynomial becomes.

$$P(x) = y_0 \frac{x-x_1}{x_0-x_1} + y_1 \frac{x-x_0}{x_1-x_0}$$

Notice that when $x = x_1$ $\quad P(x) = y_1$

$\quad\quad\quad\quad\quad\quad x = x_0$ $\quad P(x) = y_0$

Therefore the known data are fitted exactly to the Lagrange polynomial.

Example 4.5

Given the following data, approximate the output value for an input of $x = 7$.

DATA

k	0	1	2	3
x_k	1	3	4	9
$f(x_k)$	1	6	8	12

Solution

$$n+1 = 4 \text{ pieces of data}$$
$$k = 0, 1, 2, 3$$

The eqn. becomes

$$P(x) = \sum_{k=0}^{n} f(x_k)L_{3,k} = f(1)L_{3,0} + f(3)L_{3,1} + f(4)L_{3,2} + f(9)L_{3,3}$$

$$L_{n,k} = \prod_{\substack{i=0 \\ i\neq k}}^{n} \frac{(x-x_i)}{(x_k-x_i)}$$

k=0

$$L_{3,0} = \prod_{\substack{i=0 \\ i\neq 0}}^{n} \frac{(x-x_i)}{(x_k-x_i)} = \frac{(x-x_1)}{(x_0-x_1)} * \frac{(x-x_2)}{(x_0-x_2)} * \frac{(x-x_3)}{(x_0-x_3)} =$$

$$\frac{(x-3)(x-4)(x-9)}{(1-3)(1-4)(1-9)} = \frac{x^3-16x^2+75x-108}{-48}$$

<u>k=1</u>

$$L_{3,1} = \prod_{\substack{i=0 \\ i \neq 1}}^{n} \frac{(x - x_i)}{(x_k - x_i)} = \frac{(x - x_1)}{(x_1 - x_0)} * \frac{(x - x_2)}{(x_1 - x_2)} * \frac{(x - x_3)}{(x_1 - x_3)} =$$

$$\frac{(x-1)(x-4)(x-9)}{(3-1)(3-4)(3-9)} = \frac{x^3 - 14x^2 + 49x - 36}{-12}$$

<u>k=2</u>

$$L_{3,2} = \prod_{\substack{i=0 \\ i \neq 2}}^{n} \frac{(x - x_i)}{(x_k - x_i)} = \frac{(x - x_1)}{(x_2 - x_0)} * \frac{(x - x_2)}{(x_2 - x_1)} * \frac{(x - x_3)}{(x_2 - x_3)} =$$

$$\frac{(x-1)(x-4)(x-9)}{(4-1)(4-3)(4-9)} = \frac{x^3 - 13x^2 + 39x - 27}{-15}$$

<u>k=3</u>

$$L_{3,3} = \prod_{\substack{i=0 \\ i \neq 3}}^{n} \frac{(x - x_i)}{(x_k - x_i)} = \frac{(x - x_1)}{(x_3 - x_0)} * \frac{(x - x_2)}{(x_3 - x_1)} * \frac{(x - x_3)}{(x_3 - x_2)}$$

$$= \frac{(x-1)(x-3)(x-4)}{(9-1)(9-3)(9-4)} = \frac{x^3 - 8x^2 + 19x - 12}{240}$$

$$P(x) = 1\left[\frac{x^3 - 16x^2 + 75x - 108}{-48}\right] + 6\left[\frac{x^3 - 14x^2 + 49x - 368}{-12}\right]$$

$$8\left[\frac{x^3 - 13x^2 + 39x - 27}{-15}\right] + 12\left[\frac{x^3 - 8x^2 + 19x - 12}{240}\right]$$

$$P(7) = 1\left(\frac{24}{48}\right) + 6\left(\frac{-36}{12}\right) + 8\left(\frac{-48}{-15}\right) + 12\left(\frac{72}{240}\right) = 11.7$$

Ans. 11.7

CAUTION

The Lagrange polynomial is fitted to pass through the known data; therefore when approximating the output at a point close to one of these points, the result should be fairly accurate. When approximating a point that is not close to any of the data points, it is possible that the resulting value will not be a good representation of the true value, as there is the possibility of heavy oscillations between the data points, especially with polynomials of high degree.

LAGRANCE POLYNOMIAL

To define a polynomial $P(x)$ through a set of non-equally spaced data points x_i, $f(x_i)$

STEP 1   Input x_i, $f(x_i)$, $i = 0, 1, 2, \dots n$

STEP 2   Evaluate

$$L_{n,k} = \prod_{\substack{i=0 \\ i \neq k}}^{n} \frac{(x - x_i)}{(x_k - x_i)} \qquad \text{for } k = 0, 1, 2, 3 \dots n$$

STEP 3   Evaluate

$$P(x) = \sum_{k=0}^{n} f(x_k) L_{m,k}$$

STEP 4   Now plug any x in the range $x_0 \leq x \leq x_n$ to

evaluate P(x).

STEP 5   Output P(x).

4.5 CUBIC SPLINE APPROXIMATION

Previously only one polynomial was formed to describe the data over the entire range. In this section different continuous polynomials will be formed to describe the function in each interval of known points. This type of approximation is known as piecewise polynomial approximation. For n+1 sets of data, there will be n piecewise polynomials formed (see fig. 4.9).

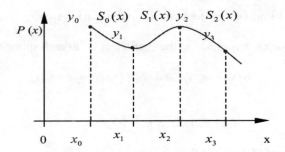

Figure 4.9: Cubic spline polynomial approximation

The advantage of such a method is that these polynomials are of a lower degree and less oscillatory, hence describing the data more accurately.

The best suited polynomial is found to be a cubic, thus the name cubic spline. With a cubic, it will also be possible to obtain an expression for the second derivative, thus describing the behavior of the data quite accurately within each interval.
The cubic polynomial will have the form:

$$S_j(x) = a_j + b_j(x-x_j) + c_j(x-x_j)^2 + d_j(x-x_j)^3 \qquad (4.5.1)$$

$$j = 0, 1, 2, \dots n-1$$

where $S_j(x)$ is valid in the interval $[x_j, x_{j+1}]$

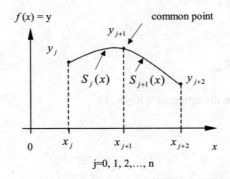

j=0, 1, 2,..., n

Figure 4.10: Common point between $s_j(x)$ and $S_{j+1}(x)$

Notice that x_{j+1} is a common point of $s_j(x)$ and $S_{j+1}(x)$.

$$S_j(x) = a_j + b_j(x - x_j) + c_j(x - x_j^2) + d_j(x - x_j)^3 \qquad (1)$$

Similarly:

$$S_{j+1}(x) = a_{j+1} + b_{j+1}(x - x_{j+1}) + c_{j+1}(x - x_{j+1})^2 + d_{j+1}(x - x_{j+1})^3 \qquad (2)$$

By letting $x = x_j$ in $S_j(x)$, or $x = x_{j+1}$ in $S_{j+1}(x)$ the coefficient 'a' in each spline equation is found.

$$S_j(x_j) = a_j = y_j \quad \text{and} \quad S_{j+1}(x_{j+1}) = a_{j+1} = y_{j+1}$$

Hence $\qquad a_j = y_j$

$$j = 0, 1, 2, \dots n$$

Since x_{j+1} is a common point of $S_j(x)$ and $S_{j+1}(x)$, then

$$S_j(x_{j+1}) = S_{j+1}(x_{j+1})$$

If $x = x_{j+1}$ in $S_j(x)$ (eqn.(1)) then;

$$S_j(x_{j+1}) = a_j + b_j(x_{j+1} - x_j) + c_j(x_{j+1} - x_j)^2 + d_j(x_{j+1} - x_j)^3$$

The expression $(x_{j+1} - x_j)$ can be referred to as an interval size h_j

$$S_j(x_{j+1}) = a_j + b_j h_j + c_j h_j^2 + d_j h_j^3$$

If $x = x_{j+1}$ in $S_{j+1}(x)$ (eqn.(2)) then

$$S_{j+1}(x_{j+1}) = a_{j+1}$$

Now equate the results of eqns (1) and (2) at $x = x_{j+1}$

$$S_{J+1}(x_{j+1}) = S_{j+1}(x_{j+1})$$
$$a_j + b_jh_j + c_jh_j^2 + d_jh_j^3 = a_{j+1} \tag{3}$$

Now take the first derivative of eqns. (1) and (2)

$$S_j'(x) = b_j + 2c_j(x - x_j) + 3d_j(x - x_j)^2 \tag{1}'$$
$$S_{j+1}'(x) = b_{j+1} + 2c_{j+1}(x - x_{j+1}) + 3d_{j+1}(x - x_{j+1})^2 \tag{2}'$$

Again the common point x_{j+1} exists between $S_j'(x)$ and $S_{j+1}'(x)$ (i.e. continuity of the slope).

Hence $\qquad S_j'(x_{j+1}) = S_{j+1}'(x_{j+1})$

Let $x = x_{j+1}$ in eqns (1)$'$ and (2) $'$;

$$S_j'(x_{j+1}) = b_j + 2c_jh_j + 3d_jh_j^2$$
$$S_{j+1}'(x_{j+1}) = b_{j+1}$$

Now equate the results of eqns (1)$'$ and (2)$'$ at $x = x_{j+1}$

$$b_j + 2c_jh_j + 3d_jh_j^2 = b_{j+1} \tag{4}'$$

Now take the second derivative of eqns (1) and (2)

$$S_j''(x) = 2c_j + 6d_j(x - x_j) \tag{1}''$$
$$S_{j+1}''(x) = 2c_{j+1} + 6d_{j+1}(x - x_{j+1}) \tag{2}''$$

Again the common point x_{j+1} exists between eqns (1)$''$ and (2)$''$ (i.e. continuity of the curvature)

Hence $S_j''(x_{j+1}) = S_{j+1}''(x_{j+1})$

Let $x = x_{j+1}$ in eqns (1)$''$ and (2)$''$

$$S_j''(x_{j+1}) = 2c_j + 6d_jh_j$$
$$S_{j+1}''(x_{j+1}) = 2c_{j+1}$$

Now equate the results of (1)$''$ and (2) $''$ at $x = x_{j+1}$

$$2c_j + 6d_jh_j = 2c_j \tag{5}$$

From (5)

$$d_j = \frac{c_{j+1} - c_j}{3h_j} \qquad j = 0, 1, 2, \dots n\text{-}1$$

From (3)

$$a_j + b_jh_j + c_jh_j^2 + d_jh_j^3 = a_{j+1}$$

$$b_j = \frac{1}{h_{j^o}} (a_{j+1} - a_j) - c_j h_j - d_j h_j^2$$

but
$$d_j = \frac{c_{j+1} - c_j}{3h_j}$$

$$b_j = \frac{1}{h_j}(a_{j+1} - a_j) - \frac{h_j}{3}(2c_j + c_{j+1}) \quad j = 0, 1, 2, \dots n-1 \tag{6}$$

From (4)
$$b_{j+1} = b_j + 2c_j h_j + 3d_j h_j^2$$

but
$$d_j = \frac{c_{j+1} - c_j}{3h_j}$$

Hence
$$b_{j+1} = b_j + 2c_j h_j + (c_{j+1} - c_j)h_j$$
$$b_{j+1} = b_j + h_j(c_j + c_{j+1})$$

The equation can be reduced by 1 to give

$$b_{j+1} = b_{j-1} + h_{j-1}(c_{j-1} + c_j) \tag{7}$$

Equation (6) can also be reduced by 1 to give

$$b_{j-1} = \frac{1}{h_{j-1}}(a_j - a_{j-1}) - \frac{h_{j-1}}{3}(2c_{j-1} + c_j)$$

Put this equation into eqn. (7) gives

$$b_j = \frac{1}{h_{j-1}}(a_j - a_{j-1}) + \frac{h_{j-1}}{3}(c_{j-1} + 2c_j) \tag{8}$$

Putting this result into eqn (6) gives an eqn with a_j's and c_j's.

$$h_{j-1}c_{j-1} + 2(h_{j-1} + h_j)c_j + h_j C_{j+1} = \frac{3}{h_j}(a_{j+1} - a_j) - \frac{3}{h_{j-1}}(a_j - a_{j-1})$$
$$j = 1, 2, \dots n-1$$

Note that j can go only upto $(n-1)$ in the above equation.

There are $(n-1)$ equations and $(n+1)$ unknowns $c_0, c_1, \dots c_n$. Hence 2 unknowns are solved by imposing boundary conditions at the two extreme end points x_0 and x_n.

Two boundary conditions normally used are given below.

(1) Natural Boundary Condition
   The second derivatives of the data at the end points x_0 and x_n are arbitrarily assumed to be zero;
$$S_0''(x_0) = S_n''(x_n) = 0$$
$$C_0 = c_n = 0$$

Because the c term represents the second derivative term of the Taylor series from which the cubic polynomial is formed.
   This condition is known as a free or natural boundary condition. This is the most common condition used to approximate the values a, b, c, d.
   The polynomials resulting from this condition are referred to as natural or free cubic splines. This may not give very accurate values close to the boundaries, but they are accurate in the interior region.

(2) Clamped Boundary Condition.
   If the first derivatives of the data are known at the end points x_0 and x_n, the boundary conditions are

$$S_0'(x_0) = f'(x_0)$$
$$S_n'(x_n) = f'(x_n)$$
$$b_0 = f'(x_0) \quad b_n = f'(x_n)$$

because the b term represents the first derivative term of the Taylor series from which the cubic polynomial is formed.

   This condition is known as a clamped boundary condition. With this type of condition, the coefficients a, b, c, d are more accurately evaluated.

   From the four expressions derived ((a), (b), (c), (d)), with the free boundary condition ($c_n = 0$ and $c_0 = 0$), it is possible to form n-1 natural cubic spline equations for n+1 sets of data. This is illustrated in example 4.5.1.   In order to evaluate clamped cubic spline equations, it becomes necessary to manipulate a few more equations. These equations will be formed after the natural cubic spline example so that the reason for their need becomes evident.

SUMMARY
Natural Cubic Spline, free boundary.

$$a_j = y_j \qquad = 0, 1, 2, \dots n$$

This is known directly from the data.

$$h_{j-1}c_{j-1} + 2(h_{j-1} + h_j)c_j + h_j C_{j+1} = \frac{3}{h_j}(a_{j+1} - a_j) - \frac{3}{h_{j-1}}(a_j - a_{j-1})$$

$$j = 1, 2, \dots n-1$$

c_j can be solved since a_j is known, using the known boundary conditions $c_0 = 0$ and $c_n = 0$.
Now d_j and b_j are found.

$$d_j = \frac{c_{j+1} - c_j}{3h_j} \quad j = 0, 1, 2, \ldots \text{n-1}$$

$$b_j = \frac{1}{h_j}(a_{j+1} - a_j) - \frac{h_j}{3}(2c_j + c_{j+1}) \quad j = 0, 1, 2, \ldots \text{n-1}$$

Boundary Condition (natural cubic spline):

$$c_0 = c_n = 0$$

Note: Coefficients a are known directly from the data. With these coefficients, coefficients c can be solved. Once coefficients c are known, coefficients b and d can be solved.

EXAMPLE 4.6
Construct a natural cubic spline for the following data.

n	0	1	2	3	4
x	3.0	4.0	5.0	6.0	7.0
y	3.7	3.9	3.9	4.2	5.7

Evaluate (a) $f''(3.4)$, (b) $f'(5.2)$, (c) $f'(5.6)$

Solution
n=4

$h_j = h_0 = h_1 = h_2 = h_3 = 1.0$
$a_j = y_j$ $j = 0, 1, 2, \ldots$ n
$a_0 = 3.7, a_1 = 3.9, a_2 = 3.9, a_3 = 4.2, a_4 = 5.7$

Now evaluate c_j values,

$$h_{j-1}c_{j-1} + 2(h_{j-1} + h_j)c_j + h_j c_{j+1} = \frac{3}{h_j}(a_{j+1} - a_j) - \frac{3}{h_{j-1}}(a_j - a_{j-1})$$

$$j = 1, 2, \ldots \text{n-1}$$
$$c_0 + 4c_1 + c_2 = 3(3.9 - 3.9) - 3(3.9 - 3.7) = -0.6$$
$$c_1 + 4c_2 + c_3 = 3(4.2 - 3.9) - 3(3.9 - 3.9) = 0.9$$
$$c_2 + 4c_3 + c_4 = 3(5.7 - 4.2) - 3(4.2 - 3.9) = 3.6$$

Enough data are available to form three equations, but there are five unknown values of c. However since it is a natural cubic spline, the boundary condition is known as

$$c_0 = c_4 = 0$$
$$0 + 4c_1 + c_2 = -0.6$$
$$c_1 + 4c_2 + c_3 = 0.9$$
$$c_2 + 4c_3 + 0 = 3.6$$

Now solve for c_1, c_2, and c_3.

$$\begin{bmatrix} 4 & 1 & 0 \\ 1 & 4 & 1 \\ 0 & 1 & 4 \end{bmatrix} \begin{bmatrix} c_1 \\ c_2 \\ c_3 \end{bmatrix} = \begin{bmatrix} -0.6 \\ 0.9 \\ 3.6 \end{bmatrix}$$

Put this into augmented form and solve by Gaussian elimination

$$\widetilde{A}^1 = \begin{bmatrix} 4 & 1 & 0 & \vdots & -0.6 \\ 1 & 4 & 1 & \vdots & 0.9 \\ 0 & 1 & 4 & \vdots & 3.6 \end{bmatrix}$$

$$\widetilde{A}^2 = \begin{bmatrix} 4 & 1 & 0 & \vdots & -0.6 \\ 0 & 3.75 & 1 & \vdots & 1.05 \\ 0 & 1 & 4 & \vdots & 3.6 \end{bmatrix}$$

$$\widetilde{A}^3 = \begin{bmatrix} 4 & 1 & 0 & \vdots & -0.6 \\ 0 & 3.75 & 1 & \vdots & 1.05 \\ 0 & 0 & 3.73 & \vdots & 3.32 \end{bmatrix}$$

Back substitution gives

$$3.73c_3 = 3.32$$
$$c_3 = 0.89$$
$$3.74c_2 + 0.89 = 1.05$$
$$c_2 = 0.043$$
$$4c_1 + 0.043 = -0.6$$
$$c_1 = -0.1607$$

Now all value of c_j have been found

$$c_0 = 0, c_1 = -0.1607, c_2 = 0.043, c_3 = 0.89, c_4 = 0$$

It is now possible to evaluate values of d_j

$$d_j = \frac{c_{j+1} - c_j}{3h_j}, \quad j = 0,1,2,\ldots n-1$$

$$d_0 = \frac{c_1 - c_0}{3} = \frac{-0.1607 - 0}{3} = -0.054$$

$$d_1 = \frac{c_2 - c_1}{3} = \frac{-0.043 - (-0.1607)}{3} = 0.0679$$

$$d_2 = \frac{c_3 - c_2}{3} = \frac{0.89 - 0.043}{3} = 0.282$$

$$d_3 = \frac{c_4 - c_3}{3} = \frac{0 - 0.89}{3} = -0.2967$$

Now b_j can be evaluated.

$$b_j = \frac{1}{h_j}(a_{j+1} - a_j) - \frac{h_j}{3}(2c_j + c_{j+1}), \quad j = 0, 1, 2, \ldots n-1$$

$$b_0 = (a_1 - a_0) - \frac{1}{3}(2c_0 + c_1) = (3.9 - 3.7) - \frac{1}{3}(2(0) - 0.1607) = 0.2536$$

$$b_1 = (a_2 - a_1) - \frac{1}{3}(2c_1 + c_2) = (3.9 - 3.9) - \frac{1}{3}(2(-0.1607) + 0.043) = 0.1214$$

$$b_2 = (a_3 - a_2) - \frac{1}{3}(2c_2 + c_3) = (4.2 - 3.9) - \frac{1}{3}(2(0.043) + 0.89) = -0.025$$

$$b_3 = (a_4 - a_3) - \frac{1}{3}(2c_3 + c_4) = (5.7 - 4.2) - \frac{1}{3}(2(0.89) + 0) = 0.907$$

Now substitute the values into eqn. 4.5.1 to evaluate the spline equations.

$$S_j(x) = a_j + b_j(x - x_j) + c_j(x - x_j)^2 + d_j(x - x_j)^3 \quad j = 0, 1, 2, \ldots n-1$$

$j = 0$ first spline eqn.
Valid for the range $x_0 \le x \le x_1$ (3.0$\le$ x $\le$ 4.0)

$$S_0(x) = 3.7 + 0.2536(x - 3.0) + 0.0536(x - 3.0)^3$$

$j = 1$ for the range $x_1 \le x \le x_2$ (4.0 $\le$ x $\le$ 5.0)
$$S_1(x) = 3.9 + 0.1214(x - 4.0) - 0.1607(x - 4.0)^2 + 0.0679(x - 4.0)^3$$

$j = 2$ for the range $x_2 \le x \le x_3$ (5.0 $\le$ x $\le$ 6.0)
$$S_2(x) = 3.9 - 0.025(x - 5.0) - 0.043(x - 5.0)^2 + 0.282(x - 5.0)^3$$

$j = 3$ for the range $x_3 \le x \le x_4$ (6.0 $\le$ x $\le$ 7.0)
$$S_3(x) = 4.2 + 0.907(x - 6.0) + 0.89(x - 6.0)^2 - 0.2967(x - 6.0)^3$$

Notice: In the free boundary case, the coefficient c is evaluated first, since the boundary condition contains information about the second derivative; however in the clamped boundary case, the coefficient c is evaluated by manipulating equations in a different fashion so that the information about the first derivative is used.

(a)   To approximate f″(3.4), the first equation is differentiated twice, since this equation is valid in
the range $3.0 \le x \le 4.0$.

$S_0(x) \doteq 3.7 + 0.2536(x - 3.0) + 0.0536(x - 3.0)^3$

$S_0'(x) = 0.2536 + 3(0.0536)(x - 3.0)^2$

$S_0''(x) = 2(3)(0.0536)(x - 3.0)$

$S_0''(3.4) = 2(3)(0.0536)(0.4) = 0.12864$

(b)   To approximate f′(5.2) the third equation is differentiated once, since this eqn. is valid in the
range $5.0 \le x \le 6.0$.

$S_2(x) = 3.9 - 0.025(x - 5.0) - 0.043(x - 5.0)^2 - 0.283(x - 5.0)^3$

$S_2'(x) = -0.025 - 2(0.43)(x - 5.0) + 3(0.282)(x - 5.0)^2$

$S_2'(x) = -0.025 - 2(0.043)(0.2) + 3(0.282)(0.2)^2 = -0.00836$

(c)   To approximate f(5.6) the same equation is valid,

$S_2(x) = 3.9 - 0.025(x - 5.0) - 0.043(x - 5.0)^2 + 0.283(x - 5.0)^3$

$S_2(5.6) = 3.9 - 0.025(0.6) - 0.043(0.6)^2 + 0.283(0.6)^3 = 3.930432$

SUMMARY: Clamped Cubic Spline

In order to construct clamped cubic spline equations, the same four equations must be used with the
clamped boundary condition. With this condition, b_0 and b_n are given, however no information about
'c' coefficients is known. The problem then arises when attempting to solve the 'c' coefficients, which
must be solved before b and d coefficients can be gotten. This problem arises because only n-1 'c'
equations can be formed, and there are n+1 unkown values of 'c'. It now becomes necessary to use the
known conditions b_0 and b_n in order to express c_0 and c_n in terms of other 'c' coefficients. This will
reduce the amount of unkown 'c' coefficients by 2, thus giving all the 'c' values to be evaluated. Then
the procedure continues in the same fashion as the natural cubic spline.

The manipulation, using b_0 and b_n, to get c_0 and c_n in terms of intermediate c values is done in the
following way;

Recall eqn (6) $b_j = \dfrac{1}{h_j}(a_{j+1} - a_j) - \dfrac{h_j}{3}(2c_j + c_{j+1})$

b_0 and b_n are known;

Let j = 0

$$b_0 = \frac{1}{h_0}(a_1 - a_0) - \frac{h_0}{3}(2c_0 + c_1)$$

Rearrange this to give;

$$\frac{h_0}{3}(2c_0 + c_1) = \frac{(a_1 - a_0)}{h_0} - b_0$$

$$2c_0 = \frac{\dfrac{(a_1 - a_0)}{h_0} - b_0}{h_0/3} - c_1$$

$$2c_0 = \frac{3(a_1 - a_0)}{(h_0)^2} - \frac{3b_0}{h_0} - c_1$$

(e) $$c_0 = \frac{3(a_1 - a_0)}{2h_0^2} - \frac{3b_0}{2h_0} - \frac{c_1}{2}$$

Now c_0 is in terms of c_1. All other
values in this equation are known.

Recall eqn (8)

$$b_j = \frac{1}{h_{j-1}}(a_j - a_{j-1}) + \frac{h_{j-1}}{3}(c_{j-1} + 2c_j)$$

Let $j = n$:

$$b_n = \frac{1}{h_{n-1}}(a_n - a_{n-1}) + \frac{h_{n-1}}{3}(c_{n-1} + 2c_n)$$

Rearrange this in terms of c_n,

$$\frac{h_{n-1}}{3}(c_{n-1} + 2c_n) = \frac{-1}{h_{n-1}}(a_n - a_{n-1}) + b_n$$

$$c_{n-1} + 2c_n = -\frac{3}{h_{n-1}^2}(a_n - a_{n-1}) + \frac{3b_n}{h_{n-1}}$$

(f) $$c_n = -\frac{3(a_n - a_{n-1})}{2h_{n-1}^2} + \frac{3b_n}{2h_{n-1}} - \frac{1}{2}c_{n-1}$$

Now c_n is in tems of c_{n-1}. All other values in this equation are known.

EXAMPLE 4.7

Construct a clamped cubic spline for the following data if it is known that the data have a slope of 0.2 at x_0 and a slope of 0.6 at x_n.

N	0	1	2	3	4
x	3.0	4.0	5.0	6.0	7.0
y	3.7	3.9	3.9	4.2	5.7

Evaluate (a) f″ (3.4), (b) f′ (5.2), (c) f(5.6)

Solution

Given boundary condition is $f'(x_0) = 0.2$ and $f'(x_n) = 0.6$

$n = 4$ $h_j = h_0 = h_1 = h_2 = h_3 = 1.0$

$$a_j = y_j \qquad\qquad j = 0, 1, 2, \dots n$$
$$a_0 = 3.7,\ a_1 = 3.9,\ a_2 = 3.9,\ a_3 = 4.2,\ a_4 = 5.7$$

Now evaluate c_j values,

$$h_{j-1}c_{j-1} + 2(h_{j-1} + h_j)c_j + h_j c_{j+1} = \frac{3}{h_j}(a_{j+1} - a_j) - \frac{3}{h_{j-1}}(a_j - a_{j-1})$$

$$j = 1, 2, \dots n-1$$

$$C_0 + 4c_1 + c_2 = 3(3.9 - 3.9) - 3(3.9 - 3.7) = -0.6$$
$$C_1 + 4c_2 + c_3 = 3(4.2 - 3.9) - 3(3.9 - 3.9) = 0.9$$
$$C_2 + 4c_3 + c_4 = 3(5.7 - 4.2) - 3(4.2 - 3.9) = 3.6$$

Enough data are available to form three equations, but there are five unknown values of c.

However since it is a clamped cubic spline, c_0 and c_4 can be expressed in terms of c_1, c_2 and c_3 by using equations (e) and (f) since b_0 and b_4 are known.

From equation (e)

$$c_0 = \frac{3}{2}\frac{(a_1 - a_0)}{(h_0)^2} - \frac{3}{2}\frac{b_0}{h_0} - \frac{c_1}{2} \qquad\qquad b_0 = 0.2$$

Hence,

$$c_0 = \frac{3}{2}(0.2) - \frac{3}{2}(0.2) - \frac{1}{2}c_1$$

$$c_0 = -\frac{1}{2}c_1$$

From equation (f)

$$c_n = \frac{3}{2}\frac{(a_n - a_{n-1})}{(h_{n-1})^2} - \frac{3}{2}\frac{b_n}{h_{n-1}} - \frac{1}{2}c_{n-1} \qquad\qquad n = 4;\quad b_4 = 0.6$$

$$c_4 = -\frac{3}{2}(5.7\text{-}4.2) + \frac{3}{2}(0.6) - \frac{1}{2}c_3$$

$$c_4 = -1.35 - 0.5c_3$$

Substitute these values into the above equations so that the values of c_1, c_2, and c_3 can be evaluated.

$$-.5c_1 + 4c_1 + c_2 = -.6$$
$$c_1 + 4c_2 + c_3 = .9$$
$$c_2 + 4c_3 - 0.5c_3 = 3.6 + 1.35$$

Put this system into an augmented matrix and solve c_1, c_2, c_3 by Gaussian elimination.

$$\widetilde{A}^1 = \begin{bmatrix} 3.5 & 1 & 0 & \vdots & -0.6 \\ 1 & 4 & 1 & \vdots & 0.9 \\ 0 & 1 & 3.5 & \vdots & 4.95 \end{bmatrix}$$

$$\widetilde{A}^2 = \begin{bmatrix} 3.5 & 1 & 0 & \vdots & -0.6 \\ 0 & 3.714 & 1 & \vdots & 1.0714 \\ 0 & 1 & 3.5 & \vdots & 4.95 \end{bmatrix}$$

$$\widetilde{A}^3 = \begin{bmatrix} 3.5 & 1 & 0 & \vdots & -0.6 \\ 1 & 3.714 & 1 & \vdots & 1.0714 \\ 0 & 0 & 3.23 & \vdots & 4.662 \end{bmatrix}$$

Back substitution gives;

$c_3 = 4.662/3.23 = 1.4432$

$c_2 = \dfrac{1.0714 - 1.4432}{3.714} = -0.1$

$c_1 = \dfrac{-0.6 - (-0.1)}{3.5} = -0.1429$

Now substitute back into eqns (e) and (f) to get c_0 and c_4.
From (e)
$$C_0 = -0.5c_1 = -0.5(-.1429) = 0.0714$$
From (f)
$$C_4 = -1.35 - 0.5c_3$$
$$C_4 = -1.35 - 0.5(1.4432)$$
$$C_4 = -2.0716$$

It is now possible to evaluate values of d_j
$$d_j = \frac{c_{j+1} - c_j}{3h_j} \qquad j = 0, 1, 2, \ldots, n-1$$

$$d_0 = \frac{c_1 - c_0}{3} = \frac{-0.1429 - 0.0714}{3} = -.0714$$

$$d_1 = \frac{c_2 - c_1}{3} = \frac{-0.1 - (-0.1429)}{3} = .0143$$

$$d_2 = \frac{c_3 - c_2}{3} = \frac{1.4432 - (-0.1)}{3} = 0.5144$$

$$d_3 = \frac{c_4 - c_3}{3} = \frac{-2.0716 - 1.4432}{3} = -1.1716$$

Now b_j can be evaluated;

$$b_j = \frac{1}{h_j}(a_{j+1} - a_j) - \frac{h_j}{3}(2c_j + c_{j+1})$$

$b_0 = 0.2$ from boundary condition

$$b_1 = (a_2 - a_1) - \frac{1}{3}(2c_1 + c_2) = (3.9 - 3.9) - \frac{1}{3}(2(-0.1429) + (-0.1)) = 0.1268$$

$$b_2 = (a_3 - a_2) - \frac{1}{3}(2c_2 + c_3) = (4.2 - 3.9) - \frac{1}{3}(2(-0.1) + (1.4432)) = -0.1144$$

$$b_3 = (a_4 - a_3) - \frac{1}{3}(2c_3 + c_4) = (5.7 - 4.2) - \frac{1}{3}(2(1.4432) + (-2.0716)) = 0.2284$$

Now substitute the values in to eqn 4.5.1 to evaluate the clamped spline equations.

$$S_j(x) = a_j + b_j(x - x_j) + c_j(x - x_j)^2 + d_j(x - x_j)^3 \quad j = 0, 1, 2, \ldots, n\text{-}1$$
$$j = 0 \text{ first spline eqn.}$$

Valid for the range $x_0 \leq x \leq x_1$ $(3.0 \leq x \leq 4.0)$
$$S_0(x) = 3.7 - 0.2(x - 3.0) + 0.0714(x\text{-}3.0)^2 - 0.0714(x - 3.0)^3$$

$j = 1$ for the range $x_1 \leq x \leq x_2$ $(4.0 \leq x \leq 5.0)$
$$S_1(x) = 3.9 + 0.1286(x - 4.0) - 0.1429(x - 4.0)^2 + 0.0143(x - 4.0)^3$$

$j = 2$ for the range $x_2 \leq x \leq x_3$ $(5.0 \leq x \leq 6.0)$
$$S_2(x) = 3.9 - 0.1144(x - 5.0) - 0.1(x - 5.0)^2 + 0.5144(x - 5.0)^3$$

$j = 3$ for the range $x_3 \leq x \leq x_4$ $(6.0 \leq x \leq 7.0)$
$$S_3(x) = 4.2 + 0.2284(x - 6.0) + 1.4432(x - 6.0)^2 - 1.1716(x - 6.0)^3$$

(a)   To approximate $f''(3.4)$, the first equation is differentiated twice, since this equation is valid in the range $3.0 \leq x \leq 4.0$.
$$S_0(x) = 3.7 + 0.2(x - 3.0) + 0.0714(x - 3.0)^2 - 0.0714(x - 3.0)^3$$
$$S'_0(x) = 0.2 + 2(0.0714)(x - 3.0) - 3(0.0714)(x - 3.0)^2$$
$$S''_0(x) = 2(0.0714) - 2(3)(0.0714)(x - 3.0)$$
$$S''_0(3.4) = 2(0.0714) - 2(3)(0.0714)(0.4) = -0.02856$$

(b)   To approximate $f'(5.2)$, the third equation is differentiated once, since this eqn is valid in the range $5.0 \leq x \leq 6.0$.
$$S_2(x) = 3.9 - 0.1144(x - 5.0) - 0.1(x - 5.0)^2 + 0.5144(x - 5.0)^3$$
$$S'_2(x) = -0.1144 - 2(0.1)(x - 5.0) + 3(0.5144)(x - 5.0)^2$$
$$S'_2(5.2) = -0.1144 - 2(0.1)(0.2) + 3(0.5144)(0.2)^2 = -0.0927$$

(c)   To approximate $f(5.6)$ the same equation is valid,
$$S_2(x) = 3.9 - 0.1144(x - 5.0) - 0.1(x\text{-}5.0)^2 + 0.5144(x - 5.0)^3$$

$$S_2(5.6) = 3.9 - 0.1144(0.6) - 0.1(0.6)^2 + 0.5144(0.6)^3 = 3.9065$$

CUBIC SPLINE APPROXIMATION

To construct cubic splines S_j for a function $y(x)$ defined by a set of points (x_j, y_j), $j = 0, 1, 2, …, n$.

Spline: $S_j = a_j + b_j(x-x_j) + c_j(x-x_j)^2 + d_j(x-x_j)^3, x_j \le x \le x_{j+1}$

(a) <u>Natural Cubic Splines</u>, $S''_0 = 0$ and $S''_n = 0$

STEP 1   Input: x_j, y_i, i $=0, 1, 2, …, n$

STEP 2   Set $a_j = y_j$ and $h_j = x_{j+1} - x_j$, j $=0, 1, 2, …, n-1$

STEP 3   For i $=1, 2, 3, …, n-1$

          Set (right side of simultaneous equation in c_1)

$$\alpha_i = \frac{3[a_{i+1}h_{i-1} - a_i(h_i + h_{i-1}) + a_{i-1}h_i]}{h_i h_{i-1}}$$

STEP 4   Set $I_0=1$, $\mu_0=0$, $z_n = 0$, $c_n = z_n$

STEP 5   For i $=1, 2, 3, …, n-1$

          Set $I_i = 2(h_i - h_{i-1}) - h_{i-1}\mu_{i-1}$

          $\mu_i = h_i/I_i$

          $z_i = (\alpha_i - h_{i-1}z_{i-1})/I_i$

STEP 6   Set $c_j = z_j - \mu_j c_{j+1}$

$$b_j = \frac{(a_{j+1} - a_j)}{h_j} - \frac{h_j}{3}(c_{j+1} + 2c_j)$$

$$d_j = \frac{1}{3h_j}(c_{j+1} - c_j)$$

STEP 7   Output: a_j, b_j, c_j, d_j, j $= 0, 1, 2, …, n-1$

(b) <u>Clamped Cubic Splines</u>

         Only steps which are different from those of the natural cubic spline   are given below.

STEP 8   Input x_i, y_i, i $=0, 1, 2, …, n$

          $f'_0 = S'_0(x_0)$, $f'_n = S'_n(x_n)$

          $\alpha_0 = 3(a_1 - a_0)/h_0 - 3f'_0$

          $\alpha_n = 3f'_n - 3(a_n - a_{n-1})/h_{n-1}$

STEP 9   Set $I_0=2h_0$, $\mu_0=0.5$, $z_0=\alpha_0/I_0$

$I_n=h_{n-1}(2-\mu_{n-1})$

$z_n=(\alpha_n-h_{n-1}z_{n-1})/I_n$

$c_n = z_n$

PROBLEMS

1.   A calibration test on a proximity pickup gave the following results:

Displacement,	X	5	10	15
Voltage,	V	19.4	18.7	18.2

Displacement is expressed in thousands of an inch and voltage is in volts.

a)   Obtain a least square approximation of the type
$$P(x) = c_0 + c_1 x + c_2 x^2$$

b)   Obtain the Lagrange Polynomial for the given data.

2.   Find the best fit in the least squares sense, to the data.

x_i	0	1	2	3	4	5	6	7	8	9
f_i	0	2	2	5	5	6	7	7	7	10

by a polynomial of degree at most 5.

3.   Obtain the least squares solution for x and y to satisfy the following set of y equations.

(i) $x + y = 1$, $2x + 2y = 0$, $-x - y = 2$

(ii) $x + y = 1$, $2x = 0$, $-x + 3y = 2$

(Hint: the squared error $E = \sum_{i=1}^{n} [f_i(x,y)]^2$, when the individual equations are of the form

$f_i(x,y) = 0$)

4.   Find the best linear fit in the least squares sense to the data given below.

$$x_1 = 1 \quad x_2 = 2 \quad x_3 = 3$$
$$f_1 = 1 \quad f_2 = 3 \quad f_3 = 1$$

5.   Find the least squares polynomial of degree 1,2,3, and 4 for the data given.

i	0	1	2	3	4	5
x_i	0	0.15	0.31	0.5	0.6	0.75
y_i	1.0	1.004	1.031	1.117	1.2223	1422

Which degree gives the best least-squares approximation?

6. Find the least squares line that fits to the following data, assuming that the x values are free of error using polynomials of degree 1,2 and 3.

x	1	2	3	4	5	6
y	2.04	4.12	5.64	7.18	9.2	12.04

7. From the following set of data construct a function of the type $f(x) = a\ e^x + b\ e^{-x}$ using the principle of least squares.

x	0.2	0.3	0.4	0.5
f(x)	2.0	5.0	3.5	3.0

8. The stress and strain are known to follow a relation of the type
$\sigma = k_1\ \varepsilon\ \exp(-k_2\varepsilon)$
Obtain the least squares fit using the below data.

Stress (σ)	Strain (ε)
1030 psi	260×10^{-6} in/in
1410	410
1720	510
2060	710
2435	960
2750	1350

9. The relationship between resistance R, velocity v and time t is given by

$$t = \int_{v_0}^{v_1} \frac{m}{R(v)}\, dv$$

where $R(v) = -v^{3/2}$ and m=1kg, $v_0 = 10$ m/sec, $v_1 = 5$ m/sec Evaluate the integrand $f(v) = m /R(v)$ at 6 equally spaced velocities between 5 and 10 m/sec and fit the best least-squares polynomial fit.

10. In modal testing procedure to obtain the modal parameters of a vibrating system, it is necessary to fit a circle to the measured mobility values. For the given data set fit a least squares circle and obtain its center point and radius. The equation of a general circle is of the form.

$$x^2 + y^2 + ax + by + c = 0$$

x	0.0789	0.3570	60.0	8.70	0.4360	0.1200
y	2.2194	4.700	0.100	-21.00	-5.200	-2.700

11. Fit polynomials of order (i) one (ii) two (iii) three, for the given data, using the principle of least squares. Compute the corresponding square errors.

x	0.0	1.0	2.0	3.0	3.5	4.0
y	1.0	2.718	7.389	20.09	33.12	54.60

12. From the compression test of a 15 x 30 cm concrete specimen, the following data are obtained relating stress and strain, in appropriate units:

stress (σ)	1025	1710	2760	2675
strain (ε)	265	500	1360	2940

Theory shows that it may be related to the following function $\sigma = k_1 \ \varepsilon \ \exp(-k_2\varepsilon)$.
By applying the method of least squares, evaluate the constants k_1 and k_2. Use the formula to extrapolate the stress of this specimen when the strain is 800 units.
(Note: you may re-arrange the function to $(\sigma/\varepsilon) = k_1 \ e^{-k_2\varepsilon}$ before proceeding.

13. Given the date:

x	1.0	1.25	1.50
y	5.1	5.8	6.5

Find a function of the form $y = be^{ax}$ that is the best approximation to the given data in the discrete least squares sense.

14. Fit a second order polynomial using the method of least square errors for the data.

x	1.0	1.3	1.6	1.9	2.2
y	0.7652	0.6201	0.4554	0.2818	0.1104

15. Fit a Lagrange polynomial to the data
a)   (1,2.8) , (2, 7.4), (3, 20.1).
b)   (1.0, 2.0) , (2.0,4.0), (3.0, 5.5).

16. Use the Lagrange interpolating polynomial to approximate cos (0.750) using the following values

   Cos (0.698) = 0.7661
   Cos (0.733) = 0.7432
   Cos (0.768) = 0.7193

17. Fit a free cubic spline to the data given below

x	0.0	1.0	2.0	3.0	4.0
y	1.0	2.718	7.389	20.09	54.60

18. Fit a polynomial of order (i) One (ii) two and (iii) three, for the data in problem 18, using the principle of least squares. Compute the corresponding squared errors.

19. Given the data

x	1.00	1.50	2.00
y	1.90	2.30	2.63

Use least squares to find the function of the type
$$f(x) = \ln (C_0 + C_1x)$$

20.   The following table gives the function values y(x) at the corresponding x values.

x	1.0	2.0	3.0	f4.0
y	11.0	4.0	-9.0	-16.0

Approximate the function with free cubic splines.

21.   The slopes at the two end points are required to fit a clamped cubic spline to the data given below:

x	1	2	3	4	5
y	-7	-16	9	128	425

Find the cubic spline functions if $b_0=-26$ and $b_n=386$.

22.   Fit a free cubic spline for the data given in problem 12.

23.   Fit free cubic splines for the following data.

x	1.0	1.5	2.0	2.5	3.0
f(x)	1.0	0.677	.50	0.40	.333

Find values for $f'(x)$ and $f''(x)$ at $x = 1.4, 2.0$ and 2.7 using the corresponding cubic spline functions that approximate f(x).

24.   Fit free cubic spline for the following data:

x:	0.15	0.76	0.89	1.07	1.73	2.11
y:	0.3945	0.2989	0.2685	0.2251	0.0893	0.0431

25.   Construct a free cubic spline to approximate $f(x) = e^{-x}$ by using the values given by f(x) at $x = 0, .0.25, 0.75, 1$.

26.   Fit clamped cubic spline functions for the measured deflections of a centrally loaded beam with clamped ends.
The measured deflections δ are given by

x	0	0.25	0.5	0.75	1
δ	0	0.0352	0.0625	0.0352	0

Using the spline functions obtain $\delta(0.35)$, $\delta(0.7)$, $\delta'(0.35)$ and $\delta'(0.7)$. The deflections and slopes are zero at the clamped ends of the beam.

27.   The following data were obtained from an experiment which described a periodic behavior

i	0	1	2	3	4	5	6
x	0	2	5	6	7	10	12
f(x)	0	2	1.5	0	-1.5	-2	

Derive the cubic spline functions for one period.

Hint: These are neither free nor clamped cubic splines. Periodicity means that the function repeats itself, that the first point is the same as the last point, so here

$$f'(0) = f'(12) \text{ and } f''(0) = f''(12)$$

28.   Construct a general purpose computer program implementing discrete least square approximations for a given set of data. Solve problem 6 using this method.

Chapter 5

NUMERICAL INTEGRATION

5.1 INTRODUCTION

Numerical integration techniques are used when the integral of a given function is too difficult to obtain analytically. Using the numerical methods explained in this chapter, such functions can be numerically integrated in a computer to give highly accurate results.

The computer can evaluate the value of a known function, y = f(x), for specified values of x; however, a computer cannot recognize integration in the same sense that it is performed analytically. Hence, the computer can only perform integration by evaluating the values of f(x) for specified values of x and by relating them to the area under the curve between two end points, (see Fig. 5.1), given by

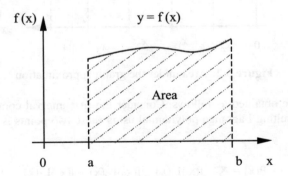

Figure 5.1: Area under the graph

$$A = \int_a^b f(x)dx$$

The curve is generated by evaluating function values at discrete values of x. In order to obtain good accuracy, it is necessary to use values of x that are very close to each other. If this is done, then function values will only change slightly between points, and a simple polynomial approximation P(x) between, say x_i and x_{i+1}, can be formed to described $f(x)$ with sufficient accuracy between these points. The area under such a polynomial will closely approximate the actual area below $f(x)$ between x_i and x_{i+1}. Computing the area between each such points for i = 0, 1, 2, ... n, and adding them will give the complete integral. The area under the curve is approximated by multiplying the function values $f(x_i)$ by an appropriate weight function C_i. For instance, the sub-interval area under the polynomial between x_i and x_{i+1} is expressed as;

$$\Delta A_i = \int_{x_i}^{x_{i+1}} f(x)dx \approx \int_{x_i}^{x_{i+1}} P(x)dx = C_i f(x_i) + C_{i+1} f(x_{i+1})$$

The weight factors C_i will depend on the degree of the polynomial approximation chosen. The complete integral is obtained by multiplying the function values $f(x_i)$ at each x_i, i = 0, 1, 2, ... n, by the corresponding weight factors C_i as shown below.

$$\int_a^b f(x)dx \approx \sum_{i=0}^{n} C_i f(x_i)$$

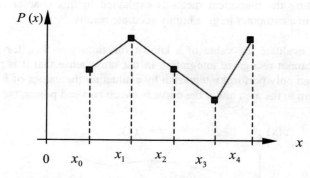

Figure 5.2: Area under the graph approximation

The weight factors C_i are obtained as follows. Consider the first interval containing the end points x_0 and x_1 (fig. 5.2). The resulting Lagrange polynomial using these two points is derived in the following way;

$$P(x) = \sum_{j=0}^{1} f(x_j)L_j(x) = f(x_0)L_0(x) + f(x_1)L_1(x)$$

where;

$$L_0(x) = \prod_{\substack{k=0 \\ k \neq j}}^{1} \frac{x - x_k}{x_j - x_k}; \quad j = 0$$

$$L_0(x) = \frac{x - x_1}{x_0 - x_1}$$

$$L_1(x) = \frac{x - x_0}{x_1 - x_0}$$

Hence $\qquad P(x) = f(x_0) \dfrac{x - x_1}{x_0 - x_1} + f(x_1) \dfrac{x - x_0}{x_1 - x_0}$

This polynomial is integrated between x_0 and x_1 to approximate the area A_1 in the first interval;

$$A_1 = \int_{x_0}^{x_1} P(x)dx = \int_{x_0}^{x_1}\left[f(x_0) \frac{x - x_1}{x_0 - x_1} + f(x) \frac{x - x_0}{x_1 - x_0} \right] dx$$

Let the denominator x_1-x_0 equal the interval h. Hence,

$$A_1 = -\frac{1}{h}\int_{x_0}^{x_1} f(x_0)(x - x_1)dx + \frac{1}{h}\int_{x_0}^{x_1} f(x_1)(x - x_0)dx$$

$$= -\frac{1}{h}\left[\frac{x^2}{2} f(x_0) - x_1 x f(x_0) \right]_{x_0}^{x_1} + \frac{1}{h}\left[\frac{x^2}{2} f(x_1) - x_0 x f(x_1) \right]_{x_0}^{x_1}$$

$$= \frac{f(x_0)}{h}\left[\left(\frac{x_1^2}{2} - x_1^2 \right) - \left(\frac{x_0^2}{2} - x_1 x_0 \right) \right] + \frac{f(x_1)}{h}\left[\left(\frac{x_1^2}{2} - x_0 x_1 \right) - \left(\frac{x_0^2}{2} - x_0^2 \right) \right]$$

$$= \frac{(x_1 - x_0)^2}{2h}\left[f(x_0) + f(x_1) \right]$$

Since $x_1 - x_0 = h$, we have

$$A_1 = \int_{x_0}^{x_1} P(x)dx = \frac{(x_1 - x_0)^2}{2h}\left[f(x_0) + f(x_1) \right] = \frac{h\left[f(x_0) + f(x_1) \right]}{2}$$

Now if this procedure is performed on the second interval between x_1 and x_2 and the interval size $x_2-x_1 = h$, the resulting area A_2 will be

$$A_2 = \int_{x_1}^{x_2} P(x)dx = \frac{h\left[f(x_1) + f(x_2) \right]}{2}$$

Continuing in this fashion, we have

$$A_i = \int_{x_{i-1}}^{x_i} P(x)dx = \frac{h\left[f(x_{i-1}) + f(x_i) \right]}{2}$$

The complete integral is given by $\displaystyle\int_a^b f(x)dx = \sum_{i=1}^{n} A_i$

In this method, the accuracy will increase when the function values are known at more points in the interval; hence there will be more sub-intervals with smaller interval size h.

When the polynomial approximation between two successive points is a line, as shown above, the method is referred to as the trapezoidal rule; when it is a quadratic expression, the method is called the Simpson's rule, and so on. These methods will be discussed in the following sections.

5.2 TRAPEZOIDAL RULE

In this method, the known function values are joined by straight lines. The area that is enclosed by these lines between the given endpoints is calculated to approximate the integral (see fig. 5.3). This is done by dividing the given range of integration into n equally spaced subintervals given by:

$$h = \frac{b-a}{n}$$

where   h is interval size
         n is number of sub-intervals,
         a and b are limits of integration with b>a.

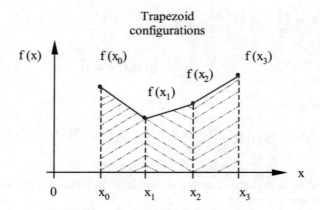

Figure 5.3: Trapezoidal rule configurations

Each subinterval with the line approximation for the function forms a trapezoid (see Fig. 5.3). The area of each trapezoid is evaluated by multiplying the interval size h by the average value of the function value in the subinterval. Once the individual trapezoidal areas are evaluated, they are added to give an overall approximation to the integral. To illustrate the method, consider three points from Figure 5.4. These points are divided into two equally sized intervals (see Fig. 5.4).

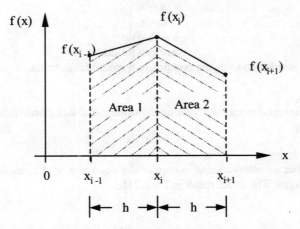

Figure 5.4: Area divided into two equal intervals

The area of each interval is approximated by;

$$\int_{x_{i-1}}^{x_i} f(x)dx \approx \frac{f(x_{i-1}) + f(x_i)}{2} * h \approx \text{ Area 1}$$

$$\int_{x_i}^{x_{i+1}} f(x)dx \approx \frac{f(x_i) + f(x_{i+1})}{2} * h \approx \text{ Area 2}$$

The total integral approximation is evaluated by summing the individual areas form x_{i-1} to x_{i+1}. Notice that the intermediate function value $f(x_i)$ is common in both area expressions.

$$\int_{x_{i-1}}^{x_{i-1}} f(x)dx \approx \frac{hf(x_{i-1})}{2} + \frac{hf(x_i)}{2} + \frac{hf(x_i)}{2} + \frac{hf(x_{i+1})}{2}$$

$$\int_{x_{i-1}}^{x_{i+1}} f(x)dx \approx \frac{h}{2}\left[f(x_{i-1}) + 2f(x_i) + f(x_{i+1})\right]$$

Since the intermediate value, $f(x_i)$, is common in both area expressions, a coefficient of 2 will appear when the terms are summed. If there are more intermediate points, they will also have coefficients of 2; however, the end points will always appear only once.

The general expression to approximate an integral for n+1 equally spaced data sets is;

$$\int_a^b f(x)dx \approx \frac{h}{2}\left[f(x_0) + 2\sum_{j=1}^{n-1} f(x_j) + f(x_n)\right] \tag{5.2.1}$$

$$a = x_0$$
$$b = x_n$$
$$h = (b-a)/n$$
a, b are end points

The above equation is referred to as the composite trapezoidal formula.

EXAMPLE 5.1

Carry out the following integral numerically using the trapezoidal composite rule.

$$I = \int_0^1 e^x dx$$

Use the following number of intervals and compare the accuracy of the result obtained in each case after rounding off to 5 digits, The exact result is I = 1.7183.

    (a) n = 1
    (b) n = 2
    (c) n = 4
    (d) n = 8

Solution

As the number of intervals used increases between 0 and 1, and interval size, h, decreases allowing for a more accurate approximation.

    (a) n = 1

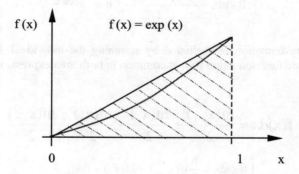

Figure 5.5: Trapezoidal approximation 1 interval

$$h = \frac{1-0}{1} = 1 \text{ and } I = \int_a^b f(x)dx \approx \frac{h}{2}\left[f(a) + f(b)\right]$$

$$I \approx \frac{1}{2}\left[e^0 + e^1\right] = 1.8591$$

Notice that for n = 1 interval, n+1=2 points are used in the approximation. The relative error is

$$\xi = \left|\frac{1.7183 - 1.8591}{1.7183}\right| = 0.0819$$

(b) n = 2

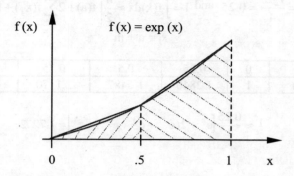

Figure 5.6: Trapezoidal approximation 2 interval

$$h = \frac{1-0}{2} = 0.5 \quad \text{and} \quad I = \int_a^b f(x)dx \approx \frac{h}{2}\left[f(a) + 2\sum_{j=1}^{1} f(x_j) + f(b)\right]$$

In this case one intermediate value exists. Form a table where values of x_j can be evaluated as

$$x_j = a + jh$$
$$x_1 = 0 + 1(0.5)$$
$$x_1 = 0.5$$

Table

x	0	0.5	1
f(x)	1	1.6487	2.7183

$$I \approx \frac{0.5}{2}\left[e^0 + 2e^{.5} + e^1\right] = 1.7539$$

The relative error is

$$\xi = \left|\frac{1.7183 - 1.7539}{1.7183}\right| = 0.0207$$

(c) n = 4

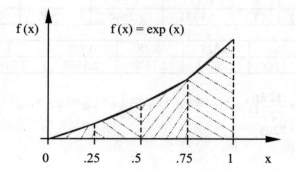

Figure 5.7 Trapezoidal approximation 4 interval

$$h = \frac{1-0}{4} = 0.25 \text{ and } I = \int_a^b f(x)dx \approx \frac{h}{2}\left[f(a) + 2\sum_{j=1}^{3} f(x_j) + f(b)\right]$$

$$x_j = a + jh$$

x	0	0.25	0.5	0.75	1
f(x)	1	1.2840	1.6487	2.1170	2.7183

$$I \approx \frac{0.25}{2}\left[e^0 + 2(e^{.25} + e^{.5} + e^{.75}) + e^1\right] = 1.7272$$

The relative error is

$$\xi = \left|\frac{1.7183 - 1.7272}{1.7183}\right| = 0.0052$$

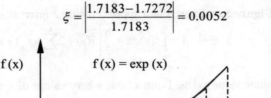

Figure 5.8: Trapezoidal approximation 8 interval

(d) n = 8

$$h = \frac{1-0}{8} = 0.125 \text{ and } I = \int_a^b f(x)dx \approx \frac{h}{2}\left[f(a) + 2\sum_{j=1}^{7} f(x_j) + f(b)\right]$$

$$x_j = a + jh$$

x	0	0.125	0.25	0.375	0.5
f(x)	1	1.1331	1.2840	1.4550	1.6487

x	0.625	0.75	0.875	1
f(x)	1.8682	2.1170	2.3989	2.7183

$$I \approx \frac{.125}{2}\left[e^0 + 2(e^{.125} + e^{.25} + e^{.375} + e^{.5} + e^{.625} + e^{.75} + e^{.875}) + e^1\right]$$

$$= 1.7205$$

The relative error is

$$\xi = \frac{1.7183 - 1.7272}{1.7183} = 0.0052$$

EXAMPLE 5.2
Evaluate the area bounded by the curve

$$f(x) = xe^{2x}$$

and the x-axis between x=0 and x=1 using the trapezoidal rule.
(a) With an interval size of h = .5.
(b) With an interval size of h = .1.

Solution
    Since the function is known, it will be possible to evaluate the function value at any point. This means that the interval size can be changed in order to increase the accuracy.
(a) The end points are a=0 and b=1.

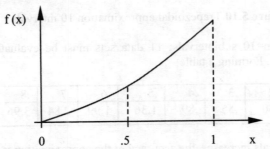

Figure 5.9 Trapezoidal approximation 2 interval

    The interval size h=0.5, hence n = (b-a)/n =(1-0)/.5 = 2 subintervals.

    Two intervals are used. Therefore three data sets are evaluated from the function, starting at a=0 and increasing by h=.5 until b=1 is reached.
with x_j = a+jh, a table is formed below.

x	0	.5	1
f(x)	0.0	1.36	7.39

Note: Number of decimal places in the answer cannot be more than those in the   data
Now applying the trapezoidal rule.

$$\int_a^b f(x)dx \approx \frac{.5}{2}[0 + 2(1.36) + 7.39] = 2.53$$

(b) The end points are a=0 and b=1.

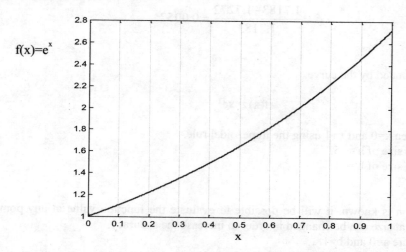

$f(x)=e^x$

Figure 5.10 Trapezoidal approximation 10 intervals

The interval size is h=0.1; n=10 subintervals; 11 data sets must be evaluated starting at a=0 and increasing by h=0.1 until b=1. Forming a table

x	0	.1	.2	.3	.4	.5	.6	.7	.8	.9	1
f (x)	0	.12	.30	.55	.89	1.36	1.99	2.84	3.96	5.44	7.39

Note: As the number of intervals increases, the accuracy of the approximation is increased.

Now applying the trapezoidal rule;

$$\int_a^b f(x)dx \approx \frac{1}{2}[0 + 2(.12) + 2(.3) + 2(.55) + 2(.89) + 2(1.36) +$$

$$2(1.99 + 2(2.84) + 2(3.96) + 2(5.44) + 7.39] = 2.12$$

The increase in accuracy produced by decreasing the interval size can be seen once the analytical solution is obtained.

$$\int_0^1 f(x)dx = \int_0^1 xe^{2x}dx$$

$$u = x \qquad dv = e^{2x}\,dx$$

This integration is performed by parts: $du = dx \qquad v = \frac{1}{2}e^{2x}$

$$\int u\,dv = uv - \int v\,du$$

$$\int_0^1 xe^{2x}\,dx = \left| x\left(\frac{1}{2}e^{2x}\right) \right|_0^1 - \int_0^1 \frac{1}{2}e^{2x}\,dx$$

$$\int_0^1 xe^{2x}\,dx = \frac{1}{2}xe^{2x} - \left(\frac{1}{4}e^{2x}\right)_0^1 = 2.09726$$

The relative error for h = 0.5 is

$$\xi = \left| \frac{2.09726 - 2.5268}{2.09726} \right| = 0.2$$

whereas the relative error for h = 0.1 is

$$\xi = \left| \frac{2.09726 - 2.1122}{2.09726} \right| = .00086$$

TRAPEZOIDAL RULE FOR NUMERICAL INTEGRATION

To obtain the integral $I = \displaystyle\int_a^b f(x)\,dx$

STEP 1   Input: a,b the limits of the integration

              n   number of intervals

STEP 2   Set h = (b-a)/n

STEP 3   Obtain $I_n = \dfrac{h}{2}\left[f(a) + f(b) + 2\displaystyle\sum_{i=1}^{n-1} f(a+ih) \right]$

STEP 4   Output I_n numerical integration value of I

5.3 SIMPSON'S RULE

In the trapezoidal rule a line approximation was made for function values between two points. In Simpson's rule, the function is approximated by a second degree polynomial between successive points. Since a second degree polynomial contains three constants, it is necessary to know three consecutive function values forming two intervals (see Fig. 5.11).

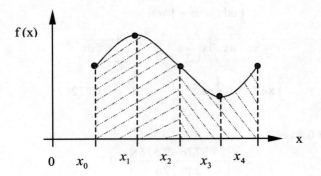

Figure 5.11: Simpson's rule approximation

From the figure it becomes obvious that an odd number of points must be used so that each parabolic arc can be fitted to a pair of intervals. The area that is enclosed by the parabolic sections between the end points of the known data is calculated to approximate the integral. The second degree polynomial in each interval pair has the form:

$$P(x) = a_0 + a_1 x + a_2 x^2$$

This polynomial is then integrated between the end points of the interval pair in order to approximate the area in the interval pair. Once the individual interval pair areas are approximated, they are summed to give an overall approximation to the integral.

Consider three equally spaced points x_0, x_1, and x_2. Since the data are equally spaced, let $h = x_{i+1} - x_i$ (see Fig. 5.12).

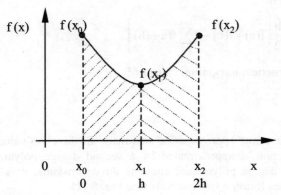

Figure 5.12: Polynomial integrated between three equally spaced points

Three equations can be formed by substituting the data sets into the quadratic polynomial form. With these three equations the coefficients a_0, a_1 and a_2 can be solved for the polynomial which will describe the curve covering two intervals $[x_0, x_1, x_2]$.
The function between x_0 and x_2 is approximated by

$$P(x) = a_0 + a_1x + a_2x^2$$

Data	**equation**	
$(0, f(x_0))$	$f(x_0) = a_0$	(1)
$(h, f(x_1))$	$f(x_1) = a_2h^2 + a_1h + a_0$	(2)
$(2h, f(x_2))$	$f(x_2) = 4a_2h^2 + 2a_1h + a_0$	(3)

Solving for a_2, a_1 and a_0 from the above three equations we get

$$a_0 = f(x_0)$$
$$a_1 = \frac{-f(x_2) + 4f(x_1) - 3f(x_0)}{2h}$$
$$a_2 = \frac{f(x_2) - 2f(x_1) + f(x_0)}{2h^2}$$

The area under the polynomial, which is an approximation to the integral is given by,

$$I = \int_{x_0}^{x_2} f(x)dx \approx \int_0^{2h} (a_2x^2 + a_1x + a_0)dx$$

$$I = \frac{8}{3}a_2h^3 + 2a_1h^2 + 2a_0h$$

Substituting for a_2, a_1, and a_0, we get

$$I \approx \frac{h}{3}[f(x_0) + 4f(x_1) + f(x_2)]$$

Notice that the interval pair is 2h units in width, and the total number of interval pairs is m=n/2. The total interval [a, b] is divided into n number of sub intervals (i.e. m number of interval pairs), where the m interval pairs are given by (x_0, x_1, x_2), (x_2, x_3, x_4), (x_4, x_5, x_6), and so on. Simpson's rule can be applied to each interval pair to get expressions for the area of each interval pair. After adding all the individual area expressions, the equation takes the form;

$$\int_a^b f(x)dx \approx \frac{h}{3}[f(x_0) + 4f(x_1) + 2f(x_2) + 4f(x_3) + 2f(x_4) + \ldots f(x_n)]$$

It can be seen that the common points x_2, x_4, x_6, ... are multiplied by a factor of 2. This happens because these points appear as the last and first in the expression for each pair of interval.

The general form of the Simpson's formula is;

$$\int_a^b f(x)dx \approx \frac{h}{3}\left[f(x_0)+4\underset{\substack{j=1\\ \text{odd}}}{\overset{n-1}{\sum}}f(x_j)+2\underset{\substack{j=2\\ \text{even}}}{\overset{n-2}{\sum}}f(x_j)+f(x_n)\right]$$

where h= (b-a)/n

EXAMPLE 5.3
Approximate the integral

$$I = \int_0^1 e^x dx$$

using Simpson's rule with n = 8 intervals. Round off the result at 4 digits.

Solution
In the case of Simpson's formula, for n intervals, there are n+1 points which must be evaluated starting at x=0 and increasing by h until x = 1.

$$h = \frac{b-a}{n} = \frac{1}{8} = .125 \qquad\qquad a = 0$$
$$b = 1$$

The data table is formed:

j	0	1	2	3	4	5	6	7	8
x	0	.125	.25	.375	.5	.625	.75	.875	1
f(x)	1	1.133	1.284	1.455	1.649	1.868	2.117	2.399	2.718

Applying Simpson's rule now gives;

$$\int_a^b f(x)dx \approx \frac{.125}{3}[1+4(1.133+1.455+1.868+2.399)$$
$$+ 2(1.284 +1.649 + 2.117) + 2.718] = 1.718$$

The analytical answer (rounded off to 4 digits) is;

$$\int_0^1 e^x dx = e^x\Big|_0^1 = e^1 - e^0 = 1.718$$

The trapezoidal rule gives; $h = \dfrac{b-a}{n} = 0.125$

$$\int_a^b f(x)dx \approx \frac{.125}{2}[1 + 2(1.133) + 2(1.284) + 2(1.455) + 2(1.649)$$
$$+ 2(1.868) + 2(2.117) + 2(2.399) + 2.718] = 1.721$$

SIMPSON'S RULE FOR NUMERICAL INTEGRATION

To obtain the integral $I = \int_a^b f(x)dx.$ Number of intervals n must be even.

STEP 1   Input: a,b the limits of the integration

n, number of intervals (n/2) interval pairs)

STEP 2   Set h = (b-a)/n

STEP 3   Obtain

$$I_n = \frac{h}{3}\left[f(a) + f(b) + 4\sum_{i=1,3,5}^{n-1} f(a+ih) + 2\sum_{i=2,4,6}^{n-2} f(a+ih) \right]$$

STEP 4   Output I_n numerical integration value of I

5.4 NEWTON-COTES FORMULAS

In the last two sections, a straight line and a parabolic polynomial approximation were made in order to approximate the integral. These polynomials could have been formed by fitting a Lagrange polynomial to the data points available in the subinterval. For instance, the trapezoidal rule uses two points for the LaGrange polynomial, Simpson's rule uses three points, and so on. In general, it is possible to fit Lagrange polynomials of any order and then integrate them to get the value of the integral. These formulas are known as the Newton-Cotes formulas.

Recall the trapezoidal rule general equation.

$$\int_a^b f(x)dx \approx \frac{h}{2}\left[f(x_0) + 2\sum_{j=1}^{n-1} f(x_j) + f(x_n) \right]$$

$$f(x_0) = f(a) \; ; \; f(x_k) = f(b)$$
$$a,b \rightarrow \text{ end points}$$

This equation could have been derived by fitting a Lagrange polynomial of degree one (straight line) to each subinterval. Each polynomial would be integrated between its respective nodes to get individual subinterval areas. These areas would then be summed to approximate the overall integral. The Lagrange polynomials are formed by using the following form of the quadrature formula;

$$P_i(x) = \sum_{j=1}^{i+1} L_j f(x_j) \qquad i = 0, 1, 2, \ldots n\text{-}1$$

Each of these polynomials is then integrated between its subinterval end points and summed to give the overall integral.

$$\int_a^b f(x)dx \approx \sum_{i=0}^{n-1} \int_{x_i}^{x_{i+1}} P_i(x)dx$$

$P_i(x) \rightarrow$ Lagrange polynomial derived by two points x_i and x_{i+1}; hence a straight line. By using more points and fitting Lagrange polynomials of higher order on them, we can derive Newton-Cotes formulas. Trapezoidal rule and Simpson's rule are Newton-Cotes formulas of first and second order, respectively.

Simpson's rule can also be derived using the Lagrange polynomial for the interval pairs. Consider the first three known data points, see Figure 5.6. Since the data are equally spaced, let $x_{i+1} - x_i = h$. Notice that interval pairs of 2h units width are used. Let there be n intervals. The Lagrange polynomials are parabolas which can be represented by;

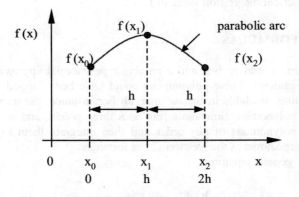

Figure 5.13: Lagrange polynomials approximation

$$P(x) = \sum_{j=1}^{i+2} f(x_j) L_j(x), i = 0,2,4,\ldots n-2$$

where;

$$L_j(x) = \prod_{\substack{k=i \\ k \neq j}}^{i+2} \frac{x - x_k}{x_j - x_k}, j = i, i+1, i+2$$

For the first polynomial; i=0

$$P(x) = \sum_{j=0}^{2} f(x_j)L_j(x) = f(x_0)L_0(x) + f(x_1)L_1(x) + f(x_2)L_2(x)$$

where; j = 0

$$L_0(x) = \prod_{\substack{k=0 \\ k \neq 0}}^{2} \frac{x - x_k}{x_j - x_k} = \frac{(x - x_1)}{(x_0 - x_1)} * \frac{(x - x_2)}{(x_0 - x_2)}$$

j = 1

$$L_1(x) = \prod_{\substack{k=0 \\ k \neq 1}}^{2} \frac{x - x_k}{x_j - x_k} = \frac{(x - x_0)}{(x_1 - x_0)} * \frac{(x - x_2)}{(x_1 - x_2)}$$

j = 2

$$L_2(x) = \prod_{\substack{k=0 \\ k \neq 2}}^{2} \frac{x - x_k}{x_j - x_k} = \frac{(x - x_0)}{(x_2 - x_0)} * \frac{(x - x_1)}{(x_2 - x_1)}$$

The first polynomial becomes;

$$P(x) = \frac{(x - x_1)}{(x_0 - x_1)} * \frac{(x - x_2)}{(x_0 - x_2)} f(x_0) + \frac{(x - x_0)}{(x_1 - x_0)} * \frac{(x - x_2)}{(x - x_2)} f(x_1) +$$

$$\frac{(x - x_0)}{(x_2 - x_0)} * \frac{(x - x_1)}{(x_2 - x_1)} f(x_2)$$

Since the data are equally spaced, the denominator in the polynomial expressions can be replaced by multiples of h. The polynomial becomes;

$$P(x) = \frac{(x - x_1)(x - x_2)}{2h^2} f(x_0) - \frac{(x - x_0)(x - x_2)}{h^2} f(x_1)$$

$$+ \frac{(x - x_0)(x - x_1)}{2h^2} f(x_2)$$

This polynomial is integrated between x_0 and x_2 to approximate the area in the first interval pair,

$$\int_{x_0}^{x_2} P(x)dx = \frac{f(x_0)}{2h^2} \int_{x_0}^{x_2} (x - x_1)(x - x_2)dx - \frac{f(x_1)}{h^2} \int_{x_1}^{x_2} (x - x_0)(x - x_2)dx$$

$$+ \frac{f(x_2)}{2h^2} \int_{x_0}^{x_2} (x - x_0)(x - x_1)dx$$

This becomes;

$$\int_{x_0}^{x_2} P(x)dx = \frac{h}{3}\left[f(x_0) + 4f(x_1) + f(x_2)\right] = A_1$$

If the area in the second interval pair is approximated, the result is;

$$\int_{x_0}^{x_4} P(x)dx = \frac{h}{3}\left[f(x_2)+4f(x_3)+f(x_4)\right] = A_2$$

and so on.

Adding the areas for all interval pairs gives an expression for the overall integral from x_0 to x_n.

$$\int_{x_0}^{x_n} f(x)dx = \frac{h}{3}\left[f(x_0)+4f(x_i)+2f(x_2)+4f(x_3)+2f(x_4)+...f(x_n)\right]$$

Designating the end point x_0 and x_n as a and b, the general form of the Simpson's formula becomes;

$$\int_{x_0}^{x_n} f(x)dx \approx \frac{h}{3}\left[f(a)+4\sum_{\substack{j=1\\odd}}^{n-1}f(x_j)+2\sum_{\substack{j=2\\even}}^{n-2}f(x_j)+f(b)\right]$$

where; $h = \dfrac{b-a}{n}$;n =number of intervals; n = even

Simpson's 3/8 rule uses four points, having three intervals, to approximate a cubic Lagrange polynomial in the interval. This polynomial is then integrated in the same fashion as outlined in the trapezoidal and Simpson's rule in order to get the following expression for three intervals.

$$\int_{x_0}^{x_3} f(x)dx = \frac{3h}{8}\left[f(x_0)+3f(x_1)+3f(x_2)+f(x_3)\right]$$

where $h = \dfrac{x_3 - x_0}{3}$

The general composite rule for the overall interval [a,b] consists of adding the area in each group having three intervals to give a general equation;

$$\int_{a}^{b} f(x)dx \approx \frac{3h}{8}\left[f(x_0)+3\sum_{\substack{j=1\\j\neq3,6,9...}}^{n-1}f(x_j)+2\sum_{j=3,6,9,...}^{n-3}f(x_j)+f(x_n)\right]$$

$$h = \frac{b-a}{n}$$

Newton-Cotes formulas can be derived by fitting a Lagrange polynomial by considering 5,6,7 or more interval groups. The accuracy of the approximation will increase as the degree of the polynomial is increased.

Error in Trapezoidal and Simpson's Rules

The error involved can be obtained by considering the Lagrange polynomial approximation for the function to be integrated given by

$$f(x) \approx P(x) + \frac{f^{n+1}(\xi\xi)}{(n+1)!} \prod_{i=0}^{n} (x - x_i)$$

$$\underbrace{\hspace{3cm}}_{\text{error term}}$$

with $x_0 \le \xi \le x_n$.

In the case of the trapezoidal rule, only two points are used to evaluate the Lagrange polynomial; hence the second derivative term of the expansion is the largest unused term. This will represent the error involved in the approximation.

Trapezoidal rule:

Let $\qquad\qquad\qquad\qquad\qquad\qquad\qquad\qquad x - x_0 = k$

$$E = \frac{f''(\xi\xi)}{2}(x - x_0)(x - x_1) \qquad \text{to simplify} \qquad \begin{aligned} dx &= dk \\ x\text{-}x_1 &= x\text{-}x_0 + x_0\text{-}x_1 \\ x\text{-}x_1 &= k\text{-}h \end{aligned}$$

Substituting values

$$E = \frac{f''(\xi\xi)}{2}(k^2 - kh)$$

The error integral becomes;

$$\int_a^b E dx = \int_0^h \frac{f''(\xi)}{2}(k^2 - kh) dk \qquad \text{at} \quad a = x_0 \to k = 0$$

$$b = x_1 \to k = h$$

$$\int_a^b E dx = \frac{f''(\xi)}{2}\left[\frac{k^3}{3} - \frac{k^2 h}{2} \right]_0^h = -\frac{f''(\xi)h^3}{12}$$

For the trapezoidal rule

$$\int_a^b E\,dx = -\frac{f''(\xi)h^3}{12}$$

For Simpson's rule

$$\int_a^b E\,dx = \frac{-h^5}{90}f^4(\xi)$$

For Simpson's 3/8 rule

$$\int_a^b E\,dx = \frac{-3h^5}{90}f^4(\xi)$$

5.5 ROMBERG INTEGRATION

Romberg integration uses a successive error reduction technique. It uses the trapezoidal rule with different interval sizes, in order to get some preliminary approximations to the integral to start with. The trapezoidal rule is the simplest to apply to obtain the integration of a function; however, the error in the trapezoidal rule is quite significant. The Romberg integration method takes the preliminary approximations obtained by the simple trapezoidal rule, and then applies the Richardson extrapolation procedure which refines these values successively to a single more accurate approximation. The initial approximations are found by applying the trapezoidal rule starting with 1, 2, 4, 8, ... 2^{k-1} intervals within the range [a,b]. The corresponding interval sizes, h_k are evaluated using the following equation.

$$h_k = \frac{b-a}{2^{k-1}} \qquad k = 1, 2, \ldots m$$

The value of m depends on the required accuracy. As the value of m increases, so does the accuracy of the integral approximation. The Richardson extrapolation procedure is carried out on the trapezoidal rule approximations at the different interval sizes in order to converge on a more accurate approximation quickly.

The procedure used to evaluate the initial approximations can be represented in the following form. Consider the case in which it is desired to integrate a function f (x) in the region [a, b].

Using the trapezoidal rule, with one interval between a and b,

$$R_{11} = h_1 \frac{[f(a)+f(b)]}{2} = (b-a)\frac{[f(a)+f(b)]}{2} \quad (k=1)$$

where;

$$h_1 = \frac{b-a}{2^0} = b-a$$

This is the crudest approximation with a straight line joining the end points a and b. The notation R_{11} denotes the first approximation of the primary value. The second approximation is more accurate since two subintervals are used. This has the form

$$R_{21} = \frac{h_2}{2}[f(a) + 2f(a+h_2) + f(b)] = \frac{(b-a)}{4}[f(a) + 2f(a+h_2) + f(b)]$$

where;

$$h_2 = \frac{b-a}{2^1} = \frac{b-a}{2} \qquad (k = 2)$$

Notice $h_2 = \frac{1}{2} h_1$. R_{21} can be put in terms of R_{11} if h_2 is replaced by $\frac{h_1}{2}$ in the R_{21} expression.

$$R_{21} = \frac{h_1}{4}\left[f(a) + 2f\left(a + \frac{1}{2}h_1\right) + f(b) \right]$$

$$R_{21} = \frac{1}{2}\left[\frac{h_1}{2}(f(a) + f(b)) + h_1 f\left(a + \frac{1}{2}h_1\right) \right]$$

$$R_{21} = \frac{1}{2}\left[R_{11} + h_1 f\left(a + \frac{1}{2}h_1\right) \right]$$

The third approximation uses four subintervals of size

$$h_3 = \frac{b-a}{2^2} = \frac{b-a}{4}$$

$$R_{31} = \frac{h_3}{2}\left[f(a) + 2f(a+h_3) + 2f(a+2h_3) + 2f(a+3h_3) + f(b) \right]$$

Similarly, R_{31} can be represented in terms of R_{21}.

$$h_3 = \frac{1}{2}h_2$$

$$R_{31} = \frac{1}{4}h_2\left[f(a) + 2f\left(a + \frac{1}{2}h_2\right) + 2f(a+h_2) + 2f\left(a + \frac{3}{2}h_2\right) + f(b) \right]$$

$$R_{31} = \frac{1}{2}\left[\frac{1}{2}h_2(f(a) + 2f(a+h_2) + f(b)) + h_2 f\left(a + \frac{1}{2}h_2\right) + h_2 f\left(a + \frac{3}{2}h_2\right) \right]$$

$$R_{31} = \frac{1}{2}\left[R_{21} + h_2\left(f\left(a + \frac{1}{2}h_2\right) + f\left(a + \frac{3}{2}h_2\right) \right) \right]$$

Similarly

$$R_{41} = \frac{1}{2}\left[R_{31} + h_3\left(f\left(a + \frac{1}{2}h_3\right) + f\left(a + \frac{3}{2}h_3\right) + f\left(a + \frac{5}{2}h_3\right) + f\left(a + \frac{7}{2}h_3\right) \right) \right]$$

The general equation which describes this procedure is:

$$R_{k1} = \frac{1}{2}\left[R_{k-1,1} + h_{k-1} \sum_{i=1}^{2^{k-2}} f(a + (i - \frac{1}{2})h_{k-1}) \right]$$

$$\text{for } k = 2, 3, \dots m$$

Notice that this equation starts at k=2. The first approximation is always evaluated as;

$$R_{11} = \frac{b-a}{2} \, [f(a)+f(b)]$$

The Richardson extrapolation method can now be carried out on these primary approximation values $R_{11}, R_{21}, R_{31}, \dots R_{n1}$.

The Richardson extrapolation method will extrapolate between the m primary approximations to give m-1 secondary approximations. These secondary values are then extrapolated to give m-2 approximations, and so on until the m-(m-1)th approximation is found. This single value is taken as the final approximation to the integral.

Richardson's Extrapolation

Richardson's extrapolation procedure identifies the nature of the error in the trapezoidal approximations and systematically eliminates these errors.

The error in the trapezoidal rule is expressed as follows. Consider the trapezoidal rule applied over the interval [-h/2, h/2]. Then

$$I = \int_{-h/2}^{h/2} f(x)dx = \frac{h}{2}\left[f(-\frac{h}{2}) + f(\frac{h}{2}) \right] + E \tag{1}$$

where E is the error. Expanding f(x) in a Taylor series about 0, we get

$$f(x) = f_0 + xf_0' + \frac{x^2}{2!} f_0'' + \dots \tag{2}$$

where $f_0 = f(O)$. Consequently,

$$f(-\frac{h}{2}) = f_0 - \frac{h}{2} f_0' + \frac{(h/2)^2}{2!} f_0'' - \frac{(h/2)^3}{3!} f_0''' + \frac{(h/2)^4}{2!} f_0'''' - \dots$$

$$f(\frac{h}{2}) = f_0 + \frac{h}{2} f_0' + \frac{(h/2)^2}{2!} f_0'' + \frac{(h/2)^3}{3!} f_0''' + \frac{(h/2)^4}{2!} f_0'''' + \dots$$

Substituting these into equation (1) we get

$$I = h\left[f_0 + \frac{(h/2)^2}{2!} f_0'' + \frac{(h/2)^4}{4!} f_0'''' + \dots \right] + E \tag{3}$$

Integrating equation (2) directly, we get

$$I = \int_{-h/2}^{h/2} (f_0 + xf_0' + \frac{x^2}{2!} f_0'' + ...) dx$$

$$= \left(xf_0 + \frac{x^2}{2!} f_0' + \frac{x^3}{3!} f_0'' + ... \right)_{-h/2}^{h/2}$$

$$= h\left[f_0 + \frac{(h/2)^2}{3!} f_0'' + \frac{(h/2)^4}{5!} f_0'''' + ... \right] \qquad (4)$$

From equations (3) and (4) we get the error as

$$E = -h\left[\frac{h^2}{12} f_0'' + \frac{h^4}{480} f_0'''' + ... \right]$$

$$= c_1 h^2 + c_2 h^4 + c_3 h^6 + c_4 h^8 + ... \qquad (5)$$

We assume that as we go on doubling the number of intervals successively the form of the error remains the same and the constants c_i remain reasonably constant. Corresponding to 2^{k-1} number of intervals, let the error be

$$E_k = c_1 h_k^2 + c_2 h_k^4 + c_3 h_k^6 + ...$$

The integral is

$$I = R_{k,1} + c_1 h_k^2 + 0(h_k^4) \qquad (6)$$

$$= R_{k-1,1} + c_1 h_{k-1}^2 + 0(h_k^4) \qquad (7)$$

where $0(h_k^4)$ indicates error of order h_k^4.

From equations (6) and (7) we get

$$C_1 = \frac{R_{k,1} - R_{k-1,1}}{h_{k-1}^2 - h_k^2}$$

Substituting this in equation (6)

$$I = R_{k,1} + \frac{R_{k,1} - R_{k-1,1}}{(h_{k-1}/h_k)^2 - 1} + 0(h_k^4)$$

Since $(h_{k-1}/h_k) = 2$, we get

$$I = R_{k,1} + \frac{R_{k,1} - R_{k-1,1}}{4-1} + 0(h_k^4)$$

$$= \frac{4R_{k-1,1} - R_{k-1,1}}{4-1} + O(h_k^4)$$

$$= R_{k,2} + O(h_k^4)$$

The extrapolation procedure can now be continued to eliminate c_2 as follows

$$I = R_{k,2} + c_2 h_k^4 + O(h_k^6) \tag{8}$$

$$= R_{k-1,2} + c_2 h_{k-1}^4 + O(h_k^6) \tag{9}$$

The constant c_2 is eliminated between equations (8) and (9) as before.
The general form of the Richardson extrapolation scheme is represented by the following.

$$R_{ij} = \frac{4^{j-1} R_{i,j-1} - R_{i-1,j-1}}{4^{j-1} - 1} \qquad i = 2, 3, \ldots m; j = 2, 3, \ldots i$$

Notice that i values start at 2 since the primary values must already be known. The following table can be formed to represent the extrapolation stages.

The results R_{11}, R_{22}, R_{33},... R_{mm} will gradually converge towards the exact result as m goes to ∞. m is chosen so as to limit the relative error.

$$\frac{|R_{mm} - R_{m-1,m-1}|}{|R_{mm}|} < \xi$$

EXAMPLE 5.4
Approximate to integral of the following equation between $x_o = 0$ and $x_n = 1$, using Romberg integration with accuracy of n=8 intervals. Round off result to 6 digits.

$$f(x) = \int_0^1 xe^{-x}\,dx$$

Solution

The maximum number of intervals using the trapezoidal rule is n=8, or n=2^{m-1}. This means that m=4. The primary values are,

$$R_{11} = \frac{b-a}{2}[f(a)+f(b)]$$

$$a = x_0 = 0$$

$$b = x_n^0 = 1$$

$$R_{11} = \frac{1}{2}\left[0e^{-0}+1.e^{-1}\right] = .183940$$

k = 2;

$$R_{21} = \frac{1}{2}\left[R_{11} + h_1 \sum_{i=1}^{1} f(a+\frac{h_1}{2})\right]$$

where;

$$h_1 = \frac{b-a}{2^0} = 1$$

$$R_{21} = \frac{1}{2}[.183940+1f(.5)] = 0.243603$$

k = 3;

$$R_{31} = \frac{1}{2}\left[R_{21} + h_2 \sum_{i=1}^{2} f(a+(i-\frac{1}{2})h_2)\right]$$

where;

$$h_2 = \frac{b-a}{2^1} = \frac{1}{2}$$

$$R_{31} = \frac{1}{2}\left[0.243603+ \frac{1}{2}\{f(0.25)+f(0.75)\}\right] = 0.259045$$

k = 4;

$$R_{41} = \frac{1}{2}\left[R_{31} + h_3 \sum_{i=1}^{4} f(a+i\frac{1}{2})h_3)\right]$$

where;

$$h_3 = \frac{b-a}{2^2} = \frac{1}{4}$$

$$R_{41} = \frac{1}{2}\left[0.259045+ \frac{1}{4}\{f(1.25)+f(0.375)+f(.625)+f(.875)\}\right]$$

$$R_{41} = 0.262940$$

The primary values are;

$$R_{11} = 0.183940$$

$$R_{21} = 0.243603$$
$$R_{31} = 0.259045$$
$$R_{41} = 0.262938$$

Now the Richardson extrapolation is carried out on these values.

The second approximation is obtained by interpolating the primary values.

$$R_{ij} = \frac{4^{j-1} R_{i,j-1} - R_{i-1,j-1}}{4^{j-1} - 1}$$

$i = 2, j = 2 \quad R_{22} = \dfrac{4^{(1)} R_{21} - R_{11}}{4^{(1)} - 1} = \dfrac{4(.243603) - 0.183940}{3} = 0.263490$

$i = 3, j = 2 \quad R_{32} = \dfrac{4^{(1)} R_{31} - R_{21}}{4^{(1)} - 1} = \dfrac{4(.259045) - 0.243603}{3} = 0.264193$

$i = 4, j = 2 \quad R_{42} = \dfrac{4^{(1)} R_{41} - R_{31}}{4^{(1)} - 1} = \dfrac{4(.262940) - 0.259045}{3} = 0.264238$

The third approximation is obtained interpolating the second approximations.

$i = 3, j = 3 \quad R_{33} = \dfrac{4^{(2)} R_{32} - R_{22}}{4^{(2)} - 1} = 0.264239$

$i = 4, j = 3 \quad R_{43} = \dfrac{4^{(2)} R_{42} - R_{32}}{4^{(2)} - 1} = \dfrac{16(0.264193) - 0.263490}{15} = 0.264241$

The fourth and final approximation is obtained by interpolating between two values from in the last approximation.

$i = 4, j = 4 \quad R_{44} = \dfrac{4^{(3)} R_{43} - R_{33}}{4^{(3)} - 1} = \dfrac{64(0.264241) - 0.264239}{63} = 0.264241$

This is the best approximation for n=4

In tale form the values are displayed in the following form;

$R_{11} = 0.183940$

$R_{21} = 0.243603 \qquad R_{22} = 0.263490$

$R_{31} = 0.259045 \qquad R_{32} = 0.264193 \qquad R_{33} = 0.264239$

$R_{41} = 0.262940 \qquad R_{42} = 0.264238 \qquad R_{43} = 0.264241 \qquad R_{44} = 0.264241$

The best answer is $R_{44} = 0.264241$

$$\xi = \text{relative error} = \left| \frac{R_{44} - R_{33}}{R_{44}} \right| = \left| \frac{0.264241 - 0.264239}{0.264241} \right| = 0.0000065$$

ROMBERG INTEGRATION

To obtain the integral $I = \int_a^b f(x)dx$

STEP 1   Input: a, b the limits of the integration

              ξ relative error

              m final approximation number

STEP 2   Set $R_{11} = \dfrac{(b-a)}{2}[f(a)+f(b)]$

STEP 3   Set k = 2

STEP 4   Set $h_{k-1} = \dfrac{(b-a)}{2^{k-2}}$

STEP 5   Set $R_{k1} = \dfrac{1}{2}\left[R_{k-1,1} + h_{k-1}\sum_{i=1}^{2^{k-2}} f(a+[i-0.5]h_{k-1})\right]$

STEP 6   For j = 2, 3 ... k

              Set $R_{kj} = \dfrac{4^{j-1}R_{k,j-1} - R_{k-1,j-1}}{4^{j-1}-1}$

STEP 7   If $\dfrac{\left|R_{k,k} - R_{k-1,k-1}\right|}{\left|R_{k,k}\right|} < \xi$ then go to 10

STEP 8   Set k = k + 1

STEP 9   If k < m then go to 4

STEP 10   Output: $R_{kk} \approx I$

5.6 GAUSS QUADRATURE

In all the previous methods, the function to be integrated was approximated by a simple polynomial that was formed using a specified amount of successive points. These successive points were equally spaced for convenience. When the simple polynomial was a line (2 points,) the method was known as the trapezoidal rule. When the polynomial was a quadratic (3 points), the method was known as the Simpson's rule. When higher order polynomials were used, the area expressions formed were known as Newton-Cotes formulas. When the degree of the polynomial approximation is higher, the accuracy of the integral approximation is better. When the interval [a, b] is subdivided into n subintervals, there are n+1 values of x_i, i = 0, 1, 2, ... n. Knowing the function values at these points, it is possible to fit a polynomial of degree n; hence a Newton-Cotes formula for an nth degree polynomial approximation can be used to obtain the best accuracy.

The trapezoidal and Simpson's rules can be written as the quadrature formula;

$$\int_a^b f(x)dx \approx \sum_{i=0}^n C_i f(x_i)$$

where the values of x_i were chosen conveniently to be equally spaced and $x_0 = a$, $x_n = b$. Once chosen, these values were used to evaluate the function values $f(x_i)$. In this case, (n+1) values of C_i are left to evaluate. In the case of Gaussian quadrature, the points x_i are not equally spaced and hence the values of x_i and C_i must be evaluated. If we choose n such points x_i within the interval [a, b] (or $a < x_i < b$), i = 1, 2, ... n, the quadrature formula is

$$\int_a^b f(x)dx = \sum_{i=1}^n C_i f(x_i) \qquad (5.6.1)$$

Hence there are 2n unknown quantities on the right hand side of the quadrature formula; namely n x_i's and n C_i's. The values of x_i are not chosen in an arbitrary fashion. They are now selected in such a way that they will minimize the error of the approximation. It becomes obvious that if these values of x_i are selectively chosen, then the function f (x) must be known in order to evaluate the function values at these specific points.

Since there are now 2n unknown quantities on the right hand side, the polynomial which approximates f (x) on the left hand side will be approximated exactly when it is of degree 2n−1 or less.

In order to select these optimum values of x_i the following scheme should be used.

This scheme is derived in the following way; notice that the approximating polynomial P (x) must be a polynomial of at most degree 2n-1 with 2n unknown coefficients;

$$f(x) \approx P(x) = \sum_{i=0}^{2n-1} a_i x^i \quad \text{2n unknown, coefficients } a_i$$

If this approximating polynomial of degree 2n-1 or less, is divided by another polynomial $P_n(x)$, of degree n, the result will be;

$$\frac{P(x)}{P_n(x)} = Q(x) + \frac{R(x)}{P_n(x)}$$

where $\dfrac{R(x)}{P_n(x)}$ is a remainder term

In the above Q(x) and R(x) are polynomials of degree (n-1) or less. Eqn. 5.6.2 can be expressed as

Eqn. 5.6.2 $P(x) = Q(x)P_n(x) + R(x)$

The approximation can now be found by;

$$\int_{-1}^{1} P(x)dx = \int_{-1}^{1} Q(x).P_n(x)dx + \int_{-1}^{1} R(x)dx$$

where $P_n(x)$ are chosen as orthogonal polynomials. Since $Q(x)$ is a polynomial of degree less than n, we can express $Q(x)$ in terms of orthogonal polynomials P_{n-1}, P_{n-2}, ... as $Q(x) = b_{n-1}P_{n-1}(x) + b_{n-2}P_{n-2}(x) + ...b_0$.

Hence the term $\int_{-1}^{1} Q(x)P_n(x)dx$ will be zero because of the following property of the orthogonal polynomials.

$$\int_{-1}^{1} P_n(x).P_m(x).w(x)dx = \beta_n; \quad \text{if } m = n$$

$$= 0; \quad \text{if } m \neq n$$

where P_n and P_m are orthogonal
This reduces the approximation to:

$$\int_{-1}^{1} P(x)dx = \int_{-1}^{1} R(x)dx$$

It is possible to get a quadrature formula for $\int R(x)dx$ in the form

$$\int_{-1}^{1} R(x)dx = \sum_{i=1}^{n} C_i R(x_i)$$

If the values of x_I are chosen as the roots of the orthogonal polynomial P_n, then from eqn. 5.6.2.

$$P(x_i) = Q(x_i)P_n(x_i) + R(x_i)$$
$$P(x_i) = R(x_i), \text{ since } P_n(x_i) = 0$$

Hence the approximation takes the form

$$\int_{-1}^{1} P(x)dx = \sum_{i=1}^{n} C_i P(x_i) \qquad\qquad (5.6.3)$$

where x_i are the roots of the orthogonal polynomial $P_n(x)$. The above approximation will be exact when the polynomial $P(x)$ is of degree 2n-1 or less.

Legendre polynomials are a set of orthogonal polynomials, which are used here. A table of roots of Legendre polynomial equations and the corresponding coefficients C_i are given in table 5.1 at the end of the section. The actual derivation of this polynomial is complicated, therefore only the formula to generate such polynomials will be given.

The first and second Legendre polynomials are:

$$P_0 = 1$$
$$P_1 = x$$

By knowing these, the following recursion formula can be used to get higher order Legendre polynomials;

$$P_{n+1}(x) = \frac{(2n+1) * x * P_n(x) - nP_{n-1}(x)}{(n+1)} \qquad n = 1, 2, \ldots \infty$$

These polynomials only exist between -1 and $+1$ with all their roots lying in this interval; (see fig. 5.7).

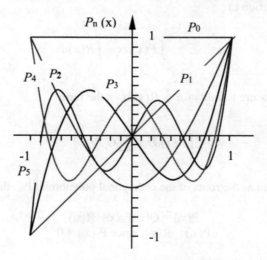

Figure5.14: Legendre polynomials

<u>Legendre polynomials</u>
$P_0(x) = 1$
$P_1(x) = x$
$P_2(x) = (3x^2-1/2$

$P_3(x) = (5x^3-3x)/2$
$P_4(x) = (35x^4-30x^2+3)/8$
$P_5(x) = (63x^5-70x^3+15x)/8$
$\vdots$

$P_\infty(x)$

The values of C_i in equation 5.6.3 are evaluated by solving the following n simultaneous equations in n unknowns $C_1, C_2, \ldots C_n$.

$$\beta_0 = \int_{-1}^{1} P_0^2(x)dx = \sum_{i=1}^{n} C_i P_{n-1}^2(x_i)$$

$$\vdots \qquad\qquad\qquad\qquad\qquad\qquad (5.6.4)$$

$$\beta_{n-1} = \int_{-1}^{1} P_{n-1}^2(x)dx = \sum_{i=1}^{n} C_i P_{n-1}^2(x_i)$$

It can be seen that eqns. (5.5) are obtained by applying eqn. (5.4) on polynomials $P(x) = P_i^2(x)$, i=0,1,2...,n-1. The coefficients C_i are solved and tabulated along with roots of Legendre Polynomials in table 5.1.

The quadrature formula can now be applied to f(x) to approximate the integral; however if the original interval of the integral is not [-1,1], then a transformation must be performed in order to convert the interval [a, b] to [-1,1]. This will be illustrated in the first of the following examples.

EXAMPLE 5.5
Transform the interval of the integral

$$\int_{0}^{1} xdx \text{ to } [-1,1].$$

Solution

The goal is to transform the function f(x)=x in such a away that the area below it in the region [0,1] will be the same as the area below the transformation in the interval [-1,1].

Any function f (x) can be transformed between −1, 1 by a linear transformation, into a new variable t as

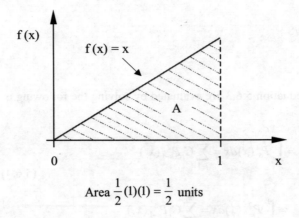

$$\text{Area } \frac{1}{2}(1)(1) = \frac{1}{2} \text{ units}$$

Figure 5.15

$$t = mx + k \qquad\qquad (5.5.5)$$

where   t = -1 for x = a and   t = 1 for x = b.

Hence, two equations can be formed, as follows:

$$-1 = ma + k$$
$$1 = mb + k$$

a and b are the old limits a=0, b=1. m and k can be solved for and replaced in equation(5.5.5) to give x in terms of a new variable k between −1 and 1.

Solving the two equations gives

$$\begin{array}{ll} -1 = ma + k & \\ \underline{1 = mb + k} & \text{(subtracting)} \\ -2 = ma - mb & \end{array}$$

$$m = \frac{-2}{a-b} \text{ and } k = 1 + \frac{2b}{a-b}$$

Putting these values in equation (5.5.5) gives

$$t = \left[\frac{1}{b-a}(2x - a - b) \right]$$

Now putting a=0 and b=1 into the equation gives;

$$t = 2x - 1 \quad \text{or} \quad x = \frac{t+1}{2}$$

The value of dx is also required.

$$dx = \frac{1}{2} dt$$

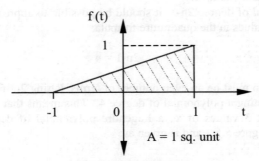

Figure 5.16: Unit area

Replacing this into the original integral gives;

$$\int_0^1 x\,dx = \int_{-1}^1 \frac{t+1}{2} \cdot \frac{1}{2} dt = \int_{-1}^1 \frac{t+1}{4} dt$$

Now the function f(x) is transformed in terms of the variable t as f(t) to exist between −1 and 1, hence the Gaussian quadrature method can now be used to approximate the integral numerically.

The transformation used to transform any function f(x) in the interval [a,b] to a suitable function f(t) in the interval [−1,1] is:

$$x = \frac{1}{2}[(b-a)t + a + b]; dx = \frac{b-a}{2} dt$$

$$t = \frac{1}{(b-a)}(2x - a - b); dt = \frac{2}{(b-a)} dx$$

Recall that if f(t) is a polynomial of degree 2n-1, then it could be approximated exactly by 2n terms on the right hand side in the Gaussian quadrature formula below:

$$\int_{-1}^1 P(t)\,dt \approx \sum_{i=1}^n C_i P(t_i)$$

EXAMPLE 5.6

Approximate the integral of the following polynomial between −1 and 1 using the Gaussian quadrature method.

$$f(x) = x^4$$

Solution
(This example is worked out in great detail. However, any given problem can be solved in a simple manner as shown in the next example.)
Since f(x) is a polynomial of degree 2n-1, it should be possible to approximate its integral between −1 and 1 exactly, using 2n values in the quadrature formula:

$$2n-1 = 4$$

and 2n=5 terms. Since 2n must be an even number of terms, letting 2n=6 or n=3 will result in an exact solution when approximating a polynomial of degree 4. This means that 3 values of x_i and 3 values of C_i are required. To get 3 values of x_i, a Legendre polynomial of degree 3 must be solved. This polynomial is shown in figure 5.5.1 and is given as;

$$n=3 \quad P_n(x) = P_3(x) = \frac{1}{2}(5x^3 - 3x)$$

The x_i values are the roots of this equation. These are stated in table 5.5.1 at the end of this section as roots:

$$x_1 = 0.774596669$$
$$x_2 = 0$$
$$x_3 = 0.774596669$$

Now following the procedure, the polynomial $P(x) = x^4$ is divided by the appropriate Legendre polynomial $P_3(x) = \frac{1}{2}(5x^3 - 3x)$ giving:

$$\frac{P(x)}{P_n(x)} = Q(x) + \frac{R(x)}{P_n(x)}$$

$$\frac{2}{5}x$$

$$\frac{5}{2}x^3 - \frac{3}{2}x \overline{\smash{\big)}\ x^4}$$

$$\underline{x^4 - \frac{3}{5}x^2}$$

$$\frac{3}{5}x^2 \quad \begin{matrix} R(x) \\ \text{Remainder term} \end{matrix}$$

$$\frac{P(x)}{P_n(x)} = \frac{2}{5}x + \frac{\frac{3}{5}x^2}{\frac{1}{2}(5x^3 - 3x)}$$

Multiplying both sides by $P_3(x)$ gives.

$$P(x) = \frac{x}{5}(5x^3 - 3x) + \frac{3}{5}x^2$$

Now using the roots of $P_n(x)$ from Table 5.4.1 as values for this function the values of $p(x_i)$ for the quadrature formula are approximated.

$$P(x_1) = P(0.774596669) = 0 + \frac{3}{5}(0.774596669)^2 = 0.36$$
$$P(x_2) = P(0) = 0$$
$$P(x_3) = P(-0.774596669) = 0 + \frac{3}{5}(-0.774596669)^2 = 0.36$$

Notice that $(0.774596669)^4 = 0.36$, therefore the approximation is exact. Using the same roots, the values of C_i can be evaluated using equation 5.4.4. These values are also given in Table 5.5.1 as

$$C_1 = 0.5555555556$$
$$C_2 = 0.8888888889$$
$$C_3 = 0.5555555556$$

The quadrature formula becomes;

$$\sum_{i=1}^{3} C_i f(x_i) = .5555555556*.36 + .8888888889*0 + .5555555556*.36 = .4$$

This is equal to the exact result given as

$$\int_{-1}^{1} x^4 dx = \left(\frac{x^5}{5}\right)_{-1}^{1} = \frac{1}{5} - \left(-\frac{1}{5}\right) = 0.4$$

This example has proven that the integral of a polynomial can be approximated exactly using this method; however, it is seldom desired to approximate such an integral due to the ease with which it can be evaluated analytically. Another example using this method is illustrated next.

EXAMPLE 5.7
Using Gaussian quadrature, approximate the following integral. Round off the result at 7 digits.

$$I = \int_0^1 e^x dx$$

Solution

The first step is to transform the function so that it will yield the same result in the interval [-1, 1]. Recall the transformation equations derived earlier

$$t = \frac{1}{(b-a)}(2x - a - b)$$

In this case substitute a = 0 and b = 1

$$t = 2x - 1 \text{ or } x = \frac{t+1}{2} \text{ and } dx = \frac{1}{d} dt$$

Substituting these parameters for x and dx and with n = 2, we get

$$\int_0^1 e^x dx = \int_{-1}^1 \frac{1}{2} e^{(\frac{t+1}{2})} dt \approx \sum_{i=1}^2 C_i f(t_i)$$

$$I \approx 1.0 * \left(\frac{1}{2} e^{(.5773503+1)/2} \right) + 1.0 * \left(\frac{1}{2} e^{(-.5773503+1)/2} \right)$$

$$I \approx 1.717896$$

It is unknown at this point whether this is a good approximation or not. In order to find out whether this approximation can be improved, use n=3. Corresponding t_i and C_i are read directly from Table 5.5.1

$$I \approx 0.5555556 * \left(\frac{1}{2} e^{(0.7745967+1)/2} \right) + 0.8888889 \left(\frac{1}{2} e^{(0+1)/2} \right)$$

$$+ 0.5555556 * \left(\frac{1}{2} e^{(-0.7745967+1)/2} \right)$$

$$I \approx 1.718281$$

The relative error is

$$\xi = \left| \frac{1.717896 - 1.718281}{1.718281} \right| = 0.00022$$

If n=4 is used

$$I \approx 0.3478548 * \left(\frac{1}{2} e^{(0.8611363+1)/2} \right) + 0.6521452 * \left(\frac{1}{2} e^{(0.3399810+1)/2} \right)$$

$$+0.6521452*\left(\frac{1}{2}e^{(-0.3399810+1)/2}\right)+0.3478548*\left(\frac{1}{2}e^{(-0.8611363+1)/2}\right)$$

$$I \approx 1.718282$$

From this last approximation it can be seen that n=3 or larger will give an excellent approximation to the integral.

Analytical solution is

$$\int_0^1 e^x dx = \left(e^x\right)_0^1 = 2.718282 - 1 = 1.718282$$

GAUSS LEGENDRE QUADRATURE METHOD

To obtain the integral $I = \int_a^b f(x)dx$.

STEP 1   Input: a, b the limits of the integration

ξ relative error

N   maximum number of points (roots) within a and b

$c_i^{(j)}, x_i^{(j)}$ coefficients and roots of Legendre

polynomials, i = 1, 2, ... j,   j = 1, 2, ... N.

STEP 2   Set $f(x) = \dfrac{(b-a)}{2} f\left[\dfrac{(b-a)t+b+a}{2}\right]$

STEP 3   Do steps until 5 for j=1,2...N

STEP 4 $I = \displaystyle\sum_{i=1}^{j} c_i^{(j)} f(x_i^{(j)})$

STEP 5   If $\dfrac{|I_j - I_{j-1}|}{|I_j|} > \xi$ then next j

STEP 6   Output: $I_j \approx I$

Table 5.1   Roots of Legendre polynomials $P_n(x)=0$ and coefficients c_i

n	Roots x_I	Coefficients c_i
2	0.5773502692	1.0000000000
	-0.5773502692	1.0000000000
3	0.7745966692	0.5555555556
	0.0000000000	0.8888888889
	-0.7745966692	0.5555555556
4	0.8611363116	0.3478548451
	0.3399810436	0.6521451549
	-0.3399810436	0.6521451549
	-0.8611363116	0.3478548451
5	0.9061798459	0.2369268850
	0.5384693101	0.4786286705
	0.0000000000	0.5688888889
	-0.5384693101	0.4786286705
	-0.9061798459	0.2369268850
6	0.9324695142	0.1713244924
	0.6612093865	0.3607615730
	0.238619161	0.4679139346
	-0.2386191861	0.4679139346
	-0.6612093865	0.3607615730
	-0.9324695142	0.1713244924
7	0.9491079123	0.1294849662
	0.7415311856	0.2797053915
	0.4058451514	0.3818300505
	0.0000000000	0.4179591837
	-0.4058451514	0.3818300505
	-0.7415311856	0.2797053915
	-0.9491079123	0.1294849662
8	0.9602898565	0.1012285363
	0.7966664774	0.2223810345
	0.5255324099	0.3137066459
	0.1834346425	0.3626837834
	-0.1834346425	0.3626837834
	-0.5255324099	0.3137066459
	-0.7966664774	0.2223810345
	-0.9602898565	0.1012285363

PROBLEMS

1.   Compute the value of definite integrals using Trapezoldal rule.

   (a) $\int_3^6 \frac{x\,dx}{4+x^2}$ using 6 sub intervals

   (b) $\int_1^9 \frac{dx}{x}$ using 8 sub intervals

   (c) $\int_0^\pi \sin x\,dx$ using 6 sub intervals

   (d) $\int_0^1 e^x dx$ in steps of 0.5

2.   Integrate $\dfrac{1}{\pi}\int_0^\pi e^{2\sin x} dx$

   using Trapezoidal rule with 4 intervals

3.   An integral was numerically evaluated using Trapezoidal rule with one and two intervals. The corresponding values are $R_{11} = 1.8591$ and $R_{21} = 1.7539$. The actual integral value is $I = R_{11} + Ch_1^2 = R_{21} = Ch_2^2$, where h_1 and h_2 are Trapezoidal interval sizes for one and two intervals. Obtain the actual integral.

4.   Apply Trapezoidal rule to integrate

   $\int_1^{1.5} \sqrt{x}\ dx$ in steps of 0.5

5.   According to Newton's 2nd law the integration equation of a particle of mass m attracted towards a fixed point p with a force of attraction that varies with the distance is derived as

   $$t = \frac{1}{\sqrt{\dfrac{2k}{m}}} \int_{0.1}^{0.9} \frac{dx}{\sqrt{\ln(1/x)}}$$

   where k/m = ½, use Trapezoidal rule with 4 intervals to approximate the time required to reach to the fixed point.

6.   Apply Trapezoidal rule to integrate with 4 intervals.

   $$I = \frac{1}{2}\int_0^4 \sqrt{x\sqrt{x}}\,dx$$

7.
   a)   Derive the Trapezoidal rule for numerical integration.

   b)   Integrate $\int_1^2 (1+x^{1/2})\, dx$ using Trapezoidal rule with 4 intervals

   c)   Integrate the above using Simpson's rule with 2 intervals.

8.   Calculate $\int_0^{\pi/2} \sqrt{1 - \dfrac{1}{2}\sin^2 x}\, dx$ using 2 intervals

   (a)   by Trapezoidal rule
   (b)   by Simpson's rule

9.   Using any of the numerical integration techniques show that

$$\int_0^{\pi/2} \frac{1}{\sin^2 t + \dfrac{1}{4}\cos^2 t}\, dt = \pi = 3.1415...$$

10.   Integrate by Simpson's Formula, using 2 and 4 intervals.

   (a) $\int_0^1 e^{-x^2}\, dx$ (b) $\int_0^{\pi} \frac{1-\cos x}{x}\, dx$

   c) $\int_0^1 \sqrt{1-x^2}\, dx$ (d) $\int_0^2 \frac{e^{-t}}{1+t^2}\, dt$

   (e) $\int_0^1 e^{-x}\, dx$ (f) $\int_0^{\pi} \sin x\, dx$

11.   Compute $\int_1^4 (2x^3 - 11x^2 + 24x)\, dx$

   using Simpson's rule with two intervals.

12.   Compute the value of I by Simpson's rule, considering 6 intervals.
$$I = \int_0^{\pi/2} \sqrt{1 - 0.162\sin^2 \theta}\, d\theta$$

13.   Using the data of the given table.

x	0.5	0.6	0.7	0.8	0.9	1.0	1.1
y	.4804	.5669	.6490	.7262	.7985	.8658	.9281

Compute the following integrals using Simpson's rule.

   (a) $\int_{0.5}^{1.1} xy\, dx$ (b) $\int_{0.5}^{1.1} y^2\, dx$

(c) $\displaystyle\int_{0.5}^{1.1} x^2 y\,dx$ (d) $\displaystyle\int_{0.5}^{1.1} y^3\,dx$

14. Using Simpson's rule, find $\displaystyle\int_{0}^{2\pi} y\,dx$, accurate to 4 decimal places, where

$$Y = y_0 + \left(\frac{y_n - y_0}{2\pi}\right) x - 0.1 \sin x, \; y_0 = 1 \text{ and } y_n = 2$$

15. A plunger of mass m moving through a fluid is subjected to a viscous resistance R, where R = R(v) = $-v^{3/2}$, in the range of interest. The relationship between the resistance R, velocity v, and time t is given by the equation

$$T = \int_{v(t)}^{v(t)} \frac{m}{R(v)}\,dv$$

where R is in Newton and v is in m/sec. If m = 1kg and v(0) =10m/sec, approximate the time required for the plunger to slow to v =5 m/sec.
(a) Use Simpson's rule with 6 intervals.
(b) Use the Trapezoidal rule with 6 intervals
(c) Compare these approximations to the actual value.

16. Solve the following by Simpson's rule, using 2 and 4 intervals.
(a) $I = \displaystyle\int_{100}^{200} \frac{1}{\log_{10}x}\,dx$

(b) $I = \displaystyle\int_{\pi/6}^{\pi/2} \log_{10}(\sin x)\,dx$

17. The average lining temperature T, of the brake pad of a disk brake may be obtained from the equation.

$$T = \frac{\displaystyle\int_{r_e}^{r_0} T(r)r\theta_p\,dr}{\displaystyle\int_{r_e}^{r_0} r\theta_p\,dr}$$

where the denominator is the total area of the pad and the numerator is the integration of the product of the elemental area with the local temperature.

For a given brake pad if $r_e = 10$ cm, $r_0 = 14$ cm, $\theta_p = 45^0$ and the temperatures given in the following table have been calculated at the various points on the disk, find an approximation for T using Simpson's rule.

r (cm)	10	11	12	13	14
T(r) (^{0}F)	640	885	1034	1114	1204

18. Evaluate $\displaystyle\int_{0}^{2} e^x\,dx$ by Romberg integration $\xi < 10^{-4}$

19.   Calculate the following integral with an error less than 10^{-4} using Romberg integration

$$\int_0^1 \frac{1}{1+x^2}\,dx$$

20.   Integrate $\dfrac{3}{\sqrt{\pi}}\displaystyle\int_0^4 e^{-x^2}\,dx$

by gauss quadrature with 3 points.

24.   Integrate $I = \displaystyle\int_0^1 \exp(x^2)\,dx$ using Gauss-Legendre quadratures. Use n =3 points.

Note: $\displaystyle\int_a^b f(x)\,dx = \int_{-1}^1 f\!\left[\frac{(b-a)t+b+a}{2}\right]\!\left(\frac{b-a}{2}\right)dt$

Roots	Coefficients
0.7746	0.5556
0.0000	0.8889
-0.7746	0.5556

25.   Apply Gauss quadrature with n=4 to

$$\int_{-1}^1 \frac{1}{1+x^2}\,dx$$

23.   Evaluate $y = \dfrac{1}{\pi}\displaystyle\int_0^\pi \cos(x\sin\theta)\,d\theta$ by Gauss quadrature with n=3 for x = 0 to 5π in steps of

$1/4\pi$. From your estimate find the smallest positive value of x for which y = 0.

24.   Approximate the following integral using Gauss quadrature with n = 2.

$$\int_{-2}^2 \frac{1}{1+x^2}\,dx$$

25.   Compute the integral by Gauss quadratures method using n = 5.

$$\int_0^1 \frac{dx}{\sqrt{x^4+1}}$$

26.   Integrate $\displaystyle\int_0^1 (e^x + x)\,dx$

(a)   Using Simpson's rule with h = 0.25
(b)   Using Gauss-Legendre quadrature with 3 points.

27.   Integrate the following using Gauss quadratures. Choose the number of points, n, until the relative error between the results using n and (n-1) points is less than 5×10^{-4}.

$$\int_{1}^{1.5} e^{-x^2} dx$$

28. Solve problem 18 using Gauss quadratures. Obtain the answer such that the relative error between answers with n points and (n-1) points is 5×10^{-4}.

29. By using numerical integration show that
$$4 \int_{0}^{1} \frac{dx}{1+x^2} = \pi = 3.1415926$$
Note: You should adopt a method which enables you to improve the accuracy progressively to at least 5 significant digits. Show all the working steps and intermediate computations.

30. Write a computer program to implement Romberg integration procedure. Using this program, obtain the integral
$$I = \int_{0}^{1} \sin x \, dx$$
to an accuracy of 5 digits.

Chapter 6

NUMERICAL DIFFERENTIATION

Numerical differentiation is used to approximate the derivatives of a function when analytical differentiation may result in a very complicated and cumbersome expression. It is also used to obtain derivatives at a point when the function is given in a tabular form and its analytical form is unknown. This is the case when experimental data points are the only information available. One way to approximate the derivative at any one of these points, is to assume that these points follow some unknown continuous function. In this case, interpolation techniques discussed in chapter 4 can be used to approximate a function form in the form of a polynomial or cubic splines which can then be differentiated analytically. Although this is a valid method for approximating the derivative, it is not the method discussed in this chapter. This chapter deals only with numerical techniques which use finite difference methods to approximate derivatives. There are three difference formulae which can be used depending on the nature of the data. In any one of these formulae, $n+1$ data points must be known in order to approximate the nth derivative. If more than $n+1$ points are available, then the approximation for the nth derivative becomes more accurate.

The three methods are known as the central, forward and backward differences.

6.1 CENTRAL DIFFERENCES

The central or midpoint differences are used when the point at which the approximation is being made has an equal amount of known points on either side. In order to get an approximation for the first derivative, two points must be known, and the point at which the approximation is being made must lie between them (see fig. 6.1).

Expressing the function at x_{i+1} and x_{i-1} by a Taylor series expansion about x_i we get the following expressions;

$$f(x_{i+1}) = f(x_i) + (x_{i+1} - x_i)\, f'(x_i) + \frac{(x_{i+1} - x_i)^2}{2!} + f''(x_i) +$$

$$\frac{(x_{i+1} - x_i)^3}{3!} f'''(x_i) + \dots\dots$$

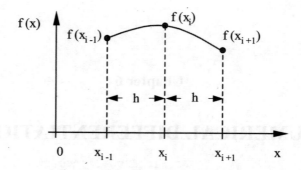

Figure 6.1: Central or midpoint differences approximation

$$f(x_{i-1}) = f(x_i) + (x_{i-1}-x_i)\,f'(x_i) + \frac{(x_{i-1}-x_i)^2}{2!}f''(x_i) +$$

$$\frac{(x_{i-1}-x_i)^3}{3!}f'''(x_i) + \ldots\ldots$$

If the data points are equally spaced; $(x_{i+1}-x_i) = h$ and $(x_{i-1}-x_i) = -h$

$$f(x_{i+1}) = f(x_i) + hf'(x_i) + \frac{h^2}{2!}f''(x_i) + \frac{h^3}{3!}f'''(x_i) + \ldots. \qquad (6.1.1)$$

$$f(x_{i-1}) = f(x_i) - hf'(x_i) + \frac{h^2}{2!}f''(x_i) - \frac{h^3}{3!}f'''(x_i) + \ldots. \qquad (6.1.2)$$

Subtracting equation 6.1.2 from 6.1.1, we get

$$f(x_{i+1}) - f(x_{i-1}) = 2hf'(x_i) + \frac{2h^3}{6}f'''(x_i) + \cdots$$

Dividing this expression throughout by 2h.

$$\frac{f(x_{i+1}) - f(x_{i-1})}{2h} = f'(x_1) + \left[\frac{h^2}{6}f'''(x_i)\right]$$

The term within the rectangular brackets is the error term.

Let all the terms to the right of the first derivative express the error involved in the approximation. The first term of this error is the most significant, and is therefore the only one retained in the approximation. In the case of the central difference formula, when using n+1 points to approximate nth derivative, the error term is of order h^2.

By adding equations 6.1.1 and 6.1.2, we get

$$f(x_{i+1}) - f(x_{i-1}) = 2hf(x_i) + h^2f'(x_i) + \frac{h^4}{12} f'''(x_i) + \cdots$$

From this expression we get the second derivative as

$$\frac{f(x_{i-1}) - 2f(x_i) + f(x_{i+1})}{h^2} = f''(x_i) + \left(\frac{h^2}{12} f'''(x_i) + \ldots\right)$$

where the last term is the error term of order h^2

The above equation can be expressed in the form,

$$\frac{1}{h}\left(\frac{f(x_{i+1}) - f(x_i)}{h} - \frac{f(x_i) - f(x_{i-1})}{h}\right) = f''(x_i) + f(E)$$

which means that the second derivative is the slope of the line tangent to the curve at $x_{i+1/2}$ minus the slope of the line at $x_{i-1/2}$ divided by h.

To obtain an expression for the third derivative, function values at four points must be expanded about x_i, and therefore expansion occurs about x_{i+2}, x_{i+1}, x_{i-1}, and x_{i-2} to give a total of 4 equations. These four equations are expanded to include the third derivative term, and then manipulated in the following way.

$$f(x_{i+1}) = f(x_i) + hf'(x_i) + \frac{h^2}{2!} f''(x_i) + \frac{h^3}{3!} f'''(x_i) + \ldots \qquad (6.1.3)$$

$$f(x_{i-1}) = f(x_i) - hf'(x_i) + \frac{h^2}{2!} f''(x_i) - \frac{h^3}{3!} f'''(x_i) + \ldots \qquad (6.1.4)$$

$$f(x_{i+2}) = f(x_i) + 2hf'(x_i) + \frac{4h^2}{2!} f''(x_i) + \frac{8h^3}{3!} f'''(x_i) + \ldots \qquad (6.1.5)$$

$$f(x_{i-2}) = f(x_i) - 2hf'(x_i) + \frac{4h^2}{2!} f''(x_i) - \frac{8h^3}{3!} f'''(x_i) + \ldots \qquad (6.1.6)$$

Eqn. 6.1.3 – eqn. 6.1.4 →

$$f(x_{i+1}) - f(x_{i-1}) = 2h f'(x_i) + \frac{2h^3}{3!} f'''(x_i) + \ldots\ldots$$

Eqn. 6.1.5 – eqn. 6.1.6 →

$$f(x_{i+2}) - f(x_{I-2}) = 4hf'(x_i) + \frac{16h^3}{3!}f'''(x_i) + ...$$

Solving these two equations simultaneously by eliminating $f'(x_i)$ between them gives

$$f'''(x_i) = \frac{(x_{i+2}) - 2f(x_{i+1}) + 2f(x_{i-1}) - f(x_{i-2})}{2h^3} + f(E)$$

In the above $f(E)$ is the error term of order h^2.

Similarly, if the expansion terms are extended to the fourth derivative term, then adding equations 6.1.3 and 6.1.4, 6.1.5 and 6.1.6, and solving the resulting eqns by eliminating the $f''(x)$ term would give;

$$f''''(x) = \frac{f(x_{i+2}) - 4f(x_{i+1}) + 6f(x_i) - 4f(x_{i-1}) + f(x_{i-2})}{h^4} + f(E)$$

where $f(E)$ is an error term of order h^2

If approximations are required for the fifth and sixth derivatives, then two more expansion equations are needed. Hence an even number of expansion equations must be available to get derivative expressions using the central differences. An equal number of points on either side of the point where the derivatives are desired must be used.

Expressions for the first four central differences are given below. In these expressions, exactly $n+1$ points were used to get the nth derivative expression. Hence the error involved is of the order h^2. In the last section of this chapter, it will be shown that the error term can be reduced by using more than $n+1$ points to obtain the nth derivative.

Central Difference Expressions with Error of order h^2

$$f'(x_i) = \frac{f(x_{i+1}) - f(x_{i-1})}{2h}$$

$$f''(x_i) = \frac{f(x_{i+1}) - 2f(x_i) + f(x_{i-1})}{h^2}$$

$$f'''(x_i) = \frac{f(x_{i+2}) - 2f(x_{i+1}) + 2f(x_{i-1}) - f(x_{i-2})}{2h^3}$$

$$f''''(x_i) = \frac{f(x_{i+2}) - 4f(x_{i+1}) + 6f(x_i) - 4f(x_{i-1}) + f(x_{i-2})}{h^4}$$

Note that for equally spaced data of interval size h, a table identical to the one described in chapter 4 can be used to approximate the derivative. This table is known as a divided difference table. Recall that the values in the divided difference table differ from the derivative approximation by an appropriate factorial term, (1! for a_1 values, 2! for a_2 values, etc.). The first divided difference values, when multiplied by 1! will give the approximation to the first derivative values; similarly when the second divided   differences are multiplied by 2!, they will give the approximation to the second derivative values, and so on. The advantage of using such a table is the ease with which the derivative value can be obtained.

The following example will illustrate that such a table will give the same derivative approximations as the central difference expressions formed earlier.

EXAMPLE 6.1
Given the following data set,

x_1	0	1	2	3	4	5	6
$f(x_1)$	0	3	8	9	16	20	23

Part I   Approximate the following quantities using the central difference equations.
   (a) $f'(x_i)$ at $x_i = 1$
   (b) $f'''(x_i)$ at $x_i = 4$
   (c) $f'''(x_i)$ at $x_i = 3$
   (d) $f''(x_i)$ at $x_i = 3$
   (e) $f''(x_i)$ at $x_i = 2$
   (f) $f'(x_i)$ at $x_i = 1$

Part II   Using a difference table confirm the above quantities. In addition to this use the table to evaluate;

   (g) $f^{vi}(x_i)$ at $x_i = 3$
   (h) $f'''(x_i)$ at $x_i = 1$

Solution
Part I
The interval size is h=1.0 and the previously derived equations are used to get the approximations for the   derivatives.
   (a) $f'(x_i)$ at $x_i = 1$
Using the central difference equation for $f'(x_i)$;

$$f'(x_i) \approx \frac{f(x_{i+1}) - f(x_{i-1})}{2h}$$

   where;

$$x_{i+1} = 2; f(x_{i+1}) = 8$$
$$x_{i+1} = 0; f(x_{i+1}) = 0$$
$$f'(x_i) \approx 4.0$$

   (b) $f'''(x_i)$ at $x_i = 4$
Using the central difference equation for $f'''(x_i)$;

$$f'''(x_i) \approx \frac{f(x_{i+2}) - 4f(x_{i+1}) + 6f(x_i) - 4f(x_{i-1}) + f(x_{i-2})}{h^4}$$

where for $x_i = 4;$ $f(x_i) = 16$

                     $x_{i+1} = 5;$ $f(x_{i+1}) = 20$

                     $x_{i-1} = 3;$ $f(x_{i+1}) = 9$

                     $x_{i+2} = 6;$ $f(x_{i+2}) = 23$

                     $x_{i-2} = 2;$ $f(x_{i-2}) = 8$

$$f'''(4) \approx \frac{23 - 4(20) + 6(16) - 4(9) + 8}{1} = 11$$

(c) Using $f''''(x_i)$ central difference with $x_i = 3$; $f(x_i) = 9$

where ; $x_{i+1} = 4;$ $f(x_{i+1}) = 16$

                     $x_{i-1} = 2;$ $f(x_{i-1}) = 8$

                     $x_{i-2} = 5;$ $f(x_{i+2}) = 20$

                     $x_{i+2} = 1;$ $f(x_{i-2}) = 3$

$$f'''(3) \approx \frac{20 - 4(16) + 6(9) - 4(8) + 3}{1} = -19$$

(d) Using the central difference equation for $f'''(x_i)$;

$$f'''(x_i) \approx \frac{f(x_{i+2}) - 2f(x_{i+1}) + 2f(x_{i-1}) - f(x_{i-2})}{2h^3}$$

where for $x_i = 3$

         $x_{i+1} = 4;$ $f(x_{i+1}) = 16$

         $x_{i-1} = 2;$ $f(x_{i-1}) = 8$

         $x_{i+2} = 5;$ $f(x_{i+2}) = 20$

         $x_{i-2} = 1;$ $f(x_{i-2}) = 3$

$$f'''(3) \approx \frac{20 - 2(16) + 2(8) - 3}{2} = \frac{1}{2} = 0.5$$

(e) Using the central difference equation for $f''(x_i)$;

$$f''(x_i) \approx \frac{f(x_{i+1}) - 2f(x_i) + f(x_{i-1})}{h^2}$$

where for $x_i = 2;$ $f(x_i) = 8$

                     $x_{i+1} = 3;$ $f(x_{i+1}) = 9$

                     $x_{i-1} = 1;$ $f(x_{i-1}) = 3$

$$f''(2) \approx \frac{9 - 2(8) + 3}{1} = -4$$

(f)   Using the $f''(x_i)$ central difference formula at $x_i = 1$;

$$\begin{array}{lll} & & f(x_i) = 3 \\ x_{i+1} = 2; & & f(x_{i+1}) = 8 \\ x_{i-1} = 0; & & f(x_{i-1}) = 0 \end{array}$$

$$f''(1) \approx \frac{8 - 2(3) + 0}{1} = 2$$

Part II

Difference Table

x_i	$f(x_i)$	$f'/1!$	$f''/2!$	$f'''/3!$	$f^{iv}/4!$	$f^v/5!$	$f^{vi}/6!$
0	0						
		3					
1	3		1				
		5		-1			
2	8		-2		.6667		
		1		1.6667		-.2917	
3	9		3		-.7917		.0903
		7		-1.5000		.2500	
4	16		-1.5000		.4583		
		4		.3333			
5	20		-.5000				
		3					
6	23						

Note: More approximations are available for the middle points than those at the top or bottom.

(a) $f'(x_i)$ for $x_i = 1$

To get the first derivative values using the central difference scheme, the average value must be taken.

$$f'(1) = \frac{3 + 5}{2} * 1! = 4 * 1! = 4$$

Average values must be taken for all odd value derivatives. It becomes obvious that the first derivative approximation and the first difference are the same.

(b) $f^{iv}(4) = 0.4583 * 4! = 11$

(c) $f^{iv}(3) = -0.7917 * 4! = -19$

(d) $f'''(3) = \dfrac{1.6667 + (-1.5)}{2} = 0.085 \; *3! = 0.5$

(e) $f''(2) = -2 * 2! = -4$

(f) $f''(1) = 1 * 2! = 2$

(g) $f^{vi}(3) = 0.1 * 6! = 72$

This is the only value with a vi^{th} derivative approximation. This is because it is the center value.

(h) $f'''(1)$ not enough points available to approximate this value.

NUMERICAL DIFFERENTIATION – CENTRAL DIFFERENCES

To obtain first, second, third, etc derivative at points for which only the function values are known.

STEP 1 Input: x_i, $f(x_i)$, $i = 0, 1, 2, \ldots n$

STEP 2 Form a divided difference table with n-1 difference terms, where;

$f[x_i]$ terms are the function values from the data

$f[x_i, x_{i+1}]$ terms are the first divided differences

$f[x_i, x_{i+1}, x_{i+2}]$ terms are the second divided differences

etc as indicated in chapter four.

STEP 3 Multiply all the terms in the first divided difference column by 1!, all the terms in the second divided difference column by 2!, and so on. This will give the derivative approximations for each point.

STEP 4 To evaluate an even number, simply read it from the table at that point $(f''(x_i), (f''''(x_i),\ldots)$. To evaluate an odd number derivative, an average value must be taken using the immediate values above and below the point in question.

STEP 5 Stop.

CAUTION

Remember that an equal number of known values must be available on both sides of the point being approximated.

To approximate the first or second derivative, at least one known point on both sides of the point where the approximation is taking place must be available. For the third or fourth, two known points must be available on both sides of the point, for the fifth and sixth, 3 known points, and so on.

6.2 FORWARD DIFFERENCES

The forward differences are used when an excess of points lie to the right of the point where the approximation is taking place. In this case more derivative expressions can be formed by forming a Taylor series expansion at the known points in a forward direction, such as f(x_0+h), f(x_0+2h)..., about x_0.

To get an expression for the first derivative value; expansion about a point x in the vicinity of x

$$f(x_{i+1}) = f(x_i) + hf'(x_i) + \frac{h^2}{2!} f''(x_i) + ...$$

All terms to the right of the first derivative term are regarded as the error in the first derivative approximation. Since the first of these terms is the most significant, it will be the only term retained to represent the error.

$$f(x_{i+1}) = f(x_i) + hf'(x_i) + \left[\frac{h^2}{2!} f''(x_i)\right]$$

The term inside the rectangular brackets is the error term.

$$f'(x_i) = \frac{f(x_{i+1}) - f(x_i)}{h} - \left[\frac{h}{2} f''(x_i)\right]$$

Error is of order h, O(h). This is larger than the central difference error because only one point was used to get the approximation.

$$f'(x_i) = \frac{f(x_{i+1}) - f(x_i)}{h} + O(h)$$

To get an expression for the second derivative, it is necessary to get expansion expressions for two points in the forward direction.

Expansion about a point x_i in the vicinity of x_{i+1} is

$$f(x_{i+1}) = f(x_i) + hf'(x_i) + \frac{h^2}{2!} f''(x_i) + \frac{h^3}{3!} f'''(x_i) + ...$$

Expansion about a point x_i in the vicinity of x_{i+2} is

$$f(x_{i+2}) = f(x_i) + 2hf'(x_i) + \frac{4h^2}{2} f''(x_i) + \frac{8h^3}{6} f'''(x_i) + ...$$

Multiply the first equation by 2 and subtract it from the second to eliminate the first derivative term $f'(x_i)$

$$f(x_{i+2}) - 2f(x_{i+1}) = -f'(x_i) + h^2 f''(x_i) + h^3 f'''(x_i) + ...$$

$$f''(x_i) = \frac{f(x_{i+2}) - 2f(x_{i+1}) + f(x_i)}{h^2} - \left[hf'''(x_i) + ... \right]$$

The error term within the rectangular brackets is of order h.

To get an expression for the third derivative, expansion equations must be formed for three points in the forward direction.

Expansion about a point x_i in the vicinity of x_{i+1} is

$$f(x_{i+1}) = f(x_i) + hf'(x_i) + \frac{h^2}{2!} f''(x_i) + \frac{h^3}{3!} f'''(x_i) + \frac{h^4}{4!} f^{IV}(x_i) + ... \quad (6.2.1)$$

Expansion about a point x_i in the vicinity of x_{i+2} is

$$f(x_{i+2}) = f(x_i) + 2hf'(x_i) + 2h^2 f''(x_i) + \frac{4}{3} h^3 f'''(x_i) + \frac{2}{3} h^4 f^{IV}(x_i) + ... \quad (6.2.2)$$

Expansion about a point x_i in the vicinity of x_{i+3} is

$$f(x_{i+3}) = f(x_i) + 3hf'(x_i) + \frac{9}{2} h^2 f''(x_i) + \frac{9}{2} h^3 f'''(x_i) + \frac{27}{8} h^4 f^{IV}(x_i) + ... \quad (6.2.3)$$

Elimination of $f''(x_i)$ from equations 6.2.1, 6.2.2, and 6.2.3 results in two equations in terms of $f''(x_i)$ and higher derivatives.

Elimination of $f''(x_i)$ from these equations gives

$$f'''(x_i) = \frac{f(x_{i+3}) - 3f(x_{i+2}) + 3f(x_{i+1}) - f(x_i)}{h^3} + O(h)$$

Expressions for the first four forward differences are given below. In these expressions, exactly n+1 points were used to get the nth derivative expression. Hence the largest error involved turned out to be of the order h. This error term will reduce if more than n+1 terms are used. This will be shown in the last section of this chapter.

Forward Difference Expressions with Error of Order h

$$f'(x_i) = \frac{f(x_{i+1}) - f(x_i)}{h}$$

$$f''(x_i) = \frac{f(x_{i+2}) - 2f(x_{i+1}) + f(x_i)}{h^2}$$

$$f'''(x_i) = \frac{f(x_{i+3}) - 3f(x_{i+2}) + 3f(x_{i+1}) - f(x_i)}{h^3}$$

$$f^{IV}(x_i) = \frac{f(x_{i+4}) - 4f(x_{i+3}) + 6f(x_{i+2}) - 4f(x_{i+1}) + f(x_i)}{h^4}$$

Again the divided difference table can be used instead of the derived expressions; however, the table must be read in a different way than the straight forward manner involved in the central divided difference method.

In order to get approximations for this method using the divided difference table, start at the point being approximated and proceed to read the table in a diagonal fashion going towards the bottom. The values being read will be the forward divided difference for the point at which we started. Consider the divided difference table of the previous example and compare results.

EXAMPLE 6.2
Given the following data set

x_i	0	1	2	3	4	5	6
$f(x_i)$	0	3	8	9	16	20	23

Part I   Approximate the following derivative approximations using the forward difference equations.
(a) $f'(x_i)$ at $x_i = 1$
(b) $f''(x_i)$ at $x_i = 1$
(c) $f'''(x_i)$ at $x_i = 1$
(d) $f^{iv}(x_i)$ at $x_i = 1$
(e) $f^{iv}(x_i)$ at $x_i = 3$
(f) $f''(x_i)$ at $x_i = 2$
Part II   Using a difference table for the data confirm the above approximations plus
(g) $f^{vi}(x_i)$ at $x_i = 0$

Solution
Part I
    Assuming the data points are following a smooth curve, then the interval size h=1.0 and the previously derived equations are used to get the approximations.
(a) $f'(x_i)$ at $x_i = 1$
The forward difference equation for $f'(x_i)$ is;

$$f'(x_i) = \frac{f(x_{i+1}) - f(x_i)}{h}$$

where;

$$x_i = 1; \qquad f(x_i) = 3$$

$$x_{i+1} = 2; \quad f(x_{i+1}) = 8$$
$$f'(1) = \frac{8-3}{1} = 5$$

(b) f''(x$_i$)   at   x$_i$ = 1
The forward difference equation for f''(x$_i$) is;

$$f''(x_i) = \frac{f(x_{i+2}) - 2f(x_{i+1}) + f(x_i)}{h^2}$$

where for

$$x_i = 1; \quad f(x_i) = 3$$
$$x_{i+1} = 1; \quad f(x_{i+1}) = 8$$
$$x_{i+2} = 3; \quad f(x_{i+2}) = 9$$

$$f''(1) = \frac{9 - 2(8) + 3}{1} = -4$$

(c) f'''(x$_i$)   at   x$_i$ = 1
The forward difference equation for f'''(x$_i$) is;

$$f'''(x_i) = \frac{f(x_{i+3}) - 3f(x_{i+2}) + 3f(x_{i+1}) - f(x_i)}{h^3}$$

where ;

$$x_i = 1; \qquad f(x_i) = 3$$
$$x_{i+1} = 2; \quad f(x_{i+1}) = 8$$
$$x_{i+2} = 3; \quad f(x_{i+2}) = 9$$
$$x_{i+3} = 1; \quad f(x_{i+3}) = 16$$

$$f'''(1) = \frac{16 - 3(9) + 3(8) - 3}{1} = 10$$

(d) f^{iv} (x$_i$)   at   x$_i$ = 1
The forward difference equation for f^{iv}(x$_i$) is;
$$f^{iv}(x_i) = \frac{f(x_{i+4}) - 4f(x_{i+3}) + 6f(x_{i+2}) - 4f(x_{i+1}) + f(x_i)}{h^4}$$

where for   x$_i$=1   ; f(x$_i$)   = 3
             x$_{i+1}$=2   ; f(x$_{i+1}$) = 8
             x$_{i+2}$=3   ; f(x$_{i+2}$) = 9
             x$_{i+3}$=4   ; f(x$_{i+3}$) = 16
             x$_{i+4}$=5   ; f(x$_{i+4}$) = 20

$$f^{iv}(1) = \frac{20 - 4(16) + 6(9) - 4(8) + 3}{1} = -19$$

(e) $f^{vi}(x_i)$ at $x_i = 3$

Using the above equation again we find that not enough information is   available for the fourth derivative approximation at $x_i = 3$.

$$
\begin{array}{ll}
x_i = 3 & ; f(x_i) \ = 9 \\
x_{i+1} = 4 & ; f(x_{i+1}) \ = 16 \\
x_{i+2} = 5 & ; f(x_{i+2}) \ = 20 \\
x_{i+3} = 6 & ; f(x_{i+3}) \ = 23 \\
x_{i+4} = 7 & ; f(x_{i+4}) \ = ?
\end{array}
$$

(f) $f''(x_i)$ at $x_i = 2$

The forward difference equation for $f''(x_i)$ is;

$$f''(x_i) = \frac{f(x_{i+2}) - 2f(x_{i+1}) + f(x_i)}{h^2}$$

$$
\begin{array}{ll}
x_i = 2 & ; f(x_i) \ = 8 \\
x_{i+1} = 3 & ; f(x_{i+1}) \ = 9 \\
x_{i+2} = 4 & ; f(x_{i+2}) \ = 16
\end{array}
$$

$$f''(2) = \frac{16 - 2(9) + (8)}{1} = 6$$

Part II

Form the difference table as in ex. 6.1.1 for all the data.

x_i	$f(x_I)$	$f'/1!$	$f''/2!$	$f'''/3!$	$f^{iv}/4!$	$f^v/5!$	$f^{vi}/6!$
0	0						
		3					
1	3		1				
		5		-1			
2	8		-2		.6667		
		1		1.6667		-.2917	
3	9		3		-.7917		.0903
		7		-1.5000		.2500	
4	16		-1.5000		.4583		
		4		.3333			
5	20		-.5000				
		3					
6	23						

Note: More approximations are available for values at the top of the table. The amount of available approximations decreases as we proceed in a forward.

Direction from $x_i=0$ to $x_i=6$.

(a) f'(x_i) at $x_i=1$

Start at $f(x_i) = f(1)$ and read in the diagonal fashion indicated on the table until the desired value is reached.

In this case;

$$\frac{f'(1)}{1!} = 5 \quad \text{Hence } f'(1) = 5$$

(b) f''(x_i) at $x_i=1$

Again start at $f(x_i)=f(1)$ and continue to the $\frac{f''(x_i)}{2!}$ value.

$$\frac{f''(x_i)}{2!} = \frac{f''(1)}{2!} = -2 \quad \text{Hence } f''(1) = -4$$

(c) $f'''(x_i)$ at $x_i=1$

Again start at $f(x_i)=f(1)$ and continue to the $\frac{f'''(x_i)}{3!}$ value which is;

$$\frac{f'''(x_i)}{3!} = \frac{f'''(1)}{3!} = 1.6667 \quad \text{Hence } f'''(1) = 10$$

(d) f^{iv} (x_i) at $x_i=1$

Start at $f(x_i)=f(1)$ and continue to the $\frac{f^{iv}}{4!}$ value which is;

$$\frac{f^{iv}(x_i)}{4!} = \frac{f^{iv}(1)}{4!} = -0.7917 \quad \text{Hence } f^{iv}(1) = -19$$

(e) f^{iv} (x_i) at $x_i = 3$

Start at $f(x_i)=f(3)$ and continue in the diagonal fashion to the $\frac{f^{iv}(x_i)}{4!}$ term. Since this term does not exist, no $f^{iv}(x_i)$ approximation can be made for this point.

(f) f'' (x_i) at $x_i = 2$

Start at $f(x_i) = f(2)$ and continue diagonally to $\frac{f''(x_i)}{2!}$ term which is:

$$\frac{f''(x_i)}{2!} = \frac{f''(2)}{2!} = 3 \quad \text{Hence } f''(2) = 6$$

(g) $f^{iv}(x_i)$ at $x_i = 0$

The only value having and n^{th} derivative approximation for $n+1$ data set in this method is the very first value.

$$\frac{f^{vi}(x_i)}{6!} = \frac{f^{vi}(0)}{6!} = 0.0903 \quad \text{Hence } f^{vi}(0) = 65.016$$

NUMERICAL DIFFERENTIATION – FORWARD DIFFERENCES

To obtain first, second, third, etc derivative at points for which only the function values are known.

STEP 1   Input: x_i, $f(x_i)$, $i = 0, 1, 2, \ldots n$

STEP 2   Form a divided difference table with n-1 difference terms, where;

   $f[x_i]$ terms are the function values from the data

   $f[x_i, x_{i+1}]$ terms are the first divided differences

   $f[x_i, x_{i+1}, x_{i+2}]$ terms are the second divided differences

   etc as indicated in chapter four.

STEP 3   Multiply all the terms in the first divided difference column by 1!, all the terms in the second divided difference column by 2!, and so on. This will give derivative approximations for each point.

STEP 4   To evaluate the derivative approximation start at the function value of the point being approximated and proceed to read the table in a diagonal fashion toward the bottom until the desired derivative value is reached.

x_i	$f(x_i)$	$f'/1!$	$f''/2!$
x_1	y_1		
		y_1'	
x_2	y_2		y_1''
		y_2'	
x_3	y_3		

STEP 5   Stop.

CAUTION

n amount of values must lie in front of the point in order to evaluate the nth derivative approximation at that point. For example since two values lie in front of x_2, then up to and including the second derivative can be approximated at x_2.

6.3 BACKWARD DIFFERENCES

The backward differences are used when an excess of points lie to the left of the point where the derivative is required. In this case more derivative expressions can be formed by expanding a Taylor series expansion at the known points in a backward direction, such as $f(x_0-h)$, $f(x_0-2h)$,…

To get an expression for the first derivative value, we get expansion about a point x_1 in the vicinity of x_{i-1}.

$$f(x_{i-1}) = f(x_i) - hf'(x_i) + \frac{h^2}{2!} f''(x_i) - ...$$

All the terms to the right of the first derivative term are regarded as the error in the first derivative approximation. Since the first of these terms is the most significant, it will be the only term retained to represent the error.

$$f(x_{i-1}) = f(x_i) - hf(x_i) + \left[\frac{h^2}{2!} f''(x_i) - ... \right]$$

The term inside rectangular brackets is the error term

$$f'(x_i) = \frac{f(x_i) - f(x_{i-1})}{h} - \left[\frac{h}{2} f''(x_i) \right]$$

The error term is of the order h.

$$f'(x_i) = \frac{f(x_i) - f(x_{i-1})}{h} + O(h)$$

In order to get an expression for the second derivative, it is necessary to get expansion expressions for two points in the backward direction.

The expansion about a point x_i in the vicinity of x_{i-1}

$$f(x_{i-1}) = f(x_i) - hf'(x_i) + \frac{h^2}{2!} f''(x_i) - \frac{h^3}{3!} f'''(x_i) + ...$$

The expansion about a point x_i in the vicinity of x_{1-2}

$$f(x_{i-2}) = f(x_i) - 2hf'(x_i) + 2h^2 f''(x_i) - \frac{4}{3}h^3 f'''(x_i) + \ldots$$

Multiply the first equation by 2 and subtract it from the second to eliminate the first derivative term $f'(x_i)$ which results in

$$f(x_{i-2}) - 2f(x_{i-1}) = -f(x_i) - h^2 f''(x_i) + h^3 f'''(x_i) + \ldots$$

$$\frac{f(x_{i-2}) - 2f(x_{i-1}) + f(x_i)}{h^2} = f''(x_i) + [hf'''(x_i)]$$

The error term is of order h

$$f''(x_i) = \frac{f(x_{i-2}) - 2f(x_{i-1}) + f(x_i)}{h^2} + O(h)$$

To get the expression for the third derivative, expansion equations must be formed for three points in the backward direction $f(x_{i-1})$, $f(x_{i-2})$, and $f(x_{i-3})$.

The expansion about a point x_i in the vicinity of x_{i-1}.

$$f(x_{i-1}) = f(x_i) - hf'(x_i) + \frac{h^2}{2!} f''(x_i) - \frac{h^3}{3!} f'''(x_i) + \frac{h^4}{4!} f^{iv}(x_i) - \ldots \qquad (6.3.1)$$

The expansion about a point x_i in the vicinity of x_{i-2}

$$f(x_{i-2}) = f(x_i) - 2hf'(x_i) + 2h^2 f''(x_i) - \frac{4h^3}{3!} f'''(x_i) + \frac{2}{3}h^4 f^{IV}(x_i) - \ldots \quad (6.3.2)$$

The expansion about a point x_i in the vicinity of x_{i-3}

$$f(x_{i-3}) = f(x_i) - 3hf'(x_i) + \frac{9}{2}h^2 f''(x_i) - \frac{9h^3}{2} f'''(x_i) + \frac{27}{8}h^4 f^{IV}(x_i) - \ldots \quad (6.3.3)$$

Elimination of $f'(x_i)$ from equations 6.3.1, 6.3.2 and 6.3.3 results in two equations in terms of $f''(x_i)$ and higher derivatives. Elimination of $f''(x_i)$ form these gives

$$f'''(x_i) = \frac{f(x_i) - 3f(x_{i-1}) + 3f(x_{i-2}) - f(x_{i-3})}{h^3} + O(h)$$

Expressions for the first four backward differences are given below. In these expressions, exactly n+1 points were used to get the nth derivative expression. The amount of error which results in these expressions is of the order h.

Backward Difference Expressions with error of Order h

$$f'(x_i) = \frac{f(x_i) - f(x_{i-1})}{h}$$

$$f''(x_i) = \frac{f(x_i) - 2f(x_{i-1}) + f(x_{i-2})}{h^2}$$

$$f'''(x_i) = \frac{f(x_i) - 3f(x_{i-1}) + 3f(x_{i-2}) - f(x_{i-3})}{h^3}$$

$$f^{iv}(x_i) = \frac{f(x_i) - 4f(x_{i-1}) + 6f(x_{i-2}) - 4f(x_{i-3}) + f(x_{i-4})}{h^4}$$

Again the divided difference table can be used instead of the derived expressions; this time the table is read starting at the function value and proceeding in a diagonal fashion going towards the top. The values being read will be the backward divided differences for the point at which we started. Consider the divided difference table of the previous example.

EXAPLE 6.3
Given the following data sets.

x_i	0	1	2	3	4	5	6
$f(x_i)$	0	3	8	9	16	20	23

Part I Approximate the following derivative approximations using the backward difference equations.

$$\begin{array}{lll}
\text{(a) } f'(x_i) & \text{at } x_i = 6 \\
\text{(b) } f''(x_i) & \text{at } x_i = 6 \\
\text{(c) } f'''(x_i) & \text{at } x_i = 6 \\
\text{(d) } f^{iv}(x_i) & \text{at } x_i = 6 \\
\text{(e) } f^{iv}(x_i) & \text{at } x_i = 1 \\
\text{(f) } f''(x_i) & \text{at } x_i = 2
\end{array}$$

Part II Using a difference table for the data confirm the above approximations plus

$$\text{(g) } f^{vi}(x_i) \qquad \text{at } x_i = 6$$

Solution
Part I

Assuming that the data points are following a smooth curve, the previously derived equations with an interval size h=1.0 can be used to get the approximations.

(a) $f'(x_i)$ at $x_i = 6$

The f'(x$_i$) equation is;

$$f'(x_i) = \frac{f(x_i) - f(x_{i-1})}{h}$$

where

$$x_i = 6 \; ; f'(x_i) = 23$$
$$x_{i-1} = 5 \; ; \; f'(x_{i-1}) = 20$$
$$f'(6) = \frac{23 - 20}{1} = 3$$

(b) f''(x$_i$)   at x$_i$ = 6

The f'(x$_i$) equation is;

$$f''(x_i) = \frac{f(x_i) - 2f(x_{i-1}) + f(x_{i-2})}{h^2}$$

where for

$$x_i = 6 \; ; \quad f(x_i) = 23$$
$$x_{i-1} = 5 \; ; \; f(x_{i-1}) = 20$$
$$x_{i-2} = 4 \; ; \; f(x_{i-2}) = 16$$

$$f''(6) = \frac{23 - 2(20) + 16}{1} = -1$$

(c) f'''(x$_i$)   at x$_i$ = 6

The f''' (x$_i$) equation is;

$$f'''(x_i) = \frac{f(x_i) - 3f(x_{i-1}) + 3f(x_{i-2}) - f(x_{i-3})}{h^3}$$

where

$$x_i = 6 \; ; \; f(x_i) = 23$$
$$x_{i-1} = 5 \; ; f(x_{i-1}) = 20$$
$$x_{i-2} = 4 \; ; f(x_{i-2}) = 16$$
$$x_{i-3} = 3 \; ; \quad f(x_{i-3}) = 9$$

$$f'''(6) = \frac{23 - 3(20) + 3(16) - 9}{1} = 2$$

(d) f^{iv} (x$_i$)   at x$_i$ = 6

The f^{iv} (x$_i$) equation is;

$$f^{iv}(x_i) = \frac{f(x_i) - 4f(x_{i-1}) + 6f(x_{i-2}) - 4f(x_{i-3}) + f(x_{i-4})}{h^4}$$

where　　for
$$x_i = 6 \; ; f(x_i) = 23$$
$$x_{i-1} = 5 \; ; f(x_{i-1}) = 20$$
$$x_{i-2} = 4 \; ; f(x_{i-2}) = 16$$
$$x_{i-3} = 3 \; ; f(x_{i-3}) = 9$$
$$x_{i-4} = 2 \; ; f(x_{i-4}) = 8$$

$$f^{iv}(6) = \frac{23 - 4(20) + 6(16) - 4(9) + 8}{1} = 11$$

(e) $f^{iv}(x_i)$　　　　at $x_i = 1$

The $f^{iv}(x_i)$ equation is used; however there is not enough information.

$$x_{i-1} = 0$$
$$x_{i-2} = ?$$

(f) $f'(x_i)$　　　　at $x_i = 2$

The $f''(x_i)$ equation is;

$$f''(x_i) = \frac{f(x_i) - 2f(x_{i-1}) + f(x_{i-2})}{h^2}$$

where for　　　　　　$x_i = 2 \; ; f(x_i) = 8$
$$x_{i-1} = 1 \; ; f(x_{i-1}) = 3$$
$$x_{i-2} = 0 \; ; f(x_{i-2}) = 0$$

$$f''(2) = \frac{8 - 2(3) + 0}{1} = 2$$

Part II

Form the divided difference table as in ex. 6.1.1 for all the data.

x_I	$f(x_I)$	$f'/1!$	$f''/2!$	$f'''/3!$	$f^{iv}/4!$	$f^v/5!$	$f^{vi}/6!$
0	0						
		3					
1	3		1				
		5		-1			
2	8		-2		.6667		
		1		1.6667		-.2917	
3	9		3		-.7917		.0903
		7		-1.5000		.2500	
4	16		-1.5000		.4583		
		4		.3333			
5	20		-.5000				
		3					
6	23						

Note: More approximations are available for values at the top of the table. The amount of available approximations decreases as we proceed in a backward direction from $x_i = 6$ to $x_i = 0$.

(a) f'(x_i) at $x_i = 6$

Start at $f(x_i) = f(6)$ and read in the diagonal fashion indicated on the table until the desired value is reached.

In this case;

$$\frac{f'(6)}{1!} = 3 \qquad\qquad \text{Hence} \quad f'(6) = 3$$

(b) f''(x_i) at $x_i = 6$

Again start at $f(6) = 23$ and proceed to the $\dfrac{f''(x_i)}{2!}$ term.

$$\frac{f''(6)}{2!} = -0.5 \text{ Hence } f''(6) = -1.0$$

(c) f''(x_i) at $x_i = 6$

Again start at $f(6) = 23$ and proceed to the $\dfrac{f'''(x_i)}{3!}$ term.

$$\frac{f'''(6)}{3!} = 0.3333 \text{ Hence } f'''(6) = 2.0$$

(d) f^{iv} (x_i) at $x_i = 6$

Start at $f(6) = 23$ and proceed to the $\dfrac{f^{iv}}{4!}$ value term.

$$\frac{f^{iv}(6)}{4!} = 0.4583 \text{ Hence } f^{iv}(6) = 11$$

(e) f^{iv} (x_i) at $x_i = 1$

Start at $f(1) = 3$ and proceed to the $\dfrac{f^{iv}(x_i)}{4!}$ term.

Since this information does not exist for this method no approximation can be made.

(f) f' (x_i) at $x_i = 2$

Start at $f(2) = 8$ and proceed to the $\dfrac{f''(x_i)}{2!}$ term.

$$\frac{f''(2)}{2!} = 1 \quad \text{Hence } f''(2) = 2$$

(g) $f^{vi}(x_i)$ at $x_i = 6$

Start at $f(6) = 23$ and proceed to the $\dfrac{f^{vi}(x_i)}{6!}$ term.

$$\frac{f^{vi}(6)}{6!} = 0.0903 \quad \text{Hence } f^{vi}(6) = 65.016$$

NUMERICAL DIFFERENTIATION – BACKWARD DIFFERENCES

To obtain first, second, third, etc derivative at points for which only the function values are known.

STEP 1   Input: x_i, $f(x_i)$, i = 0, 1, 2, … n

STEP 2   Form a divided difference table with n-1 difference terms, where;

   $f[x_i]$ terms are the function values form the data

   $f[x_i,x_{i+1}]$ terms are the first divided differences

   $f[x_i,x_{i+1},x_{i+2}]$ terms are the second divided differences etc as indicated in chapter four.

STEP 3   Multiply all the terms in the first divided difference column by 1!, all the terms in the second divided difference column by 2!, and so on. This will give the derivative approximations for each point.

STEP 4   To evaluate the derivative approximation start at the function value of the point being approximated and proceed to read the table in a diagonal fashion toward the top until the desired derivative value is reached.

x_i	$f(x_i)$	$f'/1!$	$f''/2!$
x_1	y_1		
		y_1'	
x_2	y_2		y_1''
		y_2'	
x_3	y_3		

STEP 5   Stop

CAUTION

n amount of values must lie in back of the point in order to evaluate the nth derivative approximation at that point. For example since two values lie in back of x_3, then up to and including the second derivative can be approximated at x_3.

ALTERNATE APPROACH TO DERIVING DIFFERNECE EXPRESSIONS

Let P_n be a polynomial of degree at most n that agrees with the function f(x) at points x_i i =1, 2, ...n. Then f(x) $\approx P_n$ (x) can be represented as

$$f(x) = f(x_0) + \frac{(x - x_0)f'(x_0)}{1!} + \frac{(x - x_0)(x - x_1)f''(x_0)}{2!} +$$
$$\frac{(x - x_0)(x - x_1)(x - x_2)f'''(x_0)}{3!}$$

where the derivatives of this expression can be approximated by divided differences to give a polynomial form. Recall that the difference term is just the derivative approximation divided by a factorial term. This results in the following form.

$$P(x) = f[x_0] + (x-x_0)f[x_0,x_1] + (x-x_0)(x-x_1)f[x_0,x_1,x_2] + ...$$

Consider the polynomial formed using the first three points x_0, x_1, x_2, (see Fig. 6.3).

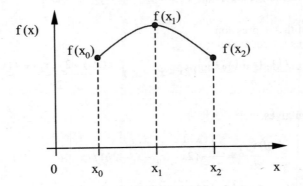

Figure 6.3: Polynomial formed using three points x_0, x_1, x_2

where; $f[x_0]=f(x_0)$

$$f[x_0, x_1] = \frac{f(x_1) - f(x_0)}{x_1 - x_0}$$

$$f[x_0, x_1, x_2] = \frac{\dfrac{f(x_2) - f(x_1)}{x_2 - x_1} - \dfrac{f(x_1) - f(x_0)}{x_1 - x_0}}{x_2 - x_0}$$

Assuming a constant interval size h, the above expressions reduce to:

$$f[x_0] = f(x_0)$$

$$f[x_0, x_1] = \frac{f(x_1) - f(x_0)}{h}$$

$$f[x_0, x_1, x_2] = \frac{f(x_2) - 2f(x_1) + f(x_0)}{2h^2}$$

Replacing these divided difference terms into the polynomial expression gives.

$$P(x) = f(x_0) + (x - x_0)\left(\frac{f(x_1) - f(x_0)}{h}\right) + (x - x_0)(x - x_1)\left(\frac{f(x_2) - 2f(x_1) + f(x_0)}{2h^2}\right)$$

$$P(x) = f(x_0) + (x - x_0)\left(\frac{f(x_1) - f(x_0)}{h}\right) +$$

$$(x^2 - x(x_0 + x_1) + x_0 x_1)\left(\frac{f(x_2) - 2f(x_1) + f(x_0)}{2h^2}\right)$$

Take the first derivative of this expression,

$$\frac{dP(x)}{dx} = \left(\frac{f(x_1) - f(x_0)}{h}\right) + (2x - x_0 - x_1)\left(\frac{f(x_2) - 2f(x_1) + f(x_0)}{2h^2}\right)$$

Simplifying this expression gives,

$$\frac{dP(x)}{dx} = \frac{1}{2h^2}(2x - x_0 - x_1 - 2h)f(x_0) +$$

$$\frac{1}{2h^2}(-4x + 2x_0 + 2x_1 + 2h)f(x_1) + \frac{1}{2h^2}(2x - x_0 - x_1)f(x_2)$$

To get the first derivative approximation, at x_0, set $x = x_0$,

$$\frac{dP(x_0)}{dx} = \frac{1}{2h^2}(x_0 - x_1 - 2h)f(x_0) + \frac{1}{2h^2}(-2x_0 + 2x_1 + 2h)f(x_1) +$$

$$\frac{1}{2h^2}(x_0 - x_1)f(x_2)$$

But $x_1 - x_0 = h$, so

$$\frac{dP(x_0)}{dx} = \frac{1}{2h}[-3f(x_0) + 4f(x_1) - f(x_2)] + O(h^2)$$

where the error term is:

$$\frac{h^2}{3} f^{(3)}(\xi)$$

Notice that this is the forward difference expression for the first derivative with error of order h^2.

Similarly $x = x_1$ will result in the central difference expression.

$$\frac{dP(x_1)}{dx} = \frac{1}{2h}(f(x_2) - f(x_0)) + O(h^2)$$

where the error term is:

$$\frac{h^2}{6} f^{(3)}(\xi)$$

Notice that the central difference expression with error of order h^2 gives the best approximation. The error term is reduced by ½ because information is available on both sides of the point being approximated, thus enabling an average value.

If $x = x_2$, the result will be the backward difference expression.

$$\frac{dP(x_2)}{dx} = \frac{1}{2h}[f(x_0) - 4f(x_1) + 3f(x_2)] + O(h^2)$$

where the error term is

$$\frac{h^2}{3} f^{(3)}(\xi)$$

Note that the in the forward and backward expressions with error of order h^2, the error is larger because the approximations use points that are farther ahead or farther behind the point being approximated.

Similarly if the second derivative were taken, the results would be the same as those derived previously; however higher order difference expressions require more points.

6.4 ERROR CONSIDERATIONS

Recall from the previous sections that a minimum or n+1 points were required to get the nth derivative approximation. The error terms involved in these approximations were of the order h for the forward or backward divided differences and of order h^2 for the central divided differences.

This means that for interval sizes h, less than one, the central divided difference expressions will have a much smaller error.

If more than n+1 points are used to get any of the difference expressions, the error will be reduced. Consider the case of using three points to get the forward expression for the first derivative at a point x_i (see fig. 6.4).

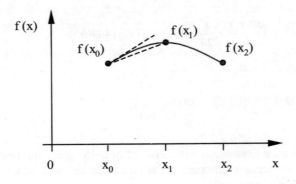

Figure 6.4: Forward expression for the first derivative

Previously, when two points were used only one expansion was required. This time the extra point is used to get another expansion expression.

The expansion about a point x_i in the vicinity of x_{i+1} is

$$f(x_{i+1}) = f(x_i) + hf'(x_i) + \frac{h^2}{2!} f''(x_i) + \frac{h^3}{3!} f'''(x_i) + \dots \qquad (6.4.1)$$

$$f(x_{i+2}) = f(x_i) + 2hf'(x_i) + 2h^2 f''(x_i) + \frac{4}{3}h^3 f'''(x_i) + \dots \qquad (6.4.2)$$

Since the first derivative is being approximated, all the terms to the right of it are considered as the error terms.

Multiply equation 6.4.1 by four and subtract equation 6.4.2 from it.

$$4f(x_{i+1}) - f(x_{i+2}) = 3f(x_i) + 2hf'(x_i) - \frac{2}{3}h^3 f'''(x_i) + ...$$

The last term on the right is the error term.

$$f'(x_i) = \frac{4f(x_{i+1}) - f(x_{i+2}) - 3f(x_i)}{2h} + \frac{1}{3}h^2 f'''(x_i) - ...$$

The error is of order h^2.

In this case n+2 points were used to give a forward difference expression for the nth derivative with an error of order h^2. If n+3 points are used, then the error involved will be of the order h^3, and so on. Similarly, the backward difference expression using n+2 points will result in an order h^2 error and so on; however when using the central difference expressions, the error is always of the order h to an even power. If n+2 points are used, then the error is of the order h^4, if n+3 points are used then the error is of the order h^6, and so on.

Some derivative expressions for the forward, backward and central differences are given using n+2 points to approximate the nth derivative.

Forward Difference Expressions with error of order h^2

$$f'(x_i) = \frac{-f(x_{i+2}) + 4f(x_{i+1}) - 3f(x_i)}{2h}$$

$$f''(x_i) = \frac{-f(x_{i+3}) + 4f(x_{i+2}) - 5f(x_{i+1}) + 2f(x_i)}{h^2}$$

$$f'''(x_i) = \frac{-3f(x_{i+4}) + 14f(x_{i+3}) - 24f(x_{i+2}) + 18f(x_{i+1}) - 5f(x_i)}{2h^3}$$

$$f^{iv}(x_i) = \frac{-2f(x_{i+5}) + 11f(x_{i+4}) - 24f(x_{i+3}) + 26f(x_{i+2}) - 14f(x_{i+1}) + 3f(x_i)}{h^4}$$

Backward Difference Expressions with error of Order h^2

$$f'(x_i) = \frac{3f(x_i) + 4f(x_{i-1}) + f(x_{i-2})}{2h}$$

$$f''(x_i) = \frac{2f(x_i) - 5f(x_{i-1}) + 4f(x_{i-2}) - f(x_{i-3})}{h^2}$$

$$f'''(x_i) = \frac{5f(x_i) - 18f(x_{i-1}) + 24f(x_{i-2}) - 14f(x_{i-3}) + 3f(x_{i-4})}{2h^3}$$

$$f^{iv}(x_i) = \frac{3f(x_i) + 14f(x_{i-1}) + 26f(x_{i-2}) - 24f(x_{i-3}) + 11(x_{i-4}) - 2f(x_{i-5})}{h^4}$$

Central Difference Expressions with error of Order h^4

$$f'(x_i) = \frac{-f(x_{i+2}) + 8f(x_{i+1}) - 8f(x_{i-1}) + f(x_{i-2})}{12h}$$

$$f''(x_i) = \frac{-f(x_{i+2}) + 16f(x_{i+1}) - 30f(x_i) + 16f(x_{i-1}) - f(x_{i-2})}{12h^2}$$

$$f'''(x_i) = \frac{-f(x_{i+3}) - 8f(x_{i+2}) - 13f(x_{i+1}) - 13f(x_{i-1}) - 8f(x_{i-2}) + f(x_{i-3})}{8h^3}$$

$$f^{iv}(x_i) = \frac{-f(x_{i+3}) + 12f(x_{i+2}) - 39f(x_{i+1}) + 56f(x_i)}{6h^4} +$$

$$\frac{-39f(x_{i-1}) + 12f(x_{i-2}) - f(x_{i-3})}{6h^4}$$

PROBLEMS

1.   Tabulate values of $y = e^x$ for $0 \le x \le 2$ at intervals of $h = 0.1$ and obtain.
     (i) y' and y'' at $x = 0.0$
     (ii) y' and y'' at $x = 0.1$
     (iii) y' and y'' at $x = 0.9$
     (iv) y' and y'' at $x = 2.0$
using appropriate numerical difference formulas.

2.   In a circuit with impressed voltage E(t) and inductance L, Kirchhoff's first law gives the relationship.

$$E(t) = L \frac{di}{dt} + Ri$$

Where R is the resistance in the circuit and i is the current. Suppose we measure the current for several values of t and obtain:

t	1.00	1.01	1.02	1.03	1.04
i	3.10	3.12	3.14	3.18	3.24

where t is measured in seconds, i in amperes, the inductance L is a constant 0.98 henries, and the resistance R is 0.142 ohms. Approximate the voltage E at the values t = 1, 1.01, 1.02, 1.03, and 1.04, using the appropriate three – point formulas.

3.   The following data have been experimentally obtained.

x	1.00	1.01	1.02	1.03	1.04	1.05
f(x)	1.27	1.32	1.38	1.41	1.47	1.52

(i)   Approximate f' (1), f''(1), f'''(1)

(ii)   Obtain a Taylor series expansion for the function using the above values, at x = 1.

4.   Given the set of data

x	1	1.3	1.6	1.9	2.2
y	0.765	0.6201	0.4554	0.2818	0.1104

Obtain a Taylor's series expansion of the function about x =1. Obtain the necessary
derivatives at 1, up to $f^{(4)}$ through numerical differentiation with error of order h.

5.   Given the data below obtain   f ' (1.30), f '' (1.30) and   f ''' (1.30) and obtain a Taylor series
expansion about x = 1.30. Approximate f (1.357) using the expansion.

x	1.2	1.3	1.4	1.5	1.6
f(x)	11.5901	13.7818	14.0428	14.3074	16.8619

Chapter 7

MATRIX EIGENVALUE PROBLEMS

Eigenvalue problems occur in several engineering situations. If the physical problem is modeled by a system of equations, then the following matrix notation can be formed;

$$Y = Ax = \lambda x$$

where a real vector $x \neq O$ exists for which the vector y is real and has the same sense and direction as x. This will happen if and only if the system has an eigenvalue λ. (λ is real and positive).

In this equation the matrix 'A' is a square matrix containing the left hand side of the system equations. The vector x is a vector which describes the system's configuration and λ is the eigenvalue which satisfies the equation when 'A' and 'x' are known. The vector x is known as the eigenvector. If the matrix 'A' is of order n×n, then n eigenvalues will exist. Eigenvalues have a significant meaning in physical engineering problems. Eigenvalues are used mostly to represent a system's natural frequencies. The following sections will discuss methods which can be used to evaluate these eigenvalues and their corresponding eigenvectors.

7.1 GERSCHGORIN CIRCLE THEOREM

The Gerschgorin circle theorem identifies the region where the eigenvalues of the given matrix are likely to occur. If a matrix is diagonal, the eigenvalues are given by the diagonal elements themselves. When the matrix is diagonally dominant, we have

$$\left| a_{ii} \right| \geq \sum_{\substack{j=1 \\ j \neq i}}^{n} \left| a_{ij} \right| \text{ for each i} = 1, 2, 3, \ldots \text{ n}$$

and hence it is reasonable to assume that the eigenvalues will not be much different from the diagonal elements.

An eigenvector associated with an eigenvalue has a definite relation among its elements. However, the eigenvector will be unchanged if all the elements are multiplied or divided by a constant value. Let

{Y} be an eigenvector associated with an eigenvalue λ. This eigenvector can also be expressed in the form.

$$\{X\} = \{Y\}/\|Y\|_\infty$$

Consequently, $\|X\|_\infty = 1$. We have

$$A x - \lambda x = 0.$$

and this can also be expressed as

$$\sum_{j=1}^{n} a_{ij}x_j = \lambda x_i, \qquad i = 1, 2, \ldots n$$

Since $\|X\|_\infty = 1$ the largest element in x is 1. Let the kth element in x be equal to 1. Then

$$\sum_{j=1}^{n} a_{kj}x_j = \lambda x_k,$$

This can be written as

$$(a_{kk} - \lambda)x_k = \sum_{\substack{j=1 \\ j \neq k}}^{n} (-a_{kj}x_j)$$

Since the modulus of a product is less than the product of the moduli, we can write.

$$|(a_{kk} - \lambda)x_k| \leq \sum_{\substack{j=1 \\ j \neq k}}^{n} |a_{kj}| \, |x_j|$$

Since $x_k = 1$ and $|x_j| \leq 1$ for $j = 1, 2, \ldots k-1, k+1, \ldots n$ we can write

$$|a_{kk} - \lambda| \leq \sum_{\substack{j=1 \\ j \neq k}}^{n} |a_{kj}|$$

This result states that the eigenvalue λ cannot differ from the diagonal element a_{kk} by an amount more than the sum of the moduli of the non-diagonal elements in the same row, $\sum_{\substack{j=1 \\ j \neq k}}^{n} |a_{kj}|$. This is stated in the Gerschgorin theorem as follows.

The eigenvalues of an n×n matrix A are contained within a region R, which is made up of the union of all the circles R_k in the complex plane whose centers are given by a_{kk} and the corresponding radii given by $\sum_{\substack{j=1 \\ j \neq k}}^{n} |a_{kj}|$. Further, union of any of these circles that do not intersect the remaining (n-m) circles

must contain exactly k number of eigenvalues including multiplicities, if any.

EXAMPLE 7.1

Consider the 2x2 matrix A = $\begin{bmatrix} 3 & 1 \\ 1 & 2 \end{bmatrix}$. Identify the region of the eigenvalues.

Solution

The eigenvalues are obtained by solving the characteristic equation.

$$(3-\lambda)(2-\lambda)-1=0$$

which gives the two eigenvalues as 1.382 and 3.618.

Using the Gerschgorin theorem the eigenvalues are contained in the union of two circles shown in the figure below.

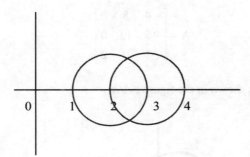

Figure 7.1: Gerschgorin circles

EXAMPLE 7.2

Locate the region of the eigenvalues of the matrix

$$A = \begin{bmatrix} -4 & 14 & 0 \\ -5 & 13 & 0 \\ -1 & 0 & 2 \end{bmatrix}$$

Solution

The region of the eigenvalues is shown in the figure below.

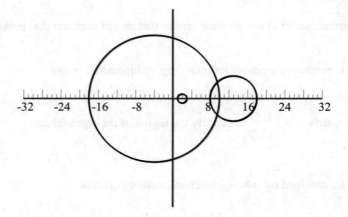

The exact eigenvalues calculated from the characteristic equation are 2,3 and 6.

EXAMPLE 7.3
Locate the region of the eigenvalues of the matrix

$$A = \begin{bmatrix} -4 & 8 & 0 \\ -5 & 13 & 0 \\ -1 & 0 & 2 \end{bmatrix}$$

<u>Solution</u>
The region of the eigenvalues is shown in the figure below.

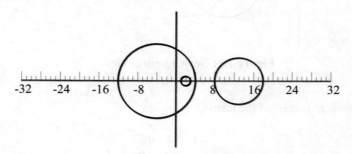

Figure 7.3: Gerschgorin circles

There are two eigenvalues in the union of the two left most circles, and one in the right most circle. The exact eigenvalues calculated from the characteristic equation are −3, 0.5969 and 13.403.

EXAMPLE 7.4
Locate the region of the eigenvalues of the matrix

$$A = \begin{bmatrix} 1 & 24 & 0 \\ -3 & 8 & 0 \\ -1 & 0 & 2 \end{bmatrix}$$

Solution
The region of the eigenvalues is shown in the figure below.

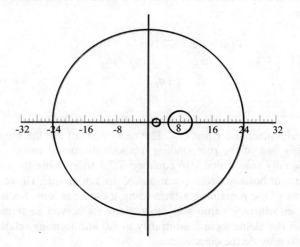

Figure 7.4: Gerschgorin circles

7.2 CHARACTERISTIC EQUATION

An eigenvalue problem appears in the form

$$Ax = \lambda x$$

This equation can be arranged in the following way.

$$(A-\lambda I)x=0 \qquad\qquad (7.2.1)$$

Note that λ is multiplied by the identity matrix so that it can be grouped on the L.H.S.

This now represents a system of homogeneous linear algebraic equations. The trivial solution would be x = 0, however the goal is to find values of λ which satisfy the equation. For a non-trivial solution, the following determinant must be zero.

$$D(\lambda) = \det(A - \lambda I) = \begin{vmatrix} a_{11} - \lambda & a_{12} & \cdots & a_{1n} \\ a_{21} & a_{22} - \lambda & \cdots & a_{2n} \\ \vdots & \vdots & \ddots & \vdots \\ a_{n1} & a_{n2} & \cdots & a_{nn} - \lambda \end{vmatrix} = 0$$

Note the values a_{ij} are known. It is desired to solve for the eigenvalues λ.

This is known as the characteristic determinant. It can be expanded using the co-factor method to give a polynomial $P(\lambda)$, given by.

$$P(\lambda) = \det(A-\lambda I) =$$

$$(a_{11} - \lambda)\begin{vmatrix} a_{22} - \lambda & \cdots & a_{2n} \\ \vdots & \ddots & \vdots \\ a_{n2} & \cdots & a_{nn} - \lambda \end{vmatrix} - a_{12}\begin{vmatrix} a_{21} & a_{23} & \cdots & a_{2n} \\ \vdots & \ddots & & \vdots \\ a_{n1} & a_{n3} & \cdots & a_{nn} - \lambda \end{vmatrix} + \cdots = 0$$

This is known as the characteristic equation. The roots of $P(\lambda)$ are the eigenvalues $\lambda_1, \lambda_2, \ldots \lambda_n$. These can be found by any one of the root finding methods discussed earlier. Once these roots are found, they can be individually substituted into equation 7.2.1 to evaluate the corresponding vector x. Since equation 7.2.1 is a set of homogeneous equations, x_i are not unique. However, (n-1) unknowns x_i can be determined in terms of the remaining x term using these equations. Such solutions of vector x, with one element given an arbitrary value and other elements solved in terms of this value is the eigenvector. Setting one of the elements in x arbitrarily to 1.0 and forming relations between the other elements based on this, will provide the eigenvector.

To find the eigenvector corresponding to the first eigenvalue, the following matrix should be solved by Gaussian elimination, where x_A is held as 1.

$$\begin{bmatrix} a_{11} - \lambda_1 & a_{12} & \cdots & a_{1n} \\ a_{21} & a_{22} - \lambda_1 & \cdots & a_{2n} \\ \vdots & \vdots & \ddots & \vdots \\ a_{n1} & a_{n2} & \cdots & a_{nn} - \lambda_1 \end{bmatrix}\begin{bmatrix} 1 \\ x_B \\ \vdots \\ x_C \end{bmatrix} = \begin{bmatrix} 0 \\ 0 \\ \vdots \\ 0 \end{bmatrix}$$

Note that all the λ values are replaced by the first root λ_1 of $P(\lambda)$.

The corresponding eigenvector $x_1 = \begin{bmatrix} 1 \\ x_B \\ x_C \end{bmatrix}$. This vector represents the systems situation. An eigenvector can be multiplied by any scalar and therefore can assume many forms. The vector is usually scaled by dividing all the elements of the vector by the element with the largest magnitude.

This method is well suited when the determinant $(A-\lambda I)$ is of size 3x3 or less; however once the size increases past this the determinants become difficult to evaluate and it is better to follow one of the numerical methods to be discussed in later sections.

EXAMPLE 7.5
Consider the following mass-spring system. This system can be described by three differential equations, and hence the system has three natural frequencies, ω_n. Find these eigenvalues and their corresponding eigenvectors.

Each spring has a stiffness of 1 N/m.

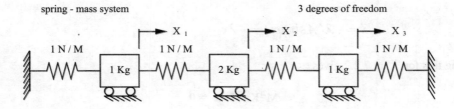

Figure 7.4: Spring mass system

Solution
In order to describe the system consider the equilibrium situation.
Force due to mass + forces by spring = 0

$$[\text{mass matrix}] \, [\ddot{x}\,] + [\text{spring stiffness matrix}][x] = [0]$$

The spring stiffness matrix is evaluated by displacing each mass to the right one unit and then evaluating what force is required to return the mass to its original position.

The equation of motion can be written as

$$[M] \, \{\ddot{x}\} + [K]\{x\} = 0$$

where $[M]$ is mass matrix, and $[K]$ is stiffness matrix.
Substituting for the mass and stiffness matrices, we get

$$\begin{bmatrix} 1 & 0 & 0 \\ 0 & 2 & 0 \\ 0 & 0 & 0 \end{bmatrix}\begin{bmatrix} \ddot{x}_1 \\ \ddot{x}_2 \\ \ddot{x}_3 \end{bmatrix} + \begin{bmatrix} 2 & -1 & 0 \\ -1 & 2 & -1 \\ 0 & -1 & 2 \end{bmatrix}\begin{bmatrix} x_1 \\ x_2 \\ x_3 \end{bmatrix} = \begin{bmatrix} 0 \\ 0 \\ 0 \end{bmatrix}$$

$$(A - \lambda I\,)x = 0$$

We have

$$M\ddot{x} + Kx = 0$$

Multiply throughout by M^{-1} to give

$$I\ddot{x} + M^{-1}Kx = 0 \qquad\qquad (.7.2.2)$$

Now if we assume that the position of the system can be described by the following equation.

Assume a sinusoidal response given by

$$x = A\sin\omega t$$

Then

$$\dot{x} = A\omega\cos\omega t$$
$$\ddot{x} = -A\omega^2\sin\omega t$$

Now let $\omega^2 = \lambda$

Hence

$$\ddot{x} = -\lambda(A\sin\omega t) = -\lambda x$$

Substituting this for $\ddot{x}$ in 7.2.2 gives

$$M^{-1}Kx = I\,\lambda x = 0$$
$$(M^{-1}K - \lambda I)x = 0$$

This is in the standard eigenvalue problem for $(A - \lambda I)x = 0$
where

$$M^{-1}K = A \text{ and } \lambda = \omega^2$$

$$A = \begin{bmatrix} 1 & 0 & 0 \\ 0 & 1/2 & 0 \\ 0 & 0 & 1/3 \end{bmatrix} \begin{bmatrix} 2 & -1 & 0 \\ -1 & 2 & -1 \\ 0 & -1 & 2 \end{bmatrix} = \begin{bmatrix} 2 & -1 & 0 \\ -1/2 & 1 & -1/2 \\ 0 & -1/3 & 2/3 \end{bmatrix}$$

$$P(\lambda) = \det|A - \lambda I|$$

$$P(\lambda) = \begin{vmatrix} 2-\lambda & -1 & 0 \\ -1/2 & 1-\lambda & -1/2 \\ 0 & -1/3 & 2/3-\lambda \end{vmatrix} = (2-\lambda)$$

$$\left[(1-\lambda)\left(\frac{2}{3}-\lambda\right) - \left(-\frac{1}{3}*\left(-\frac{1}{2}\right)\right)\right] - (-1)[-1/2(2/3-\lambda) - 0(-1/2)] = 0$$

$$P(\lambda) = \frac{1}{3}\left[-3\lambda^3 + 11\lambda^2 - 10\lambda + 2\right] = 0$$

$$P(\lambda) = \frac{1}{3}\left[(1-\lambda)\left(\lambda^2 - \frac{8}{3}\lambda + \frac{2}{3}\right)\right] = 0$$

Roots are:

eigenvalues   natural frequencies

$\lambda_1 = 2.387426 \quad \omega_1 = \sqrt{\lambda_1} = 1.54513$

$\lambda_2 = 1.0 \quad\quad\quad \omega_2 = \sqrt{\lambda_2} = 1.0$

$\lambda_3 = 0.2792408 \quad \omega_3 = \sqrt{\lambda_3} = 0.528432$

The largest of these values ω_1, is referred to as the dominant value. To find the corresponding eigenvectors we solve;

$$(A - \lambda I)x = 0$$

for the different λ values

To evaluate the eigenvector corresponding to the first eigenvalue, relationships must be formed for the homogeneous system of equations

$$\lambda_1 = 2.387426$$

$$(A - \lambda_1 I)x_1 = \begin{bmatrix} 2 - 2.387426 & -1 & 0 \\ -1/2 & 1 - 2.387426 & -1/2 \\ 0 & -1/3 & 2/3 - 2.387426 \end{bmatrix} \begin{bmatrix} x_A \\ x_B \\ x_C \end{bmatrix} = \begin{bmatrix} 0 \\ 0 \\ 0 \end{bmatrix}$$

$$(A - \lambda_1 I)x_1 = \begin{bmatrix} -0.387426 & -1 & 0 \\ -1/2 & -1.387426 & -1/2 \\ 0 & -1/3 & -1.7207594 \end{bmatrix} \begin{bmatrix} x_A \\ x_B \\ x_C \end{bmatrix} = \begin{bmatrix} 0 \\ 0 \\ 0 \end{bmatrix}$$

There are infinitely many solutions for x_1. The way to proceed is to arbitrarily choose one of the values of x to be 1.0. Say that we let $x_A = 1.0$.

Then from the first equation

$$-0.387426 \, x_A - x_B = 0$$
$$-0.387426(1.0) = x_B$$
$$x_B = -0.387426$$

From the third equation

$$-\frac{1}{3} x_B - 1.7207594 x_C = 0$$

$$-\frac{1}{3}(-0.387426) = 1.7207594 x_C$$

$$x_C = 0.075$$

The eigenvector is $\qquad$ $x_1 = \begin{bmatrix} 1.0 \\ -0.387426 \\ 0.075 \end{bmatrix}$

Note: x_A was arbitrarily chosen to be 1.0; however we could have chosen any vlaue.

x_1 is the eigenvector corresponding to λ_1. In order to get x_2 or x_3, repeat the procedure substituting λ_2 or λ_3 for λ_1.

EIGENVALUE AND ELGENVECTORS

To obtain the eigenvalues and eigenvectors of the system $Ax = \lambda x$

STEP 1 $\qquad$ Arrange the system equations in an nxn square matrix 'A' such that

$$Ax = \lambda x$$

STEP 2 $\qquad$ Multiply λ by an identity matrix I of size nxn.

STEP 3 $\qquad$ Subtract λI from A and designate the result $(A - \lambda I) X = D(\lambda) = 0$.

STEP 4 $\qquad$ Expand $D(\lambda)$ using the cofactor method to get $P(\lambda)$.

STEP 5 $\qquad$ Solve the roots of $P(\lambda)$ using incremental search with Newton Raphson to get $\lambda_1, \lambda_2, \ldots \lambda_n$.

STEP 6 $\qquad$ Substitute each value of λ back into the matrix

$$(A - \lambda_i)x_i = 0 \text{ for } i = 1, 2, \ldots n$$

and arbitrarily choose $x_A = 1.0$ to calculate and form the different values of x_i for each λ_i

STEP 7 $\qquad$ Print λ_i and x_i i = 1, 2, ... n.

STEP 8 $\qquad$ Stop

CAUTION

If the nxn matrix 'A' is symmetrical, the values of λ will be real. If 'A' is nonsymmetrical there is a chance of complex λ values.

7.3 POWER METHOD

This is an iterative technique which converges on the dominant (largest) eigenvalue and its eigenvector very quickly. In this method the assumption is made that the square n×n matrix 'A' contains n separate eigenvalues $\lambda_1, \lambda_2, \ldots \lambda_n$. It is also assumed that a dominant eigenvalue exists such that $|\lambda_1| > |\lambda_2| > \ldots |\lambda_n|$.

The governing equation of the system is given by

$$(A - \lambda I)\, x = 0 \qquad \text{i.e. } \lambda x = Ax$$

Initially, a guess vector V is assumed for the eigenvector. Since V is not the true eigenvector, it can be expressed as a linear combination of all the eigenvectors as

$$V = \sum_{i=1}^{n} c_i x_i$$

Multiplying both sides by the matrix A and replacing Ax_i by λx_i on the right hand side results in

$$AV = \sum_{i=1}^{n} c_i Ax_i = \sum_{i=1}^{n} c_i \lambda_i x_i$$

Multiplying both sides by $|m-1|^{th}$ power of the matrix A results in

$$A^m V = \sum_{i=1}^{n} c_i \lambda_i^m x_i = \lambda_i^m \sum_{i=1}^{n} c_i \left(\frac{\lambda_i}{\lambda_1} \right)^m x_i$$

Since $|\lambda_1| > |\lambda_2| > |\lambda_3| \ldots\ldots > |\lambda_n|$, the above relation becomes

$$A^m V = c_1 \lambda_1^m x_1 + \lambda_1^m \sum_{i=2}^{n} c_i \left(\frac{\lambda_i}{\lambda_1} \right)^m x_i \approx \lambda_1^m c_1 x_1$$

Hence the power method provides the largest eigenvalue λ_1 and the corresponding eigenvector x_1.

In practice, however, as soon as the guess vector V is multiplied by A several times without scaling, the result will involve very large and unwieldy numbers. To avoid this, the procedure can be set up as an iterative procedure as follows. Assume an initial guess vector as $x^{(0)}$ and obtain the product $Ax^{(0)}$, which will yield a defined value for $\lambda^{(1)} x^{(1)}$. The procedure is described by the following expression:

$$y^{(i)} = \lambda^{(i)} x^{(i)} = Ax^{(i-1)}$$

where $x^{(i-1)}$ is an initial guess at the dominant eigenvector
$x^{(i)}$ is the improved approximation to the eigenvector.

After $x^{(i-1)}$ is multiplied by A, the resulting $y^{(i)}$ vector is normalized by dividing all of its elements by the element having the largest magnitude in absolute sense (the normalizing element). This way, the normalizing element is an approximation for the eigenvalue $\lambda_1^{(i)}$.

The convergence criterion can be established on the basis of the relative error between the successive approximations for the eigenvalue being less than a specified small value,

$$\frac{\left|\lambda^{(i)} - \lambda^{(i-1)}\right|}{\left|\lambda^{(i)}\right|} < \xi$$

The power method will only converge on the dominant eigenvalue, and therefore the method has a limited application. However once one eigenvalue is known, the characteristic polynomial of the previous method can be reduced by one degree making it simpler to evaluate the remaining eigenvalues.

EXAMPLE 7.6
Evaluate the eigenvalues of the following n×n square matrix 'A', using the power method in conjunction with the characteristic polynomial. The accuracy used in the power method should be $\xi <$ 0.001.

$$A = \begin{bmatrix} 10 & -4 & 0 \\ -6 & 9 & -3 \\ 0 & -6 & 6 \end{bmatrix}$$

Solution
The general form is

$$(A - \lambda I)x = 0$$

we have

$$D(\lambda) = \det(\lambda) = \det(A - \lambda I) = \begin{bmatrix} 10-\lambda & -4 & 0 \\ -6 & 9-\lambda & -3 \\ 0 & -6 & 6-\lambda \end{bmatrix} = 0$$

This is expanded by the cofactor method to get

$$P(\lambda) = (10-\lambda)\begin{vmatrix} 9-\lambda & -3 \\ -6 & 6-\lambda \end{vmatrix} - (-4)\begin{vmatrix} -6 & -3 \\ 0 & 6-\lambda \end{vmatrix} + 0\begin{vmatrix} -6 & 9-\lambda \\ 0 & -6 \end{vmatrix} = 0$$

$$P(\lambda) = (10-\lambda)(\lambda^2 - 15\lambda + 36) + 4(6\lambda - 36) = 0$$

$$P(\lambda) = \lambda^3 - 25\lambda^2 + 162\lambda - 216 = 0$$

The roots of this polynomial are not obvious; the dominant root can be found by the power method using an iterative technique.

$$y^{(i)} = Ax^{(i-1)} \qquad i = 1, 2, \ldots$$

An initial guess at the eigenvector is required to start the iteration.

$$\text{Let } x^{(0)} = \begin{bmatrix} 1 \\ 1 \\ 1 \end{bmatrix}$$

Then

$$y^{(1)} = Ax^{(0)} = \begin{bmatrix} 10 & -4 & 0 \\ -6 & 9 & -3 \\ 0 & -6 & 6 \end{bmatrix} \begin{bmatrix} 1 \\ 1 \\ 1 \end{bmatrix} = \begin{bmatrix} 6 \\ 0 \\ 0 \end{bmatrix}$$

Now $y^{(1)}$ is normalized by dividing all of its elements by the element having the largest magnitude to get.

$$y^{(1)} = \lambda^{(1)} * x^{(1)} = 6 \begin{bmatrix} 1 \\ 0 \\ 0 \end{bmatrix}$$

where $\lambda^{(1)} = 6$ is the first approximation for the eigenvalue.
The next approximation is obtained as

$$y^{(2)} = Ax^{(1)} = \begin{bmatrix} 10 & -4 & 0 \\ -6 & 9 & -3 \\ 0 & -6 & 6 \end{bmatrix} \begin{bmatrix} 1 \\ 0 \\ 0 \end{bmatrix} = \begin{bmatrix} 10 \\ -6 \\ 0 \end{bmatrix}$$

Using the largest element in the vector $y^{(2)}$ we get $x^{(2)}$ as,

$$x^{(2)} = \begin{bmatrix} 1 \\ -0.6 \\ 0 \end{bmatrix} \text{ and } \lambda^{(2)} = 10.$$

The relative error between consecutive eigenvalue approximation is.

$$\xi = \left| \frac{10 - 6}{10} \right| = 0.4$$

This procedure is continued until a satisfactory ξ is found. The subsequent calculations are given in the form of a table.

i	$x^{(i-1)}$			$y^{(i)}$			$\lambda^{(i)}$	ξ
1	1	1	1	6	0	0	6	.4
2	1	0	0	10	-6	0	10	.194
3	1	-.6	0	12.4	-11.4	3.6	12.4	1.82
4	1	-.919	-.290	13.676	- 15.141	7.254	-15.14	2.05
5	-.903	1	-.479	-13.030	15.855	-8.874	15.855	.016
6	-.822	1	-.560	-12.220	15.612	-9.360	15.612	.007
7	-.783	1	-.600	-11.830	15.498	-9.600	15.498	.004
8	-.763	1	-.619	-11.630	15.435	-9.714	15.435	.002
9	-.753	1	-.629	-11.530	15.405	-9.774	15.405	.001
10	-.748	1	-.634	-11.480	15.390	-9.804	15.390	.0002
11	-.746	1	-.637	-11.460	15.387	-9.822	15.387	

dominant eigenvalue is 15.387

dominant eigenvector is $\qquad$ $x = \begin{bmatrix} -.746 \\ 1 \\ -.637 \end{bmatrix}$

Note that the ratio between the corresponding elements of $y^{(i)}$ and $x^{(i-1)}$ will also give the eigenvalue. In the above example the ratio between the first element of $y^{(i)}$ and $x^{(i-1)}$ is equal to 15.362 and between that of the third elements is 15.419. Hence strictly the normalization of the vector $y^{(i)}$ can be carried out using any one of its elements. However, using the element having the largest magnitude for normalization may give a better numerical result, and improve the convergency of the result.

Now that λ_i is known, $P(\lambda)$ can be reduced by one degree.

$$\frac{P(\lambda)}{(\lambda - \lambda_i)} = \frac{\lambda^3 - 25\lambda^2 + 162\lambda - 216}{(\lambda - 15.38)}$$

$$= (\lambda - 15.38) \overline{\smash{)}\lambda^3 - 25\lambda^2 + 162\lambda - 216}$$
$$\overline{\lambda^2 - 9.62\lambda + 14}$$

$$\underline{\lambda^3 - 15.38\lambda^2}$$
$$-9.62\lambda^2 + 162\lambda$$
$$\underline{-9.62\lambda^2 + 147.9\lambda}$$
$$14\lambda - 216$$
$$\underline{14\lambda - 216}$$
$$0$$

Now $P(\lambda) = (\lambda - 15.38)(\lambda^2 - 9.62\lambda + 14) = 0$

The remaining eigenvalues are evaluated by simply applying the quadratic equation.

$$\lambda_2, \lambda_3 = \frac{-b \pm \sqrt{b^2 - 4ac}}{2a} = \frac{9.62 \pm \sqrt{92.54 - 4(14)}}{2} = 7.83, 1.79$$

POWER METHOD

To find the largest eigenvalue and the associated eigenvector of an n×n matrix A.

STEP 1   Input: a_{ij}, i = 1, 2, ... n, j =1, 2, ... n, elements of matrix A

              $x^{(0)}$ initial guess vector for eigenvalue

              ξ relative error

              N maximum number of iterations

STEP 2   Set k = 1, $\lambda_0 = 0$

STEP 3   Set y=Ax

STEP 4   Find the element in y having the largest magnitude, such that

              $1 \le p \le n, \qquad \left| y_p \right| = \| y \|_\infty$

STEP 5   Set $\lambda_p = y_p$

STEP 6   Set $x = \dfrac{y}{y_p}$

STEP 7   If $\dfrac{\left| \lambda_k - \lambda_{k-1} \right|}{\left| \lambda_k \right|} < \xi$ go to step 10

STEP 8   Set k=k+1

STEP 9   If k < N go to 3

STEP 10   Output: λ_k and x

7.4 INVERSE POWER METHOD

The eigenvalue problem is stated as   Ax = λx.

If this is reformulated as $A^{-1}x = \dfrac{1}{\lambda} x,$

We see that the eigenvalue of the matrix A^{-1} obtained by the power method will be the largest $(1/\lambda)$ or the smallest λ. Hence the power method applied to the matrix A^{-1} is called the inverse power method and yields the lowest eigenvalue of matrix A. This method can be conveniently used to converge on any eigenvalue by writing.

$$(A\text{-}qI)x = (\lambda - q)x$$

which can be put in the form

$$(A\text{-}qI)^{-1} x = \left(\frac{1}{\lambda - q}\right)x$$

By choosing q very close to an eigenvalue and applying the power method on the matrix $(A\text{-}qI)^{-1}$ we will converge on the largest value of $\dfrac{1}{\lambda - q}$ which means that we have converged on the eigenvalue closest to q.

The iterative procedure can be described in the following way:

$$(A\text{-}qI)^{-1}x^{(i-1)} = y^{(i)}$$

or

$$(A\text{-}qI)y^{(i)} = x^{(i-1)}$$

where $x^{(i-1)}$ is an initial guess for the eigenvector of the eigenvalue in the vicinity of the value q; and $y^{(i)}$ is the improved approximation to the eigenvector. $y^{(i)}$ is evaluated by performing Gaussian elimination on the square matrix $(A\text{-}qI)$ using the initial guess vector $x^{(i-1)}$ as the right hand side of the system. Once $y^{(i)}$ is evaluated, it is normalized by dividing all its elements by the element having the largest magnitude. This normalized vector is now used as the new $x^{(i-1)}$ and the procedure is repeated to find a better approximation $y^{(i)}$. The element having the largest magnitude in $y^{(i)}$ which is used to normalize it, is also an approximation to the eigenvalue of $(A\text{-}qI)^{-1}$. This is equated to $\dfrac{1}{\lambda_i - q}$ to evaluate the eigenvalue. This procedure is continued for consecutive eigenvalue approximations, until the relative error between two consecutive values satisfies a relative error criterion.

Appropriate values of q are chosen by applying the Gerschgorin theorem to the n×n square matrix 'A'. This is done by denoting the diagonal element A_{ii}, i=1,2,...n as center points of circles with radii equivalent to the sum of the absolute values of non-diagonal elements in the respective rows. The circles are then graphed on the x-y plane. If k number of circles intersect each other but not the remaining (n-k) circles, then the union of the k circles will contain exactly k eigenvalues. Note that if the lowest eigenvalue is to be found the corresponding q=0.

The inverse power method is more powerful than the power method, because it allows all the eigenvalues to be found by appropriately choosing q. However, the method is extremely tedious since Gaussian elimination must be performed for each iteration.

EXAMPLE 7.7
Evaluate the smallest eigenvalue of the following system.

$$A = \begin{bmatrix} -4 & 14 & 0 \\ -5 & 13 & 0 \\ -1 & 0 & 2 \end{bmatrix}$$

Solution
The first step is to evaluate the appropriate value of q using the Gerschgorin theorem. The values of A_{ii} are –4, 13 and 2, hence the circles formed are;

Center = -4   radius = $|14| + |0|$ = 14 units

Center = 13   radius = $|5| + |0|$ = 5 units

Center = 2   radius = $|-1| + |0|$ = 1 units

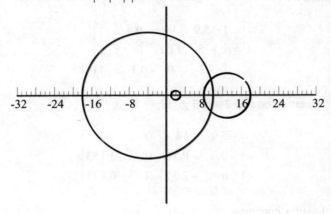

Figure 7.6: Gerschgorin circles

From Gerschgorin circles we can only asset that the eigenvalues fall in the range $-18 \le \lambda \le 18$. For the smallest eigenvalue q can be chosen as zero, or a small number. In this example let us choose q=1.9.

Then (A-qI) becomes;

$$(A - qI) = \begin{bmatrix} -5.9 & 14 & 0 \\ -5 & 11.1 & 0 \\ -1 & 0 & .1 \end{bmatrix}$$

To start the iterative procedure, choose an eigenvector,

$$x^{(0)} = \begin{bmatrix} 1 \\ 1 \\ 1 \end{bmatrix}$$

Recall the iterative procedure

$$(A-qI)y^{(i)}=x^{(i-1)}$$

i = 1 (first iteration)

$$\begin{bmatrix} -5.9 & 14 & 0 \\ -5 & 11.1 & 0 \\ -1 & 0 & .1 \end{bmatrix} \begin{bmatrix} y_A \\ y_B \\ y_C \end{bmatrix} = \begin{bmatrix} 1 \\ 1 \\ 1 \end{bmatrix}$$

Putting this into augmented form;

$$\widetilde{A}^1 = \begin{bmatrix} -5.9 & 14 & 0 & \vdots & 1 \\ -5 & 11.1 & 0 & \vdots & 1 \\ -1 & 0 & 0.1 & \vdots & 1 \end{bmatrix}$$

Performing Gaussian elimination on row 2 and 3 gives;

$$\widetilde{A}^2 = \begin{bmatrix} -5.9 & 14 & 0 & \vdots & 1 \\ 0 & -0.76 & 0 & \vdots & 0.153 \\ 0 & -2.37 & .1 & \vdots & 0.831 \end{bmatrix}$$

Now the third augmented matrix becomes;

$$\widetilde{A}^3 = \begin{bmatrix} -5.9 & 14 & 0 & \vdots & 1 \\ 0 & -0.76 & 0 & \vdots & 0.153 \\ 0 & 0 & 0.1 & \vdots & 0.357 \end{bmatrix}$$

Now back substitution yields $y^{(1)}$;

$$y^{(1)} = \begin{bmatrix} -0.643 \\ -0.200 \\ 3.57 \end{bmatrix}$$

Using the largest element in $y^{(i)}$ (in the absolute sense) consistently to get an approximation for $\dfrac{1}{\lambda_n - q}$ in the ensuing calculations we have.

$$y_m^{(i)} = \frac{1}{\lambda_i - q}$$

In this case;

$$3.57 = \frac{1}{\lambda_i - 1.9}$$
$$3.57\,(\lambda_i - 1.9) = 1$$
$$\text{hence } \lambda_i = 2.18$$

Note that this is a crude approximation; however it gives us something to compare with in the next iteration.

To get the next approximation, it is necessary to normalize $y^{(i)}$ and equate it to $x^{(i)}$. $y^{(i)}$ is normalized by dividing all of its elements by the third element as shown below,

$$x^{(1)} = \frac{1}{3.57}\, y^{(1)} = \begin{bmatrix} -0.180 \\ -0.056 \\ 1.00 \end{bmatrix}$$

Now i = 2 (second iteration)

$$(A - qI)\, y^{(2)} = x^{(1)}$$

$$\begin{bmatrix} -5.9 & 14 & 0 \\ -5 & 11.1 & 0 \\ -1 & 0 & 0.1 \end{bmatrix} \begin{bmatrix} y_A \\ y_B \\ y_C \end{bmatrix} = \begin{bmatrix} -0.180 \\ -0.056 \\ 1.00 \end{bmatrix}$$

Putting this into augmented form.

$$A \sim^1 = \begin{bmatrix} -5.9 & 14 & 0 & \vdots & -0.180 \\ -5 & 11.1 & 0 & \vdots & -0.056 \\ -1 & 0 & 0.1 & \vdots & 1.00 \end{bmatrix}$$

Notice that the L.H.S. of the augmented matrix is reduced exactly the same way as in the first iteration. Performing Gaussian elimination on rows 2 and 3 gives;

$$\tilde{A}^2 = \begin{bmatrix} -5.9 & 14 & 0 & \vdots & -.180 \\ 0 & -.76 & 0 & \vdots & .097 \\ 0 & -2.37 & .1 & \vdots & 1.030 \end{bmatrix}$$

Now the third augmented matrix becomes;

$$\tilde{A}^3 = \begin{bmatrix} -5.9 & 14 & 0 & \vdots & -.180 \\ 0 & -.76 & 0 & \vdots & .097 \\ 0 & 0 & .1 & \vdots & .730 \end{bmatrix}$$

Now back substitution yields $y^{(2)}$

$$y^{(2)} = \begin{bmatrix} -.270 \\ -.127 \\ 7.30 \end{bmatrix}$$

The approximation for the eigenvalue is

$$7.3 = \frac{1}{\lambda_i - 1.9}$$
$$7.3(\lambda_i - 1.9) = 1$$
$$\lambda_i = 2.04$$

The relative error is;

$$\xi = \left| \frac{2.04 - 2.18}{2.04} \right| = 0.070$$

Normalizing

$$x^{(2)} = \begin{bmatrix} -0.04 \\ -0.02 \\ 1.0 \end{bmatrix}$$

Now i =3 (third iteration)

$$(A-qI)y^{(3)}=x^{(2)}$$

Subsequent calculations are given in the form of a table.

i	$x^{(i-1)}$			$y^{(i)}$			λ_i	ξ
1	1	1	1	-.643	-.200	3.57	2.18	.070
2	-.180	-.056	1	-.270	-.127	7.30	2.04	.017
3	-.037	-.017	1	-.037	-.018	9.63	2.00	.002
4	-.004	-.002	1	-.004	-.002	9.96	2.00	
5	-.000	-.000	1					

Answer: $\lambda = 2.00$ with vector $x = \begin{bmatrix} -.000 \\ -.000 \\ 1 \end{bmatrix}$

INVERSE POWER METHOD

To find the eigenvalue close to q and the corresponding eigenvector of an n×n matrix A.

STEP 1   Input: a_{ij} , i = 1, 2, ... n, j = 1, 2, ... n, elements of matrix A

                q: a number close to the eigenvalue to be found; may be obtained from Gerschgorin circle
                theorem.

                ξ relative error

                N maximum number of iterations

STEP 2   Set k = 1, $\lambda_0 = 0$

STEP 3   Solve the system (A-qi)y = x using Gaussian elimination

STEP 4   Find the element in y having the largest magnitude such that

                $1 \le p \le n, |y_p| = \|y\|_\infty$

STEP 5   Set $\dfrac{1}{\lambda_k - q} = y_p$

STEP 6   Set $\lambda_k = q + \dfrac{1}{y_p}$

STEP 7   Set $x = \dfrac{y}{y_p}$

STEP 8   If $\dfrac{|\lambda_k - \lambda_{k-1}|}{|\lambda_k|} < \xi$ go to step 11

STEP 9   Set k = k+1

STEP 10   If k < N go to 3

STEP 11   Output: λ_k and x

DEFLATION TECHNIQUES

Deflation techniques are used to obtain the remaining eigenvalues and eigenvectors of a matrix, after the dominant eigenvalue is obtained using the Power Method. They involve forming a new matrix B from the original matrix A whose eigenvalues are the same as those of A with the exception that the dominant eigenvlaue of A is replaced by the eigenvalue zero in B.

Suppose that $\lambda_1, \lambda_2, \ldots \lambda_n$ are eigenvalues of A with associated eigenvectors $v_1, v_2, \ldots v_n$ and that λ_1 has multiplicity of one. If x is any vector with the property $x^T . v_1 = 1$, then the matrix.

$$B = A - \lambda_1 v_1 x^T$$

has eigenvalues 0, $\lambda_2, \lambda_3, \ldots \lambda_n$ with associated eigenvectors $v_1, w_2, w_3 \ldots w_n$ where v_i and w_i are related by

$$v_i = (\lambda_i - \lambda_1) w_i + \lambda_1 (x^T w_i) v_i$$

for each i = 2, 3, 4, ...n

Wielandt's deflation technique is the most commonly used deflation technique. Here the vector x is defined as

$$x = \frac{1}{\lambda_1 v_{1i}} \left\{ \begin{array}{c} a_{i1} \\ a_{i2} \\ \vdots \\ a_{in} \end{array} \right\}$$

where v_{1i} is the i-th element of vector v_i which is non-zero and $a_{i1}, a_{i2}, \ldots a_{in}$ are the elements from the i-th row of A. With this definition we have

$$x^T v_1 = \frac{1}{\lambda_1 v_{1i}} \sum_{j=1}^{n} a_{ij} v_{1j}$$

The right hand side is equal to the product of A with v_1. Since $Av_1 = \lambda_1 v_1$, we have the result

$$x^T v_1 = 1$$

With this x, if we formulate

$$B = A - \lambda_1 v_1 x^T$$

then the i-th row of B consists of all zeroes. If $\lambda \neq 0$ is any eigenvalue with associated eigenvector v, then $Bv = \lambda v$, implies that the i-th coordinate or element of vector v must also be zero. Consequently, the i-th column of matrix B makes no contribution to the product $Bv = \lambda v$. Hence, B can be replaced by B', which is obtained by deflating the i-th row and column from B. B' will have eigenvalues $\lambda_2, \lambda_3, .. \lambda_n$. If $|\lambda_2| > |\lambda_3|$, the power method can be reapplied to B' to determine the new dominant eigenvalue, and the associated eigenvector w_2'. The eigenvector w_2 of matrix B is obtained by inserting a zero between the $w_{2,i-1}$ and $w_{2,i}$ elements. The corresponding eigenvector of matrix A is obtained from the relation

$$v_i = (\lambda_i - \lambda_1)w_i + \lambda_1(x^T w_i)v_1$$

EXAMPLE 7.8
Obtain the eigenvalues and associated eigenvectors of the matrix

$$A = \begin{bmatrix} 4 & -1 & 1 \\ -1 & 3 & -2 \\ 1 & -2 & 3 \end{bmatrix}$$

Solution
The dominant eigenvalue can be found by applying the power method as $\lambda_1 = 6$ with the associated eigenvector $v_1 = (1,-1,1)^T$. To obtain λ_2 and v_2 form the x vector as

$$x = \frac{1}{6x_1}\begin{Bmatrix} 4 \\ -1 \\ 1 \end{Bmatrix} = \left(\frac{2}{3}, \frac{-1}{6}, \frac{1}{6}\right)^T$$

where the first column of A is used along with the first element of v_1. The matrix B is

$$B = A - \lambda_1 v_1 x^T = \begin{bmatrix} 0 & 0 & 0 \\ 3 & 2 & -1 \\ -3 & -1 & 2 \end{bmatrix}$$

Hence,

$$B' = \begin{bmatrix} 2 & -1 \\ -1 & 2 \end{bmatrix}$$

whose eigenvalues are 3 and 1. The eigenvector w_2' corresponding to 3 can be obtained as

$$w_2' = (1,-1)^T$$

Hence $w_2 = (0,-1,1)^T$

and $v_2 = (3-6)\ w_2 + 6\ (x^T w_2)\ v_1 = (-2,-1,1)^T$

Deflation techniques are prone to round off errors. When all the eigenvalues and eigenvectors are required it is more convenient to use other methods based on similarity transformations, which will be discussed later.

Matrix Similarity Transformations

Use of similarity transformations are made in the following methods to transform matrices so as to suppress the non-diagonal terms while the eigenvalues of the transformed matrices remain identical to the original matrix. In order to perform similarity transformations, it is essential to understand the following properties of matrices.

Diagonal matrix: A diagonal matrix is a square matrix whose elements $A_{j,k}=0$ for all $j \neq k$.

$$\text{diagonal matrix} \quad A = \begin{bmatrix} A_{11} & 0 & \cdots & 0 \\ 0 & A_{22} & & \vdots \\ \vdots & \vdots & \ddots & \vdots \\ 0 & 0 & \cdots & A_{nn} \end{bmatrix}$$

Similar matrix:
A and B are similar matrices if there is an invertible matrix P such that

$$B = P^{-1}AP$$

Also if A and B are similar matrices, and λ is an eigenvalue of A, then it is also an eigenvalue of B.

Matrix transpose (A^T):
The transpose A^T of an (mxn) matrix 'A' is the (nxm) matrix which will result by interchanging the rows and columns in A. In other words the kth row of A becomes the kth column of A^T.

$$A = \begin{bmatrix} A_{11} \\ A_{21} \\ A_{31} \end{bmatrix} \text{ then } A^T = \begin{bmatrix} A_{11}, & A_{21}, & A_{31} \end{bmatrix}$$

Orthogonal matrix:
A matrix 'A' is orthogonal if A is invertible and, has the property $A^{-1} = A^T$.

7.5 JACOBI'S METHOD

This method uses an iterative procedure to reduce a square symmetric matrix 'A' into a similar diagonal matrix. A series of suitable transformations is carried out on 'A' so that all the off diagonal terms are eventually reduced to values so close to zero that they can be neglected, leaving the main diagonal only. If these transformations are performed in such a way that they produce a similar matrix to 'A' each time, then the eigenvalues of the reduced matrix will be identical to those of the original matrix 'A'. The final reduced matrix will yield these eigenvalues along its main diagonal. These similarity transformations of 'A' are performed by multiplying 'A' by an invertible transformation matrix S in the following way.

$$A_{new} = S^{-1} * A_{old} * S$$

The transformation matrix S is chosen so that it is orthogonal. This property allows us to work with the transpose of S, rather than its inverse.

Since S is orthogonal; $S^{-1} = S^T$ and the iterative procedure can be described by the equation:

$$A_{new} = S^T * A_{old} * S$$

By multiplying $S^T * A_{old} * S$, the coordinate system which describes the matrix 'A' is shifted by an angle θ, so that some of the elements on 'A' become zero. This is similar to the way in which a stress element is rotated to find its principal stresses using Mohr's circle.

The transformation matrix S in the two dimensional situation has the following form.

$$S = \begin{bmatrix} \cos\theta & -\sin\theta \\ \sin\theta & \cos\theta \end{bmatrix} \text{ Hence } S^T = \begin{bmatrix} \cos\theta & \sin\theta \\ -\sin\theta & \cos\theta \end{bmatrix}$$

This is identical to the transformation used to transform from one cartesian coordinate system to another, inclined at an angle θ to the former system.

The quantities $\cos\theta$ and $\sin\theta$ can be expressed in terms of elements in the matrix 'A'. To get these standard relationships consider a 2x2 matrix 'A'.

$$A_{old} = \begin{bmatrix} A_{pp} & A_{pq} \\ A_{qp} & A_{qq} \end{bmatrix} \qquad \text{A is square symmetrical}$$

Then

$$A_{new} - S^T A_{old} S$$

or

$$\begin{bmatrix} A'_{pp} & A'_{pq} \\ A'_{qp} & A'_{qq} \end{bmatrix} = \begin{bmatrix} \cos\theta & \sin\theta \\ -\sin\theta & \cos\theta \end{bmatrix} \begin{bmatrix} A_{pp} & A_{pq} \\ A_{qp} & A_{qq} \end{bmatrix} \begin{bmatrix} \cos\theta & -\sin\theta \\ \sin\theta & \cos\theta \end{bmatrix}$$

$$\quad A_{new} \qquad\qquad S^T \qquad\qquad A_{old} \qquad\qquad S$$

Due to the symmetric property of A, $A'_{pq} = A'_{qp}$

Carrying out the multiplication results in the following equations.

$$A'_{pq} = A'_{qp} = (\cos^2\theta - \sin^2\theta)A_{pq} + \cos\theta\sin\theta(A_{qq} - A_{pp}) \qquad (7.5.1A)$$

$$A'_{pp} = \cos^2\theta A_{pp} + \sin^2\theta A_{qq} + 2\cos\theta\sin\theta A_{pq} \qquad (7.5.1B)$$
$$A'_{qq} = \cos^2\theta A_{qq} + \sin^2\theta A_{pp} - 2\cos\theta\sin\theta A_{pq} \qquad (7.5.1C)$$

The following trigonometirc identities

$$\cos^2\theta - \sin^2\theta = \cos 2\theta$$
$$\sin\theta\cos\theta = \frac{1}{2}\sin 2\theta$$

are substituted into 7.5.1(A) to yield

$$\cos 2\theta\, A_{pq} + \frac{1}{2}\sin 2\theta(A_{qq} - A_{pp}) = A'_{pq} = A'_{qp}$$

The angle of rotation θ is chosen in such a way as to reduce A'_{pq} and A'_{qp} to zero. Hence,

$$\cos 2\theta\, A_{pq} + \frac{1}{2}\sin 2\theta(A_{qq} - A_{pp}) = 0$$

Rearranging this gives,

$$\frac{\sin 2\theta}{\cos 2\theta} = \frac{A_{pq}}{\frac{1}{2}(A_{pp} - A_{qq})} = \tan 2\theta$$

Consider the right angle triangle shown. We have

$$\tan\theta = \frac{a}{b}$$
$$\sin\theta = \frac{a}{\sqrt{a^2+b^2}}$$
$$\cos\theta = \frac{b}{\sqrt{a^2+b^2}}$$

using this relation, from the second triangle we get

$$\sin 2\theta = \frac{A_{pq}}{\sqrt{A_{pq}^2 + \frac{1}{4}(A_{pp} - A_{qq})^2}} = \alpha$$

Let the $\sin 2\theta$ expression equal α for simplicity. Again let

$$\cos 2\theta = \frac{\frac{1}{2}(A_{pp} - A_{qq})}{\sqrt{A_{pq}^2 + \frac{1}{4}(A_{pp} - A_{qq})^2}} = \beta$$

We have

$$\sin^2 \theta = \frac{1 - \cos 2\theta}{2} \qquad \text{and} \qquad \cos^2 \theta = \frac{1 + \cos 2\theta}{2}$$

Hence

$$\cos^2 \theta = \left(\frac{1}{2} + \frac{\beta}{2}\right) \qquad \text{or} \qquad \cos\theta = \left(\frac{1}{2} + \frac{\beta}{2}\right)^{1/2}$$

$$\sin^2 \theta = \left(\frac{1}{2} - \frac{\beta}{2}\right)$$

or

$$\sin\theta = \left(\frac{1}{2} - \frac{\beta}{2}\right)^{1/2} \qquad \text{if} \qquad A_{pq} \geq 0$$

and

$$\sin\theta = -\left(\frac{1}{2} - \frac{\beta}{2}\right)^{1/2} \qquad \text{if} \qquad A_{pq} < 0$$

In order to eliminate a particular element in 'A', say A_{pq}, the elements of the corresponding 2x2 transformation matrix are positioned in the S matrix so that their S_{pq} elements are in a corresponding position to the A_{pq} element which is being reduced. In the case that the S matrix is larger than 2x2, all the off diagonal positions which are not occupied by transformation elements are assigned a zero value, while those unoccupied values on the main diagonal are assigned values of 1. For example if the A_{12} value of the 3x3 matrix 'A' given below is to be reduced, then p=1, q=2 and the S matrix will assume the following form.

$$S = \begin{bmatrix} \cos\theta & -\sin\theta & 0 \\ \sin\theta & \cos\theta & 0 \\ 0 & 0 & 1 \end{bmatrix}$$

Notice that the off diagonal terms of the transformation matrix are in the same position as those elements in 'A' which are being reduced (A_{12} and A_{21}).

The resulting matrix A_new, will have the form;

$$
\underbrace{\begin{bmatrix} \cos\theta & \sin\theta & 0 \\ -\sin\theta & \cos\theta & 0 \\ 0 & 0 & 1 \end{bmatrix}}_{S^T}
\underbrace{\begin{bmatrix} A_{11} & A_{12} & A_{13} \\ A_{21} & A_{22} & A_{23} \\ A_{31} & A_{32} & A_{33} \end{bmatrix}}_{A_{old}}
\underbrace{\begin{bmatrix} \cos\theta & -\sin\theta & 0 \\ \sin\theta & \cos\theta & 0 \\ 0 & 0 & 1 \end{bmatrix}}_{S}
= \underbrace{\begin{bmatrix} A'_{11} & 0 & A'_{13} \\ 0 & A'_{22} & A'_{23} \\ A'_{31} & A'_{32} & A'_{33} \end{bmatrix}}_{A_{new1}}
$$

Continuing the elimination process let $A_{old} = A_{new}$ and position the A_{pq} elements of the 2x2 transformation matrix so that they correspond with the symmetrical elements in matrix 'A' being eliminated. The normal practice is to eliminate the highest off diagonal element in A_{new}. For instance, say that this value is A'_{23}; then p = 2, q = 3.

The resulting matrix A_new2 will have the form;

$$
\begin{bmatrix} 1 & 0 & 0 \\ 0 & \cos\theta & \sin\theta \\ 0 & -\sin\theta & \cos\theta \end{bmatrix}
\begin{bmatrix} A'_{11} & 0 & A'_{13} \\ 0 & A'_{22} & A'_{23} \\ A'_{31} & A'_{32} & A'_{33} \end{bmatrix}
\begin{bmatrix} 1 & 0 & 0 \\ 0 & \cos\theta & -\sin\theta \\ 0 & \sin\theta & \cos\theta \end{bmatrix}
= \begin{bmatrix} A''_{11} & A''_{12} & A''_{13} \\ A''_{21} & A''_{22} & 0 \\ A''_{31} & 0 & A''_{33} \end{bmatrix}
$$

Notice that the previous values which were reduced to zero now have a new value. This new value is much smaller than the original value in 'A'. This procedure is continued until all the off-diagonal values are reduced to very small values; then the eigenvalues of the system will be the elements on the main diagonal.

The element a_{pq} is called the pivot element for the plane rotation in the (p,q) plane. Plane rotation in the (p,q) plane reduces the sum of the squares of the off-diagonal elements by $2a_{pq}^2$ after rotation. The rotations must be continued on the matrix A until the sum of the squares of the off-diagonal elements is less than or equal to a small quantity ξ.

The pivot elements can be chosen in the following way, a_{21}, a_{31}, a_{32}, a_{41}, a_{42}, a_{43}, and so on, considering all the off diagonal terms in a row up to the main diagonal and then proceeding to the next row.

Using a small pivot is wasteful, because if a_{pq} is small, then the rotation will reduce the sum of the squares of the resulting off-diagonal terms only by $2a_{pq}^2$. Hence, it is useful to fix a threshold value and during each sweep through the off-diagonal elements consider only those pivots which are above the threshold. Another way is to search for the off-diagonal term with the largest magnitude each time and use it as the pivot.

EXAMPLE 7.9
    Find the eigenvalues of the following system matrix 'A'

$$A = \begin{bmatrix} 4 & 2 & 0 \\ 2 & 3 & 1 \\ 0 & 1 & 2 \end{bmatrix}$$

Solution

'A' is a symmetrical, square matrix.

Begin by eliminating A_{12} (A_{21}); hence $p=1$, $q=2$. This means that the off diagonal terms of the 2x2 transformation matrix will occupy the same positions in the S matrix.

$$A_{new} = \underbrace{\begin{bmatrix} \cos\theta & \sin\theta & 0 \\ -\sin\theta & \cos\theta & 0 \\ 0 & 0 & 1 \end{bmatrix}}_{S^T} \underbrace{\begin{bmatrix} 4 & 2 & 0 \\ 2 & 3 & 1 \\ 0 & 1 & 2 \end{bmatrix}}_{A_{old}} \underbrace{\begin{bmatrix} \cos\theta & -\sin\theta & 0 \\ \sin\theta & \cos\theta & 0 \\ 0 & 0 & 1 \end{bmatrix}}_{S}$$

$p = 1, q = 2$

$$\cos\theta = \left(\frac{1}{2} + \frac{\beta}{2}\right)^{1/2} ; \qquad \sin\theta = \pm\left(\frac{1}{2} - \frac{\beta}{2}\right)^{1/2}$$

$$\beta = \frac{\frac{1}{2}\left(A_{pp} - A_{qq}\right)}{\sqrt{A_{pq}^2 + \frac{1}{4}\left(A_{pp} - A_{qp}\right)^2}} = \frac{\frac{1}{2}(4-3)}{\sqrt{4 + \frac{1}{4}(4-3)^2}} = 0.243$$

$$\cos\theta = \left(\frac{1}{2} + \frac{0.243}{2}\right)^{1/2} = 0.788$$

$A_{pq} \geq 0$, so

$$\sin\theta = \left(\frac{1}{2} - \frac{0.243}{2}\right)^{1/2} = 0.615$$

$$A_{new} = \begin{bmatrix} .788 & .615 & 0 \\ -.615 & .788 & 0 \\ 0 & 0 & 1 \end{bmatrix} \begin{bmatrix} 4 & 2 & 0 \\ 2 & 3 & 1 \\ 0 & 1 & 2 \end{bmatrix} \begin{bmatrix} .788 & -.615 & 0 \\ .615 & .788 & 0 \\ 0 & 0 & 1 \end{bmatrix}$$

$$A_{new} = \begin{bmatrix} 4.382 & 3.421 & .615 \\ -.884 & 1.134 & .788 \\ 0 & 1 & 2 \end{bmatrix} \begin{bmatrix} .788 & -.615 & 0 \\ .615 & .788 & 0 \\ 0 & 0 & 1 \end{bmatrix} = \begin{bmatrix} 5.56 & 0 & .615 \\ 0 & 1.437 & .788 \\ .615 & .788 & 2.0 \end{bmatrix}$$

Note that in A_{new}, the elements A_{12} and A_{21} have been reduced to 0.
Also A_{new} is symmetrical.

Next iteration

Let
$$A_{new} = = A_{old}$$

The method will converge faster if the highest off diagonal value is reduced after each iteration; however if a computer is being used, it will be easier to proceed by eliminating off diagonal elements row after row.

In this case the largest off diagonal is 0.788 at $p = 2$, $q = 3$.

Hence

$$A_{new} = \underbrace{\begin{bmatrix} 1 & 0 & 0 \\ 0 & \cos\theta & \sin\theta \\ 0 & -\sin\theta & \cos\theta \end{bmatrix}}_{S^T} \underbrace{\begin{bmatrix} 5.56 & 0 & .615 \\ 0 & 1.437 & .788 \\ 615 & .788 & 2.0 \end{bmatrix}}_{A_{old}} \underbrace{\begin{bmatrix} 1 & 0 & 0 \\ 0 & \cos\theta & -\sin\theta \\ 0 & \sin\theta & \cos\theta \end{bmatrix}}_{S}$$

$p = 2$, $q = 3$

$$\beta = \frac{\frac{1}{2}\left(A_{pp} - A_{qq}\right)}{\sqrt{A_{pq}^2 + \frac{1}{4}\left(A_{pp} - A_{qq}\right)^2}} = \frac{\frac{1}{2}(1.437 - 2.0)}{\sqrt{(0.788)^2 + \frac{1}{4}(1.437 - 2.0)^2}} = -0.336$$

$$\cos\theta = \left(\frac{1}{2} + \frac{-.336}{2}\right)^{1/2} = 0.576$$

$$A_{pq} \geq 0, \text{ so} \quad \sin\theta = \left(\frac{1}{2} - \frac{-.336}{2}\right)^{1/2} = 0.817$$

$$A_{new} = \begin{bmatrix} 1 & 0 & 0 \\ 0 & .576 & .817 \\ 0 & -.817 & .576 \end{bmatrix} \begin{bmatrix} 5.56 & 0 & .615 \\ 0 & 1.437 & .788 \\ .615 & .788 & 2.0 \end{bmatrix} \begin{bmatrix} 1 & 0 & 0 \\ 0 & .576 & -.817 \\ 0 & .817 & .576 \end{bmatrix}$$

$$A_{new} = \begin{bmatrix} 5.56 & 0 & .615 \\ .502 & 1.472 & 2.088 \\ .354 & -.720 & .508 \end{bmatrix} \begin{bmatrix} 1 & 0 & 0 \\ 0 & .576 & -.817 \\ 0 & .817 & .576 \end{bmatrix} = \begin{bmatrix} 5.56 & .502 & .354 \\ .502 & 2.55 & 0 \\ .354 & 0 & .881 \end{bmatrix}$$

Note that the previously deflated value A_{12}, A_{21} assume a new value much lower than their original one.

The largest off diagonal values are A_{12}, A_{21} or $p = 1$, $q = 2$.

Let
$$A_{new} = A_{old}$$

Hence

$$A_{new} = \begin{bmatrix} \cos\theta & \sin\vartheta & 0 \\ -\sin\theta & \cos\theta & 0 \\ 0 & 0 & 1 \end{bmatrix} \begin{bmatrix} 5.56 & .502 & .354 \\ .502 & 2.55 & 0 \\ .354 & 0 & .881 \end{bmatrix} \begin{bmatrix} \cos\theta & -\sin\theta & 0 \\ \sin\theta & \cos\theta & 0 \\ 0 & 0 & 1 \end{bmatrix}$$

$$\phantom{A_{new} =} \underbrace{\phantom{\begin{bmatrix} \cos\theta & \sin\vartheta \end{bmatrix}}}_{S^T} \qquad \underbrace{\phantom{\begin{bmatrix} 5.56 & .502 \end{bmatrix}}}_{A_{old}} \qquad \underbrace{\phantom{\begin{bmatrix} \cos\theta & -\sin \end{bmatrix}}}_{S}$$

$$\beta = \frac{\dfrac{1}{2}\left(A_{pp} - A_{qq}\right)}{\sqrt{A_{pq}^2 + \dfrac{1}{4}\left(A_{pp} - A_{qq}\right)^2}} = \frac{\dfrac{1}{2}(5.56 - 2.55)}{\sqrt{(0.502)^2 + \dfrac{1}{4}(5.56 - 2.55)^2}} = 0.949$$

$$\cos\theta = \left(\frac{1}{2} + \frac{.949}{2}\right)^{1/2} = 0.987$$

$$A_{pq} \geq 0, \text{ so} \qquad \sin\theta = \left(\frac{1}{2} - \frac{.949}{2}\right)^{1/2} = 0.160$$

$$A_{new} = \begin{bmatrix} .987 & .160 & 0 \\ -.160 & .980 & 0 \\ 0 & 0 & 1 \end{bmatrix} \begin{bmatrix} 5.56 & .502 & .354 \\ .502 & 2.55 & 0 \\ .354 & 0 & .881 \end{bmatrix} \begin{bmatrix} .987 & -.106 & 0 \\ .160 & .987 & 0 \\ 0 & 0 & 1 \end{bmatrix}$$

$$A_{new} = \begin{bmatrix} 5.57 & .903 & .349 \\ -.394 & 2.44 & -.057 \\ .354 & 0 & .881 \end{bmatrix} \begin{bmatrix} .987 & -.160 & 0 \\ .160 & .987 & 0 \\ 0 & 0 & 1 \end{bmatrix} = \begin{bmatrix} 5.64 & 0 & .349 \\ 0 & 2.47 & -.057 \\ .349 & -.057 & .881 \end{bmatrix}$$

Now the largest off diagonal value at $A_{pq} = A_{13}$ hence $p=1$, $q=3$.

Let
$$A_{old} = A_{new}$$

$$A_{new} = \begin{bmatrix} \cos\theta & 0 & \sin\theta \\ 0 & 1 & 0 \\ -\sin\theta & 0 & \cos\theta \end{bmatrix} \begin{bmatrix} 5.64 & 0 & .349 \\ 0 & 2.47 & -.057 \\ .349 & -.057 & .881 \end{bmatrix} \begin{bmatrix} \cos\theta & 0 & -\sin\theta \\ 0 & 1 & 0 \\ \sin\theta & 0 & \cos\theta \end{bmatrix}$$

$$\beta = \frac{\frac{1}{2}\left(A_{pp} - A_{qq}\right)}{\sqrt{A_{pq}^{2} + \frac{1}{4}\left(A_{pp} - A_{qq}\right)^{2}}} = \frac{\frac{1}{2}(5.64 - .881)}{\sqrt{(0.349)^{2} + \frac{1}{4}(5.64 - .881)^{2}}} = 0.989$$

$$\cos\theta = \left(\frac{1}{2} + \frac{\beta}{2}\right)^{1/2} = \left(\frac{1}{2} + \frac{.989}{2}\right)^{1/2} = 0.997$$

$$A_{pq} \geq 0, \text{ so} \qquad \sin\theta = \left(\frac{1}{2} - \frac{\beta}{2}\right)^{1/2} = \left(\frac{1}{2} - \frac{.989}{2}\right)^{1/2} = .074$$

$$A_{new} = \begin{bmatrix} .997 & 0 & .074 \\ 0 & 1 & 0 \\ -.074 & 0 & .997 \end{bmatrix} \begin{bmatrix} 5.64 & 0 & .349 \\ 0 & 2.47 & -.057 \\ .349 & -.057 & .881 \end{bmatrix} \begin{bmatrix} .997 & 0 & -.074 \\ 0 & 1 & 0 \\ .074 & 0 & .997 \end{bmatrix}$$

$$A_{new} = \begin{bmatrix} 5.65 & -.004 & .413 \\ 0 & 2.47 & -.057 \\ -.069 & -.057 & .852 \end{bmatrix} \begin{bmatrix} .997 & 0 & -.074 \\ 0 & 1 & 0 \\ .074 & 0 & .997 \end{bmatrix} = \begin{bmatrix} 5.66 & -.004 & -.006 \\ -.004 & 2.47 & -.057 \\ -.006 & -.057 & .855 \end{bmatrix}$$

At this point the off diagonal elements are close enough to zero so that they could be neglected. The eigenvalues are the elements in the main diagonal of the matrix A_{new}.

They are $\lambda_1 = 5.66$; $\lambda_2 = 2.47$ and $\lambda_3 = 0.855$.

JACOBI METHOD

To obtain all the eigenvalues of an n×n symmetric matrix A by similarity transformation.

STEP 1   Input a_{ij}, I = 1, 2, … n, j = 1, 2, … n, elements of matrix A

ξ relative error

STEP 2   Set k = 1

STEP 3   Search for the largest element below the main diagonal,

A_{pq}, p = 1, 2, … n, q =1, 2, … p

STEP 4   Form an n×n matrix S with

$$S_{pp} = S_{qq} = \left[\frac{1}{2} \left(1 + \frac{\frac{1}{2}\left(A_{pp} - A_{qq}\right)}{\sqrt{A_{pq}^2 + \frac{1}{4}\left(A_{pp} - A_{qq}\right)^2}} \right) \right]^{0.5}$$

$$S_{pq} = \left[\frac{1}{2} \left(1 - \frac{\frac{1}{2}\left(A_{pp} - A_{qq}\right)}{\sqrt{A_{pq}^2 + \frac{1}{4}\left(A_{pp} - A_{qq}\right)^2}} \right) \right]^{0.5} \text{, if } A_{pq} \geq 0$$

$$S_{pq} = \left[-\frac{1}{2} \left(1 - \frac{\frac{1}{2}\left(A_{pp} - A_{qq}\right)}{\sqrt{A_{pq}^2 + \frac{1}{4}\left(A_{pp} - A_{qq}\right)^2}} \right) \right]^{0.5} \text{, if } A_{pq} < 0$$

$S_{qp} = -S_{pq}$

$S_{ii} = 1$, and $S_{ij} = 0$, $i = 1, 2, \ldots n$, $j = 1, 2, \ldots n$, $ii \neq pp$, $ii \neq pp$,

$ij \neq pq$, $ij \neq qp$.

STEP 5   Find $A_{new} = S^T A_{old} S$

STEP 6   Obtain $\displaystyle\sum_{\substack{i=1 \\ i \neq j}}^{n} \sum_{j=1}^{n} A_{ij}^2 = E_k$

STEP 7   If $\dfrac{|E_k - E_{k-1}|}{|E_k|} < \xi$ go to step 10

STEP 8   Set $k = k+1$

STEP 9   If $k < n$ go to step 3

STEP 10   Output: A_{ii}

Definition: A tridiagonal matrix is a matrix in which one off diagonal element exists on the immediate top and bottom of each diagonal element. The remaining off diagonal elements are zero. A tridiagonal matrix is shown below.

$$A = \begin{bmatrix} A_{11} & A_{12} & 0 & \cdots & \cdots & 0 \\ A_{21} & A_{22} & \ddots & & & \vdots \\ 0 & A_{32} & & & & 0 \\ \vdots & & \ddots & & & A_{n-1,n} \\ 0 & \cdots & 0 & A_{n,n-1} & & A_{n,n} \end{bmatrix}$$

7.6 HOUSEHOLDER – Q,L METHOD

The Q,L method is an iterative process which is more stable than the Jacobi method, however it can only be applied to a tridiagonal matrix. The symmetric matrix 'A' is reduced to a similar tridiagonal form using a technique known as the Householder method. Once in this form, the Q,L method will converge on the system's eigenvalues by reducing the tridiagonal matrix to a diagonal form. The diagonal elements of this diagonal matrix are the eigenvalues.

HOUSEHOLDER METHOD

This method will only result in a tridiagonal form when applied to a square symmetric matrix. The original matrix 'A' is multiplied by a symmetric orthogonal matrix P in the following fashion.

$$A_{new} = P^T A_{old} P$$

The matrix P is formed in the following way,

$$P = I - \frac{2ww^T}{\|w\|_2^2}$$

where w is a vector containing the following elements;

$$w = \left\{ A_{k1}, A_{k2}, \dots A_{k,k-1} - s, 0, \dots \right\}^T$$

k is the row in matrix 'A' being reduced.

Let $V_k = \{A_{k1}, A_{k2}, \dots A_{kn}\}^T$

which is the row vector in 'A' being reduced.

and $s = \pm\sqrt{A_{k1}^2 + A_{k2}^2 + \dots A_{k,k-1}^2}$

The sign of s is chosen so that the value $\left| A_{kmk-1} - s \right|$ in w will result in the largest absolute value. The procedure is to begin with k=n and decrease in value as each successive matrix P is formed until k=2. This is best illustrated by an example.

EXAMPLE 7.10
Use Householder's method to reduce the following symmetric matrix 'A'.

$$A = \begin{bmatrix} 4 & -2 & 1 & 2 \\ -2 & 3 & 0 & -2 \\ 1 & 0 & 2 & 1 \\ 2 & -2 & 1 & -1 \end{bmatrix} \text{ n×n matrix, n=4}$$

Solution
From the first orthogonal matrix P starting with k=n=4.

$$P_1 = I - \frac{\|2w_1 w_1^T\|}{\|w_1\|_2^2}$$

$$w_1 = \{A_{41}, A_{42}, A_{43} - s, 0\}^T$$
$$V_4 = \{A_{41}, A_{42}, A_{43}, A_{44}\}^T$$
$$s = \pm\sqrt{A_{41}^2 + A_{42}^2 + A_{43}^2}$$
$$w_1 = \{2, -2, 1, -s, O\}^T$$
$$V_4 = \{2, -2, 1, -1\}^T$$

$$s = \pm\sqrt{(2)^2 + (-2)^2 + (1)^2} = \pm 3$$

take s = -3 so that 1-(-3)=4. This is larger than 1-3=-2

Hence $\qquad\qquad w_1 = \{2, -2, 4, 0\}^T$

$$P_1 = I - \frac{2w_1 w_1^T}{\|w_1\|_2^2}$$

$$w_1 w_1^T = \begin{bmatrix} 2 \\ -2 \\ 4 \\ 0 \end{bmatrix} [2, -2, 4, 0] = \begin{bmatrix} 4 & -4 & 8 & 0 \\ -4 & 4 & -8 & 0 \\ 8 & -8 & 16 & 0 \\ 0 & 0 & 0 & 0 \end{bmatrix}$$

$$\|w_2^2\| = 2^2 + (-2)^2 + 4^2 + 0^2 = 24$$

$$P_1 = \begin{bmatrix} 1 & 0 & 0 & 0 \\ 0 & 1 & 0 & 0 \\ 0 & 0 & 1 & 0 \\ 0 & 0 & 0 & 1 \end{bmatrix} - \frac{1}{12} \begin{bmatrix} 4 & -4 & 8 & 0 \\ -4 & 4 & -8 & 0 \\ 8 & -8 & 16 & 0 \\ 0 & 0 & 0 & 0 \end{bmatrix}$$

To simplify this calculation multiply both sides by 12, perform the subtraction, then divide the result by 12.

$$P_1 = \frac{1}{12} \begin{bmatrix} 8 & 4 & -8 & 0 \\ 4 & 8 & 8 & 0 \\ -8 & 8 & -4 & 0 \\ 0 & 0 & 0 & 12 \end{bmatrix}$$

$$A_{new} = P^T A_{old} P$$

$$A_{new} = \frac{1}{12} \begin{bmatrix} 8 & 4 & -8 & 0 \\ 4 & 8 & 8 & 0 \\ -8 & 8 & -4 & 0 \\ 0 & 0 & 0 & 12 \end{bmatrix} \begin{bmatrix} 4 & -2 & 1 & 2 \\ -2 & 3 & 0 & -2 \\ 1 & 0 & 2 & 1 \\ 2 & -2 & 1 & -1 \end{bmatrix} \frac{1}{12} \begin{bmatrix} 8 & 4 & -8 & 0 \\ 4 & 8 & 8 & 0 \\ -8 & 8 & -4 & 0 \\ 0 & 0 & 0 & 12 \end{bmatrix}$$

$$P_1^T A_{old} = \frac{1}{12} \begin{bmatrix} 16 & -4 & -8 & 0 \\ 8 & 16 & 20 & 0 \\ -52 & 40 & -16 & -36 \\ 24 & -24 & 12 & -12 \end{bmatrix}$$

$$P_1^T A_{old} P_1 =$$

$$\frac{1}{144} \begin{bmatrix} 176 & -32 & -128 & 0 \\ -32 & 320 & -16 & 0 \\ -128 & -16 & 800 & -432 \\ 0 & 0 & -432 & -144 \end{bmatrix} = \begin{bmatrix} 1.222 & -.2222 & -.8889 & 0 \\ -.222 & 2.222 & -.1111 & 0 \\ -.8889 & -.1111 & 5.5556 & -3 \\ 0 & 0 & -3 & -1 \end{bmatrix} = A_{new}$$

Next iteration k = 3
Let

$$A_{new} = A_{old}$$
$$V_3 = \{A_{31}, A_{32}, A_{33}, A_{34}\}^T$$
$$w_2 = \{A_{31}, A_{32} - s, 0, 0\}^T$$
$$s = \pm \sqrt{A_{31}^2 + A_{32}^2}$$
$$V_3 = \{-.8889, -.1111, 5.556, -3\}^T$$

$$s = \pm\sqrt{(-.8889)^2 + (-.1111)^2} = \pm 0.8958$$

Hence $\quad\quad w = w_2 = \{-.8889, -1.0069, 0, 0\}^T$

Hence

$$ww^T = \begin{bmatrix} -.8889 \\ -1.0069 \\ 0 \\ 0 \end{bmatrix} [-.8889, -1.0069, 0, 0] = \begin{bmatrix} .79 & .895 & 0 & 0 \\ .895 & 1.014 & 0 & 0 \\ 0 & 0 & 0 & 0 \\ 0 & 0 & 0 & 0 \end{bmatrix}$$

$$\|w\|_2^2 = (-.8889)^2 + (-1.0069)^2 = 1.804$$

$$P_2 = \begin{bmatrix} 1 & 0 & 0 & 0 \\ 0 & 1 & 0 & 0 \\ 0 & 0 & 1 & 0 \\ 0 & 0 & 0 & 1 \end{bmatrix} - \frac{2}{1.804} \begin{bmatrix} .79 & .895 & 0 & 0 \\ .895 & 1.014 & 0 & 0 \\ 0 & 0 & 0 & 0 \\ 0 & 0 & 0 & 0 \end{bmatrix} - \frac{2}{1.804} \begin{bmatrix} .79 & .895 & 0 & 0 \\ .895 & 1.014 & 0 & 0 \\ 0 & 0 & 0 & 0 \\ 0 & 0 & 0 & 0 \end{bmatrix}$$

$$P_2^T A_{old} = \frac{2}{1.804} \begin{bmatrix} .112 & -.895 & 0 & 0 \\ -.895 & -.112 & 0 & 0 \\ 0 & 0 & .902 & 0 \\ 0 & 0 & 0 & .902 \end{bmatrix} \begin{bmatrix} 1.222 & -.222 & -.8889 & 0 \\ -.222 & 2.222 & -.1111 & 0 \\ -.8889 & -.1111 & 5.556 & -3 \\ 0 & 0 & -3 & -1 \end{bmatrix}$$

$$= \frac{2}{1.804} \begin{bmatrix} .336 & -2.01 & 0 & 0 \\ -1.069 & -.050 & .808 & 0 \\ -.8018 & -.1002 & 5.02 & -2.7 \\ 0 & 0 & -2.7 & .902 \end{bmatrix}$$

$$P_2^T A_{old} P_2 = \frac{2}{1.804} \begin{bmatrix} .336 & -2.01 & 0 & 0 \\ -1.069 & -.050 & .808 & 0 \\ -.8018 & -.1002 & 5.02 & -2.7 \\ 0 & 0 & -2.7 & -.902 \end{bmatrix} * \frac{2}{1.804} \begin{bmatrix} .112 & -.895 & 0 & 0 \\ -.895 & -.112 & 0 & 0 \\ 0 & 0 & .902 & 0 \\ 0 & 0 & 0 & .902 \end{bmatrix}$$

$$= 1.229 \begin{bmatrix} 1.837 & -.076 & 0 & 0 \\ -.076 & .962 & .73 & 0 \\ 0 & .73 & 4.53 & -2.44 \\ 0 & 0 & -2.44 & .813 \end{bmatrix}$$

$$A_{new} = \begin{bmatrix} 2.26 & -.093 & 0 & 0 \\ -.093 & 1.182 & .90 & 0 \\ 0 & .90 & 5.56 & -3.0 \\ 0 & 0 & -3.0 & -1.0 \end{bmatrix}$$

This is the similar tridiagonal form of A

Now that 'A' is in a similar tridiagonal form, the Q,L method can be applied in order to reduce the matrix further to a diagonal form from which the eigenvalues can be read.

Q,L METHOD

This method can only be applied to symmetric tridiagonal matrices such as those formed by the Householder method. These tridiagonal matrices can be factorized into an orthogonal matrix and a lower triangular matrix, hence the name Q,L. Once this decomposition is completed, a reverse multiplication of L.Q is performed to get the next reduced matrix. This matrix is then factorized and the procedure is repeated until a diagonal matrix is formed. This procedure is known as a similar orthogonal transformation. This iterative transformation process is outlined below. Start by factorizing the tridiagonal matrix T_1.

$$T_1 = Q_1 L_1 \rightarrow L_1 = Q_1^{-1} T_1 = Q_1^T T_1$$
$$T_2 = L_1 Q_1 = Q_1^T T_1 Q_1$$

where T_1 is the tridiagonal symmetric matrix formed from the Householder method.

$Q_1 \rightarrow$ orthogonal matrix.

$L_1 \rightarrow$ lower triangular matrix.

T_2 is formed from an orthogonal transformation of T_1.

T_2 is then factorized to give

$T_2 = Q_2 L_2$

$T_3 = L_2 Q_2 = Q_2^T T_2 Q_2$

$T_3 = Q_3 L_3$

Until T_n is reached. T_n is a diagonal matrix with diagonal elements equal to the system's eignevalues. The factorization of the matrix T into Q and L is performed by making use of an orthogonal transformation matrix similar to that used in the Jacobi method. The transformation matrix consists of the same submatrix as in the Jacobi method. However, here the purpose of the transformation is to annihilate the terms above the diagonal. This matrix is denoted by the letter S.

$$S_k = \begin{bmatrix} \cos\theta_{k-1} & -\sin\theta_k \\ \sin\theta_k & \cos\theta_k \end{bmatrix}$$

Note: S must be the same size as the matrix A. If the size of S is larger than 2x2, then the elements on the main diagonal not occupied by $\cos\theta$ are assigned values of 1. Those off diagonal elements not occupied by $\sin\theta$ or $-\sin\theta$ are assigned values of 0. The angle θ_k is evaluated so as to annihilate the terms above the diagonal.

The matrix S_k is used to factorize the matrix T into a lower triangular matrix and an orthogonal matrix in the following way.

$$L_1 = S_2 * S_3 \ldots * S_n T_1 \qquad k = 2, 3, \ldots n;$$
$$\qquad\qquad\qquad\qquad\qquad n \text{ is the size}$$
$$Q_1 = S_n^T * S_{n-1}^T \ldots * S_2^T \qquad \text{of the matrix}$$

This is performed on each T matrix until a diagonal form is reached.

The subscript k in matrix S_k corresponds to a specific position which can be found in the following way. Consider a tridiagonal matrix with the main diagonal elements equal to a_k where k=1,2,…n, and off diagonal elements above and to left of 'a_k' values equal to b_k with k = 2,…n. The resulting matrix will have the following form.

$$T = \begin{bmatrix} a_1 & b_2 & 0 & 0 & \cdots & \cdots & \cdots & 0 \\ b_2 & a_2 & b_3 & \ddots & & & & \vdots \\ 0 & b_3 & a_3 & b_4 & \ddots & & & \vdots \\ \vdots & & \ddots & b_4 & a_4 & \ddots & & 0 \\ \vdots & & & \ddots & & \ddots & \ddots & b_n \\ 0 & \cdots & & \cdots & 0 & b_n & & a_n \end{bmatrix}$$

The subscript of these elements corresponds to the k value of S_k, thus enabling the positioning of the transformation elements correctly in the P matrices.

The transformation elements can be expressed in terms of values in the appropriate T matrix in the following way, so as to annihilate elements above the main diagonal,

$$\cos\theta_k = \cos\theta_{k-1} = \frac{a_k}{\sqrt{b_k^2 + a_k^2}}$$

$$\sin\theta_k = \frac{b_k}{\sqrt{b_k^2 + a_k^2}}$$

The procedure used to find the lower triangular matrix is the following.
Recall:

$$L_1 = S_2* S_3*\ldots S_n* T$$

First evalaute $S_n* T$ to get T^1. Use values of a_k and b_k from T to evaluate the transformation elements in S_n. Next evaluate $S_{n-1} * T^1$ using values of a_k and b_k from T^1. Since T^1 is not symmetric, the values of b_k will not be the same; therefore the appropriate b_k value to use when evaluating the transformation elements will be the b_k value above the pivot element a_k. This procedure is continued until the last multiplication is performed involving S_2. After each multiplication involving the kth matrix S, the element above the kth pivot element is annihilated from the original T matrix.

The orthogonal matrices are easily found by using the transpose of the S matrices found when evaluating L. This method is best illustrated by the following example.

EXALE 7.11
Evaluate the eigenvalues of the following tridiagonal matrix using the Q,L method.

$$T_1 = \begin{bmatrix} 4 & 2 & 0 \\ 2 & 3 & 1 \\ 0 & 1 & 2 \end{bmatrix}$$

Note: T_1 was formed by applying Householders method to some symmetric matrix 'A'.

Solution
The first step is to evaluate L_1.

$L_1 = S_2 S_3 T_1$ Note n=3   two S matrices

since k=2,3,…n

First evaluate $S_3 T_1 = T_1'$

for k=3

$$S_3 = \begin{bmatrix} 1 & 0 & 0 \\ 0 & \cos\theta_2 & -\sin\theta_3 \\ 0 & \sin\theta_3 & \cos\theta_3 \end{bmatrix}$$

where a_3 and b_3 are taken from T_1

$$\cos\theta_2 = \cos\theta_3 = \frac{a_3}{\sqrt{b_3^2 + a_3^2}} = \frac{2}{\sqrt{(1)^2 + (2)^2}} = .894$$

$$\sin\theta_3 = \frac{b_3}{\sqrt{b_3^2 + a_3^2}} = \frac{1}{\sqrt{(1)^2 + (2)^2}} = .447$$

$$\text{Hence } S_3 = \begin{bmatrix} 1 & 0 & 0 \\ 0 & .894 & -.447 \\ 0 & .447 & .894 \end{bmatrix}$$

$$S_3 T_1 = T_1' = \begin{bmatrix} 1 & 0 & 0 \\ 0 & .894 & -.447 \\ 0 & .447 & .894 \end{bmatrix} \begin{bmatrix} 4 & 2 & 0 \\ 2 & 3 & 1 \\ 0 & 1 & 2 \end{bmatrix} = \begin{bmatrix} 4 & 2 & 0 \\ 1.788 & 2.235 & 0 \\ .894 & 2.235 & 2.235 \end{bmatrix}$$

Notice that the elements above the nth diagonal element or $(n-1,n)$-th element are now reduced to zero. The $(n, n-2)$ element has become non zero.

The next step is to multiply $S_2 T_1' = T_1'' = L_1$

For $k = 2$

$$S_2 = \begin{bmatrix} \cos\theta_1 & -\sin\theta_2 & 0 \\ \sin\theta_2 & \cos\theta_2 & 0 \\ 0 & 0 & 1 \end{bmatrix}$$

$$\cos\theta_1 = \cos\theta_2 = \frac{a_2}{\sqrt{b_2^2 + a_2^2}} = \frac{2.236}{\sqrt{(2)^2 + (2.236)^2}} = .745$$

$$\sin\theta_2 = \frac{b_2}{\sqrt{b_2^2 + a_2^2}} = \frac{2}{\sqrt{(2)^2 + (2.236)^2}} = .667$$

Hence

$$S_2 = \begin{bmatrix} .745 & -.667 & 0 \\ .667 & .745 & 0 \\ 0 & 0 & 1 \end{bmatrix}$$

$$L_1 = S_2 T_1' = T_1'' =$$

$$\begin{bmatrix} .745 & -.667 & 0 \\ .667 & .745 & 0 \\ 0 & 0 & 1 \end{bmatrix} \begin{bmatrix} 4 & 2 & 0 \\ 1.788 & 2.235 & 0 \\ .894 & 2.235 & 2.235 \end{bmatrix} = \begin{bmatrix} 1.788 & 0 & 0 \\ 4.0 & 3.0 & 0 \\ .894 & 2.235 & 2.235 \end{bmatrix}$$

$$L_1 = \begin{bmatrix} 1.788 & 0 & 0 \\ 4.0 & 3.0 & 0 \\ .894 & 2.235 & 2.235 \end{bmatrix}$$

The orthogonal matrix Q_1 is formed by multiplying the transpose of the S matrices.

$$Q_1 = S_3^T * S_2^T$$

$$Q_1 = \begin{bmatrix} 1 & 0 & 0 \\ 0 & .894 & .447 \\ 0 & -.447 & .894 \end{bmatrix} \begin{bmatrix} .745 & .667 & 0 \\ -.667 & .745 & 0 \\ 0 & 0 & 1 \end{bmatrix} = \begin{bmatrix} .745 & .667 & 0 \\ -.596 & .666 & .447 \\ .298 & -.333 & .894 \end{bmatrix}$$

Now a check is performed to verify

$T_1 = Q_1 L_1$

$$T_1 = \begin{bmatrix} .745 & .667 & 0 \\ -.596 & .666 & .447 \\ .298 & -.333 & .894 \end{bmatrix} \begin{bmatrix} 1.788 & 0 & 0 \\ 4.0 & 3.0 & 0 \\ .894 & 2.235 & 2.235 \end{bmatrix} = \begin{bmatrix} 4.0 & 2.0 & 0 \\ 2.0 & 3.0 & 1.0 \\ 0 & 1.0 & 2.0 \end{bmatrix}$$

Now the first transformation can be performed.

$T_2 = L_1 Q_1$

$$T_2 = \begin{bmatrix} 1.788 & 0 & 0 \\ 4.0 & 3.0 & 0 \\ .894 & 2.235 & 2.235 \end{bmatrix} \begin{bmatrix} .745 & .667 & 0 \\ -.596 & .666 & .447 \\ .298 & -.333 & .894 \end{bmatrix}$$

$$= \begin{bmatrix} 1.33 & 1.19 & 0 \\ 1.19 & 4.667 & 1.341 \\ 0 & 1.341 & 3.0 \end{bmatrix}$$

Note that the off diagonal elements of T_2 are reduced compared to those in T_1. T_2 is now factorized into L_2 and Q_2 in the following way.

$$L_2 = S_2 * S_3 * T_2$$

Evaluate $S_3 T_2 = T_2'$ first.

For $k = 3$

$$S_3 = \begin{bmatrix} 1 & 0 & 0 \\ 0 & \cos\theta_2 & -\sin\theta_3 \\ 0 & \sin\theta_3 & \cos\theta_3 \end{bmatrix}$$

where a_3 and b_3 are taken from T_2

$$\cos\theta_2 = \cos\theta_3 \frac{a_3}{\sqrt{b_3^2 + a_3^2}} = \frac{3.0}{\sqrt{(1.341)^2 + (3)^2}} = .913$$

$$\sin\theta_3 = \frac{b_3}{\sqrt{b_3^2 + a_3^2}} = \frac{1.341}{\sqrt{(1.341)^2 + (3)^2}} = .408$$

Hence

$$S_3 = \begin{bmatrix} 1 & 0 & 0 \\ 0 & .913 & -.408 \\ 0 & .408 & .913 \end{bmatrix}$$

$$S_3 T_2 = T_2' =$$

$$\begin{bmatrix} 1 & 0 & 0 \\ 0 & .913 & -.408 \\ 0 & .408 & .913 \end{bmatrix} \begin{bmatrix} 1.33 & 1.19 & 0 \\ 1.19 & 4.67 & 1.34 \\ 0 & 1.34 & 3.0 \end{bmatrix} = \begin{bmatrix} 1.33 & 1.19 & 0 \\ 1.090 & 3.72 & 0 \\ .490 & 3.13 & 3.29 \end{bmatrix}$$

Notice that elements in T_2' are reduced to zero above the kth diagonal element. The next step is to multiply $S_2 T_2' = T_2'' = L_2$.

For k=2

$$S_2 = \begin{bmatrix} \cos\theta_1 & -\sin\theta_2 & 0 \\ \sin\theta_2 & \cos\theta_2 & 0 \\ 0 & 0 & 1 \end{bmatrix}$$

where a_2 and b_2 are taken from T_2'

$$\cos\theta_1 = \cos\theta_2 = \frac{a_2}{\sqrt{b_2^2 + a_2^2}} = \frac{3.72}{\sqrt{(1.19)^2 + (3.72)^2}} = .952$$

$$\sin\theta_2 = \frac{b_2}{\sqrt{b_2^2 + a_2^2}} = \frac{1.19}{\sqrt{(1.19)^2 + (3.72)^2}} = .305$$

Hence

$$S_2 = \begin{bmatrix} .952 & -.305 & 0 \\ .305 & .952 & 0 \\ 0 & 0 & 1 \end{bmatrix}$$

$$L_2 = T_2'' = S_2 T_2' =$$

$$\begin{bmatrix} .952 & -.305 & 0 \\ .305 & .952 & 0 \\ 0 & 0 & 1 \end{bmatrix} \begin{bmatrix} 1.33 & 1.19 & 0 \\ 1.090 & 3.71 & 0 \\ .490 & 3.13 & 3.29 \end{bmatrix} = \begin{bmatrix} .93 & 0 & 0 \\ 1.44 & 3.89 & 0 \\ .490 & 3.13 & 3.29 \end{bmatrix}$$

$$L_2 = \begin{bmatrix} .93 & 0 & 0 \\ 1.44 & 3.89 & 0 \\ .490 & 3.13 & 3.29 \end{bmatrix}$$

The orthogonal matrix Q_2 is formed by multiplying the transpose of the S matrices.

$$Q_2 = S_3^T * S_2^T$$

$$Q_2 = \begin{bmatrix} 1 & 0 & 0 \\ 0 & .913 & .408 \\ 0 & -.408 & .913 \end{bmatrix} \begin{bmatrix} .952 & .305 & 0 \\ -.305 & .952 & 0 \\ 0 & 0 & 1 \end{bmatrix} = \begin{bmatrix} .952 & .305 & 0 \\ -.278 & .869 & .408 \\ .124 & -.388 & .913 \end{bmatrix}$$

Now a check is performed.

$$T_2 = Q_2 L_2 =$$

$$\begin{bmatrix} .952 & .305 & 0 \\ -.278 & .869 & .408 \\ .124 & -.388 & .913 \end{bmatrix} \begin{bmatrix} .93 & 0 & 0 \\ 1.44 & 3.89 & 0 \\ .490 & 3.13 & 3.29 \end{bmatrix} = \begin{bmatrix} 1.33 & 1.19 & 0 \\ 1.19 & 4.66 & 1.34 \\ 0 & 1.34 & 3.0 \end{bmatrix}$$

Now the next transformation is performed.

$$T_3 = L_2 Q_2 =$$

$$\begin{bmatrix} .93 & 0 & 0 \\ 1.44 & 3.89 & 0 \\ .490 & 3.13 & 3.29 \end{bmatrix} \begin{bmatrix} .952 & .305 & 0 \\ -.278 & .869 & .408 \\ .124 & -.388 & .913 \end{bmatrix} = \begin{bmatrix} .89 & .28 & 0 \\ .28 & 3.83 & 1.59 \\ 0 & 1.59 & 4.28 \end{bmatrix}$$

This procedure is continued until the off diagonal elements are smaller than a certain specified amount, and then the diagonal elements give the eigenvalues. After 3 iterations we see that.

$$\lambda_1 \simeq 0.89, \ \lambda_2 \simeq 3.83, \ \lambda_3 \simeq 4.28$$

$$\lambda_1$$

However, the off diagonal elements are still quite large. Therefore more iterations are needed to find a more accurate answer. After 7 iterations we find that

$$\lambda_1 = 0.86, \ \lambda_2 = 2.48, \ \lambda_3 = 5.65$$

The true values are

$$\lambda_1 = 0.85, \ \lambda_2 = 2.48 \text{ and } \lambda_3 = 5.67$$

HOUSEHOLDER QL-METHOD

To obtain all the eigenvalues of an nxn symmetric matrix A.

STEP 1   Put the system in the form: $(A - \lambda I)x = 0$

STEP 2   If matrix 'A' is not in the tridiagonal form, then continue. If 'A' is tridiagonal, go to step 4.

STEP 3   Apply Householder's method to matrix 'A' in order to get a tridiagonal form. This is an iterative technique of the form.

$$A_{new} = P^T A_{old} P$$

where matrix P is formed after each iteration by using elements from A_{old} in the following way.

$$P = I - \frac{2ww^T}{\|w\|_2^2}$$

$$w = \{A_{k1}, A_{k2}, \ldots A_{k,k-1} - s, 0 \ldots\}^T$$

$$V_k = \{A_{k1}, A_{k2}, \ldots A_{kn}\}^T$$

and $s = \pm\sqrt{A_{k1}^2 + A_{k2}^2 + \ldots A_{k,k-1}^2}$

this procedure is continued up to k=n-2 columns are reduced. Let the tridiagonal form of $A_{new} = T_{old}$.

STEP 4   Once the matrix is the tridiagonal from, the next step is to apply the Q,L method in order to reduce the tridiagonal form to a main diagonal. The diagonal elements of this diagonal matrix will represent the eigenvalues. The matrix T_{old} is factored into a lower triangular and an orthogonal matrix by the following procedure.

$$L_{old} = S_2 * S_{3=} * \ldots S_n T_{old}$$

$$Q_{old} = S_n^T * S_{n-1}^T * \ldots S_2^T$$

L → lower triangular matrix

Q → orthogonal matrix

S → transformation matrix

Transformation elements are formed using elements in matrix T.

STEP5   Once the matrix T_{old} is factored into a Q,L form a reverse multiplication is carried out to get T_{new}.

$$T_{new} = L_{old} * Q_{old}$$

STEP 6   The off diagonal elements of T_{new} are checked to see whether they are below a specified tolerance. If they are, go to step 7. If they are not, let $T_{new} = T_{old}$, go to step 4.

STEP 7   Print the diagonal elements of T_{new}, these are the eigenvalues.

7.7 HOUSEHOLDER-Q,R, METHOD

This method is used to find the eigenvalues of the system when the square matrix 'A', which describes the system, is unsymmetrical. Before this method can be used the matrix 'A' must be reduced to a combination of tridiagonal and an upper triangular matrix with one more diagonal below the main diagonal.

$$A = \begin{bmatrix} A_{11} & A_{12} & A_{13} & A_{14} \\ A_{21} & A_{22} & A_{23} & A_{24} \\ 0 & A_{32} & A_{33} & A_{34} \\ 0 & 0 & A_{43} & A_{44} \end{bmatrix}$$

This form is referred to as the upper Hessenberg configuration, and is achieved by applying Householders method to a non symmetric square matrix. For an upper Hessenberg matrix all elements $A_{ij} = 0$ for $i \geq j + 2$. Once in the upper Hessenberg form, the matrix 'A' is factored into an upper triangular and orthogonal matrix by using a similar transformation matrix to that used in the Q,L method. When the factorization is completed into the Q (orthogonal) * R (upper triangular) matrix form, a reverse multiplication (R*Q) is performed in order to get a reduced matrix which is similar to the original matrix 'A'. The procedure is repeated on the reduced matrix until a matrix is produced with sufficiently low off diagonal terms. The main diagonal of such a matrix represents the eigenvalues of the system. The iterative transformation can be described by the following equations.

$$A_{old} = Q_{old} * R_{old} \qquad \text{(factorization)}$$
$$A_{new} = R_{old} * Q_{old} = Q_{old}^{-1} * A_{old} * Q_{old}$$

After each iteration let $A_{new} = A_{old}$ and repeat the procedure until the off diagonal elements are satisfactorily low.

The factorization procedure is performed in the following way.

R (upper triangular matrix) = $S_{n-1}^T * S_{n-2}^T ... S_1^T * A$

Q^T (orthogonal matrix) = $S_{n-1}^T * S_{n-2}^T ... S_1^T$

Note: A is an upper Hessenberg matrix.

The matrix S is a transformation matrix which annihilates the extra element below the main diagonal in the upper Hessenberg matrix in order to get an upper triangular matrix. For matrix of size n, the number of transformation matrices, S, required is n-1.

The transformation matrix is formed in an identical manner to the Q,L method with the exception that the b_k value used when evaluating $\cos\theta_k$ and $\sin\theta_k$ is the value that is below a_k. This is the actual value being annihilated. The values of $\cos\theta_k$ and $\sin\theta_k$ are evaluated by the following equations.

$$\cos\theta_k = \frac{a_k}{\sqrt{b_k^2 + a_k^2}}$$

$$\sin\theta_k = \frac{b_k}{\sqrt{b_k^2 + a_k^2}}$$

where a_k is a diagonal element and b_k is the element directly below a_k, as is seen in the following matrix A.

$$A = \begin{bmatrix} a_{11} & a_{12} & a_{13} & \cdots & a_{1n} \\ a_{21} & a_{22} & a_{23} & \cdots & a_{2n} \\ 0 & a_{32} & a_{33} & \cdots & \cdot \\ \cdot & & & a_k & \cdot \\ 0 & & & b_k & \cdot \\ 0 & 0 & 0 & \cdots & a_{n,n-1}\ a_{nn} \end{bmatrix}$$

The procedure used to formulate the upper triangular matrix is to multiply $S_1^T A$, using elements from A to evaluate $\cos\theta_1$ and $\sin\theta_1$. This result is denoted as A^1 and is then multiplied by S_2^T. The elements from A^1 are used to evaluate $\cos\theta_2$ and $\sin\theta_2$ in S_2^T. This procedure is continued until an upper triangular matrix is formed.

The orthogonal matrix results from the multiplication of the S^T matrices formed when evaluating the upper triangular matrix.

After each reverse multiplication of R*Q, the new matrix formed must be factorized into a new upper triangular and orthogonal matrix using the procedure outlined above.

EXAMPLE 7.10
Evaluate the eigenvalues of a system given the matrix 'A' which describes the system equations. Use the Q.R. method.

$$A = \begin{bmatrix} 1 & 2 & 3 \\ 1 & 4 & 9 \\ 1 & 8 & 27 \end{bmatrix}$$

Solution

The first step is to apply Householders method to get 'A' into an upper Hessenberg configuration. n-2 iterations will be required to perform this.

We have

$$P = I - \frac{2ww^T}{\|w\|_2^2}$$

where w is a vector containing the following elements;

$$w = \{A_{k1}, A_{k2}, \dots A_{k,k-1} - s, O, \dots\}^T$$

$$s = \pm\sqrt{A_{k1}^2 + A_{k2}^2 + \dots A_{k,k-1}^2}$$

$$V_k = \{A_{k1}, A_{k2}, \dots A_{kn}\}^T$$

K   is the row in matrix 'A' being deflated

V_k is the row vector in 'A' being deflated

Start with k=n and decrease until k=2.

k = 3

$$V_3 = \{1, 8, 27\}^T$$

$$s = \pm\sqrt{(1)^2 + (8)^2} = \pm 8.06$$

$$w_1 = \{1, 8 \pm 8.06, 0\}$$

Note: + 8.06 is chosen so that the absolute value of the appropriate value in w will be a maximum.

$$P = I - \frac{2w_1 w_1^T}{\|w_1\|_2^2}$$

$$w_1 w_1^T = \begin{bmatrix} 1 \\ 16.06 \\ 0 \end{bmatrix} * [1 \quad 16.06 \quad 0] = \begin{bmatrix} 1 & 16.06 & 0 \\ 16.06 & 257.9 & 0 \\ 0 & 0 & 0 \end{bmatrix}$$

$$\|w_1\|_2^2 = (1)^2 + (16.06)^2 + (0)^2 = 258.9$$

$$P_1 = \begin{bmatrix} 1 & 0 & 0 \\ 0 & 1 & 0 \\ 0 & 0 & 1 \end{bmatrix} - \frac{2}{258.9} \begin{bmatrix} 1 & 16.06 & 0 \\ 16.06 & 257.9 & 0 \\ 0 & 0 & 0 \end{bmatrix}$$

$$= \frac{2}{258.9} \begin{bmatrix} 128.45 & -16.06 & 0 \\ -16.06 & -128.45 & 0 \\ 0 & 0 & 129.45 \end{bmatrix}$$

Now the first iteration can be performed;

$$A_{new} = P_1^T A_{old} P_1$$

$$A_{new} = \frac{2}{258.9} * \begin{bmatrix} 128.45 & -16.06 & 0 \\ -16.06 & -128.45 & 0 \\ 0 & 0 & 129.45 \end{bmatrix} * \begin{bmatrix} 1 & 2 & 3 \\ 1 & 4 & 9 \\ 1 & 8 & 27 \end{bmatrix} *$$

$$\frac{2}{258.9} * \begin{bmatrix} 128.45 & -16.06 & 0 \\ -16.06 & -128.45 & 0 \\ 0 & 0 & 129.45 \end{bmatrix}$$

$$A_{new} = \frac{2}{258.9} * \begin{bmatrix} 112.39 & 192.66 & 240.81 \\ -144.51 & -545.92 & -1204.23 \\ 129.45 & 1035.6 & 3495.15 \end{bmatrix} *$$

$$\frac{2}{258.9} * \begin{bmatrix} 128.45 & -16.06 & 0 \\ -16.06 & -128.45 & 0 \\ 0 & 0 & 129.45 \end{bmatrix}$$

$$A_{new} = \left(\frac{2}{258.9} \right)^2 * \begin{bmatrix} 11342.4 & -26552.2 & 31172.8 \\ -9794.8 & 72444.25 & -155887.6 \\ -3.88 & -135101.8 & 452447.2 \end{bmatrix}$$

$$A_{new} = \begin{bmatrix} .68 & -1.58 & 1.86 \\ -.58 & 4.32 & -9.3 \\ 0 & -8.06 & 27 \end{bmatrix}$$

Note: this is now the upper Hessenberg form.
Now the Q,R method can be applied.

The matrix 'A' must now be factored into an upper triangular and orthogonal matrix. This can only be done if 'A' is in the upper Hessenberg form. For clarity, let this form of A be denoted by H.
The upper triangular matrix is formed first, using the following procedure.

$$R = \text{(upper triangular)} = S_{n-1}^T * S_{n-2}^T * .... * S_1^T * H \text{, since } n = 3$$
$$R = S_2^T * S_1^T * H$$

For k=1

$$S_1 = \begin{bmatrix} \cos\theta_1 & -\sin\theta_1 & 0 \\ \sin\theta_1 & \cos\theta_2 & 0 \\ 0 & 0 & 1 \end{bmatrix}$$

where

$$\cos\theta_1 = \cos\theta_2 = \frac{a_1}{\sqrt{b_1^2 + a_1^2}} = \frac{0.68}{\sqrt{(-0.58)^2 + (0.68)^2}} = 0.761$$

$$\sin\theta_1 = \frac{b_1}{\sqrt{b_1^2 + a_1^2}} = \frac{-0.58}{\sqrt{(-0.58)^2 + (0.68)^2}} = -0.649$$

The values of a_k and b_k are taken from matrix H.

Hence

$$S_1^T = \begin{bmatrix} .761 & -.649 & 0 \\ .649 & .761 & 0 \\ 0 & 0 & 1 \end{bmatrix}$$

$$S_1^T * H = H' =$$

$$H' = \begin{bmatrix} .761 & -.649 & 0 \\ .649 & .761 & 0 \\ 0 & 0 & 1 \end{bmatrix} \begin{bmatrix} .68 & -1.58 & 1.86 \\ -.58 & 4.32 & -9.3 \\ 0 & -8.06 & 27 \end{bmatrix} = \begin{bmatrix} .894 & -4.01 & 7.45 \\ 0 & 2.26 & -5.87 \\ 0 & -8.06 & 27 \end{bmatrix}$$

$$H' = \begin{bmatrix} .894 & -4.01 & 7.45 \\ 0 & 2.26 & -5.87 \\ 0 & -8.06 & 27 \end{bmatrix}$$

For k = 2

$$S_2 = \begin{bmatrix} 1 & 0 & 0 \\ 0 & \cos\theta_2 & -\sin\theta_2 \\ 0 & \sin\theta_2 & \cos\theta_3 \end{bmatrix}$$

where $\qquad \cos\theta_2 = \cos\theta_2 = \dfrac{a_2}{\sqrt{b_2^2 + a_2^2}} = \dfrac{2.26}{\sqrt{(-8.06)^2 + (2.26)^2}} = .269$

$$\sin\theta_2 = \dfrac{b_2}{\sqrt{b_2^2 + a_2^2}} = \dfrac{-8.06}{\sqrt{(-8.06)^2 + (2.26)^2}} = -.963$$

The values of a_k and b_k are taken from matrix H'.

Hence $\qquad\qquad S_2^T = \begin{bmatrix} 1 & 0 & 0 \\ 0 & .269 & -.963 \\ 0 & .963 & .269 \end{bmatrix}$

Now R can be found by;

$$S_2^T * H' = R =$$

$$\begin{bmatrix} 1 & 0 & 0 \\ 0 & .269 & -.963 \\ 0 & .963 & .269 \end{bmatrix} \begin{bmatrix} .894 & -4.01 & 7.45 \\ 0 & 2.26 & -5.87 \\ 0 & -8.06 & 27 \end{bmatrix} = \begin{bmatrix} .894 & -4.01 & 7.45 \\ 0 & 8.37 & -27.6 \\ 0 & 0 & 1.61 \end{bmatrix}$$

Now Q is found by.

$$Q^T = S_n^T * S_{n-1}^T * ... S_1^T$$

In this case

$$Q^T = s_2^T * S_1^T =$$

$$\begin{bmatrix} 1 & 0 & 0 \\ 0 & .269 & -.963 \\ 0 & .963 & .269 \end{bmatrix} \begin{bmatrix} .761 & -.649 & 0 \\ .649 & .761 & 0 \\ 0 & 0 & 1 \end{bmatrix} = \begin{bmatrix} .761 & -.649 & 0 \\ .175 & .205 & -.963 \\ .625 & .733 & .269 \end{bmatrix}$$

Hence $\qquad\qquad Q = \begin{bmatrix} .761 & .175 & .625 \\ -.649 & .205 & .733 \\ 0 & -.963 & .269 \end{bmatrix}$

Now a check is performed to verify that Q and R are the correct factors of H.

$$H = Q * R = \begin{bmatrix} .761 & .175 & .625 \\ -.649 & .205 & .733 \\ 0 & -.963 & .269 \end{bmatrix} \begin{bmatrix} .894 & -4.01 & 7.45 \\ 0 & 8.37 & -27.6 \\ 0 & 0 & 1.61 \end{bmatrix}$$

$$H = \begin{bmatrix} .68 & -1.58 & 1.85 \\ -.58 & 4.32 & -9.3 \\ 0 & -8.06 & 27 \end{bmatrix}$$

This is true, thus Q and R are valid.

Now the reduction process can be performed.

$$H_{new} = R_{old} * Q_{old}$$

$$H_{new} = \begin{bmatrix} .894 & -4.01 & 7.45 \\ 0 & 8.37 & -27.6 \\ 0 & 0 & 1.61 \end{bmatrix} \begin{bmatrix} .761 & .175 & .625 \\ -.649 & .205 & .733 \\ 0 & -.963 & .269 \end{bmatrix}$$

$$= \begin{bmatrix} 3.28 & -7.84 & -.38 \\ -5.43 & 28.3 & -1.28 \\ 0 & -1.55 & .433 \end{bmatrix}$$

Now let $H_{new} = H_{old}$, and repeat the factorization process to get the next Q_{old} and R_{old}; then multiply R_{old} *Q_{old} to get the next H_{new}. When a H_{new} is formed with sufficiently low off diagonal elements stop the iterative process. The elements along the diagonal of this H_{new} represent a close approximation to the systems eigenvalues. After four iterations $\lambda_1 = 0.22, \lambda_2 = 1.84$ and $\lambda_3 = 29.94$

CAUTION
The Q,R method will be assured to converge on the systems eigenvalues if the sum of the squares of the elements below the main diagonal of the matrix $H_{new}^{(i)}$ is smaller than that of its predecessor $H_{new}^{(i-1)}$. For instance in the above example $(5.93)^2 + (-4.8)^2 < (0.58)^2 + (-8.06)^2$. Hence the method is converging.

HOUSEHOLDER QR-METHOD

To obtain all eignevalues of a non symmetric nxn matrix A.

STEP 1   Put the system into the form

$$(A - \lambda I)x = 0$$

STEP 2   Apply Householder's transformation method in order to get a similar matrix 'A' in the upper Hessenberg form. Let this form of matrix 'A' be denoted by the letter H.

STEP 3   Once in the upper Hessenberg form, the Q,R method can be applied. Let n= size of matrix 'A'. Then factorize H into an orthogonal * an upper triangular matrix using a transformation matrix S in the following way.

$$R_{old} = (\text{upper triangular}) = S^T_{n-1} * S^T_{n-2} * ... S^T_i * H_{old}$$

$$Q^T_{old} = (\text{orthogonal triangular}) = S^T_{n-1} * S^T_{n-2} * .... S^T_1$$

where values of a_k and b_k which are used to evaluate S are taken from the current H matrix.

Check

$$H_{old} = R_{old} * Q_{old}$$

STEP 4   Now the H matrix can be reduced by a reverse multiplication.

$$H_{new} = Q_{old} * R_{old}$$

STEP 5   Check to see that the sum of the squares of those elements below the main diagonal of H_{new} is less than that of H_{old}. If they are continue. If not go to step 9.

STEP 6   Search the matrix H_{new} for the smallest off diagonal term. If this term is smaller than some pre-specified tolerance, go to step 8. If not continue.

STEP 7   Let $H_{new} = H_{old}$. Go to step 3.

STEP 8   Print the main diagonal values of H_{new}. These are the system eigen-values stop.

STEP 9   Print "THIS METHOD WILL NOT CONVERGE ON THE SYSTEM'S EIGENVALUES".

PROBLEMS

1.   Use Power method to find the larger eigenvalue and the corresponding eigenvector using 5 iterations.

(a) $\begin{bmatrix} 1 & 2 \\ 3 & 4 \end{bmatrix}$ (b) $\begin{bmatrix} 1 & 1 \\ 1 & -1 \end{bmatrix}$

2.   Perform 5 iterations of the following matrix using power method and obtain the eigenvalue and egienvector.

$$\begin{bmatrix} 1 & 3 & 2 \\ 3 & -1 & 1 \\ 2 & 1 & -2 \end{bmatrix}$$

3.   Obtain the largest eigenvalue and corresponding eignevector of matrix A given by

$$A = \begin{bmatrix} 3 & 7 & 9 \\ 7 & 4 & 3 \\ 9 & 3 & 8 \end{bmatrix}$$

Starting with a trial vector of $(1,1,1)$. Perform two iterations.

4.   Apply Power method to evaluate the largest eigenvalue and eigenvector of the given 3x3 system.

$$\begin{bmatrix} 4 & 2 & 0 \\ 0 & 5 & 3 \\ 2 & 1 & 6 \end{bmatrix} \begin{bmatrix} x_1 \\ x_2 \\ x_3 \end{bmatrix} = \lambda \begin{bmatrix} x_1 \\ x_2 \\ x_3 \end{bmatrix}$$

5.   A network is described by the following set of equations

$$(12-\lambda)x_1 \qquad - 4x_2 + 2\,x_3 = 0$$

$$-2x_1 + (12-\lambda)x_2 - 2x_3 = 0$$

$$2x_1 - 2x_2 + (9-\lambda)x_3 = 0$$

Determine the eigenvalues using Inverse Power method.

6.   Using Gerschgorin theoren give the regions of eigenvalues for the system.

$$A = \begin{bmatrix} 1 & -1 & 0 \\ -1 & 4 & -2 \\ 0 & -2 & 2 \end{bmatrix}$$

7.   Using the Power method check the largest eigenvalue of matrix A using 7 iterations.

$$A = \begin{bmatrix} 0 & 2 & 3 \\ -10 & -1 & 2 \\ -2 & 4 & 7 \end{bmatrix}$$

8.   Find the largest eigenvalue and eigenvector of the following matrix using Power method within 0.001 error.

$$A = \begin{bmatrix} 4 & 0 & 0 \\ -1 & 2 & 0 \\ -2 & -1 & -3 \end{bmatrix}$$

9.   Evaluate the dominant eigenvalue and the eigenvector of the given matrix by Power method. Use 4 iterations.

$$A = \begin{bmatrix} 2 & -2 & 0 \\ -1 & 2 & -1 \\ 0 & -4 & 2 \end{bmatrix}$$

10.   (a)   Check whether the following matrix is positive definite

$$A = \begin{bmatrix} 4 & 1 & 2 \\ 1 & 5 & 3 \\ 2 & 3 & 7 \end{bmatrix}$$

        (b)   Find the largest eigenvalue of A using the Power method.
        (c)   Find the other two eigenvalues by substituting the above calculated eigenvalue in $|A - \lambda I| = 0$

11.   The vibration of a system of 3 masses connected by springs can be described by the following 3 equations.

$$\begin{bmatrix} (2-\lambda) & -1 & 0 \\ 1 & (2-\lambda) & -1 \\ 0 & -1 & (2-\lambda) \end{bmatrix}$$

        Apply Power method with the characteristic equation to find the eigenvalues and eigenvectors of the system.

12.   Find the largest eigenvalue and corresponding eigenvector of matrix A to three digits.

$$A = \begin{bmatrix} 1 & 1 & 2 \\ 0 & 1 & 3 \\ 1 & 1 & 1 \end{bmatrix}$$

13.   A mechanical system of masses and springs can be descried by the following equation.

$$M [\ddot{y}] + K [y] = 0$$

        mass motion stiffness motion

    (i)   write down the equation of motion of the system assuming $m_1 = 1, m_2 = 2, k_1 = 2$ and $k_2 = 1$

    (ii)   show that the system can be written in the form

$$A \underline{x} = \lambda \underline{x}; \text{ assuming } y = \underline{x}e^{j\omega t}, \text{ where } j = \sqrt{-1}$$

            find matrix A, the largest eigenvalue and the corresponding eigenvector using the power method

14.   Determine the smallest eigenvalue of the given system in Problem 13 using inverse Power method.

15.   Show that $a_{11}, a_{22}, \ldots, a_{nn}$ are the eigenvalues of the triangular matrix

$$\begin{vmatrix} a_{11} & a_{12} & \cdots & a_{1n} \\ 0 & a_{22} & \cdots & a_{1n} \\ \vdots & \vdots & & \\ 0 & 0 & & a_{nn} \end{vmatrix}$$

16.   The vertical motion of a mechanical system is given in a matrix form

$$A = \begin{bmatrix} 2 & -1 & 0 \\ -1 & 2 & -1 \\ 0 & -1 & 2 \end{bmatrix}$$

    Use Jacobi method to evaluate the eigenvalues and the normalized eigenvectors. Use 5 iterations.

17.   Given

$$A = \begin{bmatrix} 4 & 3 & 0 \\ 3 & 4 & -1 \\ 0 & -1 & 4 \end{bmatrix}$$

    (a)   Is the matrix A positive definite and tridiagonal ?

    (b)   Find the eigenvalues of the matrix A using Jacobi's method.

18.   Apply Jacobi's method using 3 iterations to matrix B

$$B = \begin{bmatrix} 10 & -4 & 4 \\ -4 & 16 & -4 \\ 4 & -4 & 10 \end{bmatrix}$$

19.   Reduce the given matrix to a tridiagonal form using Householder technique.

$$\begin{bmatrix} 6 & 5 & 2 \\ 5 & 4 & -2.5 \\ 2 & -2.5 & 1.5 \end{bmatrix}$$

20.   Apply Householder's method to reduce the matrix A. Solve using QR method and use 2 iterations.

$$A = \begin{bmatrix} 6 & 2 & 0 \\ 2 & 8 & 4 \\ 0 & 4 & 2 \end{bmatrix}$$

21.   Use the Q,L method to solve the following matrix using 3 iterations.

$$\begin{bmatrix} 4 & -2 & 0 \\ -2 & -2 & -4 \\ 0 & -4 & 6 \end{bmatrix}$$

22.   Find the eigenvalues of the matrix B using Jacobi method. Perform the annihilation operation 3 times.

$$B = \begin{bmatrix} 3 & 2 & 1 \\ 2 & 3 & 0 \\ 1 & 0 & 3 \end{bmatrix}$$

23.   Using Gerschgorin theorem identify the regions of eignevalues of matrix A in problem 9. Using the inverse power method, find the eigenvalue and corresponding eigenvector in the neighbourhood of 2. Perform 3 iterations.

24.   Write a computer program to obtain the largest eigenvalue of the matrix $S^{-1} B S$

$$\text{Where } B = \begin{bmatrix} 3 & 2 & 1 \\ 2 & 3 & 0 \\ 1 & 0 & 3 \end{bmatrix} S = \begin{bmatrix} 2 & 1 & 0 \\ 1 & 2 & 1 \\ 0 & 1 & 2 \end{bmatrix}$$

$S^{-1} B S$ is termed a similarity transformation of B. Eigenvalues of B and $S^{-1} BS$ are identical.

25.   Write a computer program for power method and solve problem 2 with absolute convergence criterion of 10^{-5}.

Chapter 8

SOLUTION OF EQUATIONS FOR ENGINEERING DESIGN AND ANALYSIS

8.1 INTRODUCTION

There are several instances in engineering where it is required to solve a system of n non-linear equations with n unknowns. This system of equations cannot be solved using the methods employed for a linear system of equations. In the case of a single non-linear equation with one unknown the situation is described by a one dimensional curve. This case can easily be handled by the Newton Raphson method discussed earlier. When there are n non-linear equations with n unknowns, the system can be solved using Newton's method.

8.2 NEWTON'S METHOD

This method is a matrix form of the Newton Raphson iterative method. It will converge on a vector solution for a system of n non-linear equations with n unknowns. The system of non-linear equations must be put in the following matrix form:

$$F(x) = \begin{bmatrix} f_1(x_1, x_2, x_3, \cdots x_n) \\ f_2(x_1, x_2, x_3, \cdots x_n) \\ \vdots \\ f_n(x_1, x_2, x_3, \cdots x_n) \end{bmatrix} = \begin{bmatrix} 0 \\ 0 \\ \vdots \\ 0 \end{bmatrix} \qquad (8.1.1)$$

Note that F(x) is a column vector with the following form

$$F(x) = \begin{Bmatrix} f_1(x) \\ f_2(x) \\ f_3(x) \end{Bmatrix} \quad \text{where } (x) = (x_1, x_2, x_3, \cdots x_n)$$

Recall that the idea used in the Newton Raphson method for a single non-linear equation was to linearize the function f(x) at x_0 by expanding f(x) into a Taylor series and keeping only the first two terms.

For a single equation:

$$f(x) = f(x_0) + f'(x_0)(x - x_0) + \cdots = 0$$

This can be arranged in the familiar form:

$$x = x_0 - \frac{f(x_0)}{f'(x_0)} \quad \text{or} \quad x_{k+1} = x_k - \frac{f(x_k)}{f'(x_k)}$$

This method can be extended to a system of nonlinear equation as follows. Let vector

$$x^{(0)} = (x_1^{(0)}, x_2^{(0)}, x_3^{(0)}, \cdots\cdots, x_n^{(0)})$$

be an approximation to the actual solution. Expanding the function $f_1(x)$ about the value we have:

$$f_1(x) = f_1(x^{(0)}) + \left[\frac{\partial f_1}{\partial x_1}, \frac{\partial f_1}{\partial x_2}, \cdots\cdots, \frac{\partial f_1}{\partial x_n}\right]\{x - x^{(0)}\}$$

$$+ \text{high order terms} \qquad (8.1.2)$$

Neglecting higher order terms in equation 8.1.2 we can write equation 8.1.1 as

$$\{F(x)\} = \{F(x^{(0)})\} + [J]\{x - x^{(0)}\} = 0 \qquad (8.1.3)$$

where [J] is the Jacobian matrix given by:

$$[J(x)] = \begin{bmatrix} \dfrac{\partial f_1}{\partial x_1} & \dfrac{\partial f_1}{\partial x_2} & \dfrac{\partial f_1}{\partial x_3} & \cdots & \dfrac{\partial f_1}{\partial x_n} \\ \dfrac{\partial f_2}{\partial x_1} & \dfrac{\partial f_2}{\partial x_2} & \dfrac{\partial f_2}{\partial x_3} & \cdots & \dfrac{\partial f_2}{\partial x_n} \\ \vdots & \vdots & \vdots & & \vdots \\ \dfrac{\partial f_n}{\partial x_1} & \dfrac{\partial f_n}{\partial x_2} & \dfrac{\partial f_n}{\partial x_3} & \cdots & \dfrac{\partial f_n}{\partial x_n} \end{bmatrix}$$

Equation 8.1.3 can be arranged in two ways so that the iterative procedure can be performed on the matrix forms. Both of these equation forms will be explained in more detail by examples at the end of the following explanation.

The first form involves the inverse of the Jacobian matrix. Assuming that (x_k) is the vector of solutions $\{x\}$ at the kth iteration, equation 8.1.2 can be written in the form

$$[x_{k+1}] = [x_k] - [J(x_k)]^{-1} F(x_k) \qquad (8.1.4)$$

where

$$[J(x_k)]^{-1} = \frac{1}{\det\ J(x_k)} \begin{bmatrix} \text{confator} \\ \text{matrix of} \\ J(x_k) \end{bmatrix}^T$$

In this form x_{k+1} is obtained by substituting x_k in the right hand side of equation 8.1.4.

The inverse can also be obtained using Gauss-Jordan technic discussed in Chapter 3.

The second form of equation (8.1.3) involves Gaussian elimination. This equation is given below.

$$J(x_k)(y_k) = J(x_k)(x_{k+1} - x_k) = -F(x_k)$$

In this form y_k must be solved using Gaussian elimination after substituting for x_k on the right hand side. x_{k+1} is obtained from y_k as $x_{k+1} = y_k + x_k$.

In either equation an initial guess vector, x_k, is required to start the procedure. Once the vector x_{k+1} is evaluated, a check is performed between the norms of vectors x_{k+1} and x_k to find the relative error. This is represented by the following equation.

$$\xi = \frac{\left\| x_{k+1} - x_k \right\|_p}{\left\| x_{k+1} \right\|_p}$$

The type of the norm is decided depending on the nature of the problem.

If the value ξ is less than or equal to some prespecified tolerance, then x_{k+1} is the solution vector to the system. If it is not, then let $x_{k+1} = x_k$ and repeat the procedure. The rate of convergence for this method is of the order 2.

CAUTION:
It is impossible to monitor the relative error, ξ, so that it can be determined whether the method is converging or diverging. If this value is increasing after each iteration then the method is diverging and a better initial guess vector x_k must be chosen.

Another condition is that $J(x_k)$ must not be singular.

EXAMPLE 8.1
Solve the following system of non-linear equations using
    (a) using the 8.1.4 equation

(b) using the 8.1.5 equation

<u>system of equations</u>

$$x_1^2 + x_2 - 37 = 0 \rightarrow f_1(x)$$
$$x_1 + x_2^2 - 5 = 0 \rightarrow f_2(x)$$
$$x_1 + x_2 + x_3 - 3 = 0 \rightarrow f_3(x)$$

Solution

(a) This is the form of equation (8.1.3) in which the inverse of $J(x_k)$ is used.

Recall (8.1.4) $\rightarrow [x_{k+1}] = [x_k] - [J(x_k)]^{-1} F(x_k)$

The first step is to choose an appropriate guess vector, x_0, which is believed to be in the vicinity of the solution vector of the system.

Let $\quad x_0 = \begin{bmatrix} 1 \\ 1 \\ 1 \end{bmatrix} \quad$ Hence $\quad F(x_0) = \begin{bmatrix} 1^2 + 1 - 37 \\ 1 - 1^2 - 5 \\ 1 + 1 + 1 - 3 \end{bmatrix} = \begin{bmatrix} -35 \\ -5 \\ 0 \end{bmatrix}$

Now the Jacobian matrix is formed.

$$J_{ij} = \frac{\partial f_i(x)}{\partial x_j} \qquad \begin{matrix} i = 1, 2, 3 \\ j = 1, 2, 3 \end{matrix}$$

$$J_{11} = \frac{\partial f_1(x)}{\partial x_1} = \frac{\partial(x_1^2 + x_2 - 37)}{\partial x_1} = 2x_1$$

$$J_{12} = \frac{\partial f_1(x)}{\partial x_2} = \frac{\partial(x_1^2 + x_2 - 37)}{\partial x_2} = 1$$

$$J_{13} = \frac{\partial f_1(x)}{\partial x_3} = \frac{\partial(x_1^2 + x_2 - 37)}{\partial x_3} = 0$$

$$J_{21} = \frac{\partial f_2(x)}{\partial x_1} = \frac{\partial(x_1 - x_2^2 - 5)}{\partial x_1} = 1$$

$$J_{22} = \frac{\partial f_2(x)}{\partial x_2} = \frac{\partial(x_1 - x_2^2 - 5)}{\partial x_2} = -2x_2$$

$$J_{23} = \frac{\partial f_2(x)}{\partial x_3} = \frac{\partial(x_1 - x_2^2 - 5)}{\partial x_3} = 0$$

$$J_{31} = \frac{\partial f_3(x)}{\partial x_1} = \frac{\partial(x_1 + x_2 + x_3 - 3)}{\partial x_1} = 1$$

$$J_{32} = \frac{\partial f_3(x)}{\partial x_2} = \frac{\partial(x_1 + x_2 + x_3 - 3)}{\partial x_2} = 1$$

$$J_{33} = \frac{\partial f_3(x)}{\partial x_3} = \frac{\partial(x_1 + x_2 + x_3 - 3)}{\partial x_3} = 1$$

Hence, the Jacobian matrix is

$$J(x) = \begin{bmatrix} 2x_1 & 1 & 0 \\ 1 & -2x_2 & 0 \\ 1 & 1 & 1 \end{bmatrix}$$

Now to evaluate $J(x_0)$

$$x_0 = \begin{bmatrix} 1 \\ 1 \\ 1 \end{bmatrix}, \qquad J(x_0) = \begin{bmatrix} 2 & 1 & 0 \\ 1 & -2 & 0 \\ 1 & 1 & 1 \end{bmatrix}$$

This is $J(x_0)$ but the inverse of this is required in order to use eqn 8.1.4.

The inverse is found using the cofactor method.

$$[J(x_0)]^{-1} = \frac{1}{\det\ J(x_0)} \begin{bmatrix} \text{confators} \\ \text{of} \\ J(x_0) \end{bmatrix}^T$$

The cofactors are found by solving the sub-determinants formed by blocking out the row and column in a position where the row and column are both even or both odd values then the result is multiplied by a positive one. If the element position is a combination of an odd and even value, then multiply the value of the sub-determinant by minus one.

To illustrate this, consider the cofactor of position $J(x_0)_{11}$.

$$J(x_0) = \begin{bmatrix} 2 & \cdots & 1 & \cdots & 0 \\ \vdots & & & & \\ 1 & & -2 & & 0 \\ \vdots & & & & \\ 1 & & 1 & & 1 \end{bmatrix} \begin{matrix} + \\ \\ - \\ \\ + \end{matrix}$$

For position J_{11}, both row and column values are odd, so the sub-determinant will be multiplied by +1.

The sub-determinant is found by placing a line through the row and column of the desired position. In this case row 1 and column 1. (see fig. 8.1.1)

The sub-determinant consists of those values that are not crossed out.

Hence
$$\begin{vmatrix} -2 & 0 \\ 1 & 1 \end{vmatrix} = -2(1) - 0 = -2$$

Therefore, position 1, 1 of the cofactor matrix is $-2(+1) = -2$.

The remaining cofactor values are shown below.

$$\text{cofactor matrix} = \begin{bmatrix} -2 & -1 & 3 \\ -1 & 2 & -1 \\ 0 & 0 & -5 \end{bmatrix}$$

The determinant of the overall matrix $J(x_0)$ is

$$\det |J(x_0)| = 2 \begin{vmatrix} -2 & 0 \\ 0 & 1 \end{vmatrix} - 1 \begin{vmatrix} 1 & 0 \\ 1 & 1 \end{vmatrix} + 0 \begin{vmatrix} 1 & -2 \\ 1 & 1 \end{vmatrix} = -5$$

Now

$$[J(x_0)]^{-1} = \frac{1}{\det\ J(x_0)} \begin{bmatrix} \text{confators} \\ \text{matrix of} \\ J(x_0) \end{bmatrix}^{T} = \frac{1}{-5} \begin{bmatrix} -2 & -1 & 0 \\ -1 & 2 & 0 \\ 3 & -1 & -5 \end{bmatrix}$$

The first iteration can now be performed
$$k = 0$$
$$[x_{k+1}] = [x_k] - [J(x_k)]^{-1} F(x_k)$$

$$x_1 = \begin{bmatrix} 1 \\ 1 \\ 1 \end{bmatrix} - \left(\frac{1}{-5}\right) \begin{bmatrix} -2 & -1 & 0 \\ -1 & 2 & 0 \\ 3 & -1 & -5 \end{bmatrix} \begin{bmatrix} -35 \\ -5 \\ 0 \end{bmatrix}$$

$$x_1 = \begin{bmatrix} 16 \\ 6 \\ -19 \end{bmatrix}$$

To check the relative error, use the second norm $\|x\|_2$ for convenience.

$$\xi = \frac{\|x_{k+1} - x_k\|_2}{\|x_{k+1}\|_2} = \frac{\sqrt{15^2 + 5^2 + 20^2}}{\sqrt{16^2 + 6^2 + (-19)^2}} = .998$$

This is not satisfactory and hence, let $k = 1$ and perform the next iteration.

$$x_1 = \begin{bmatrix} 16 \\ 6 \\ -19 \end{bmatrix} \quad \text{Hence} \quad F(x_0) \begin{bmatrix} 16^2 + 6 - 37 \\ 16 - 6^2 - 5 \\ 16 + 6 - 19 - 3 \end{bmatrix} = \begin{bmatrix} 225 \\ -25 \\ 0 \end{bmatrix}$$

$$\text{For} \quad x_1 = \begin{bmatrix} 16 \\ 6 \\ -19 \end{bmatrix}, \quad J(x_1) = \begin{bmatrix} 32 & 1 & 0 \\ 1 & -12 & 0 \\ 1 & 1 & 1 \end{bmatrix},$$

$$\det|J(x_1)| = 32 \begin{vmatrix} -12 & 0 \\ 1 & 1 \end{vmatrix} - 1 \begin{vmatrix} 1 & 0 \\ 1 & 1 \end{vmatrix} + 0 \begin{vmatrix} 1 & -12 \\ 1 & 1 \end{vmatrix} = -385$$

$$\text{cofactor matrix} = \begin{bmatrix} -12 & -1 & 13 \\ -1 & 32 & -31 \\ 0 & 0 & -385 \end{bmatrix}$$

Now the second iteration can be done.

$$x_2 = \begin{bmatrix} 16 \\ 6 \\ -19 \end{bmatrix} - \left(\frac{-1}{-385} \right) \begin{bmatrix} -12 & -1 & 0 \\ -1 & 32 & -31 \\ 13 & -31 & -385 \end{bmatrix} \begin{bmatrix} 225 \\ -25 \\ 0 \end{bmatrix} = \begin{bmatrix} 16 \\ 6 \\ -19 \end{bmatrix} + \frac{1}{385} \begin{bmatrix} -2675 \\ -1025 \\ 3700 \end{bmatrix}$$

$$x_2 = \begin{bmatrix} 9.05 \\ 3.34 \\ -9.39 \end{bmatrix}$$

Relative error

$$\xi = \frac{\sqrt{6.95^2 + 2.66^2 + 9.61^2}}{\sqrt{9.05^2 + 3.34^2 + (-9.39)^2}} = .903$$

After applying five iterations, we obtain the following table.

k	0	1	2	3	4	5
	1	16	9.05	6.46	6	6
x_k	1	6	3.34	1.89	1.21	1
	1	-19	-9.39	-5.35	-4.21	-4
ξ	.998	.903	.583	.189	.041	

To check the validity of the solution vector, evaluate $F(x_5)$.

$$F(x_5) = \begin{bmatrix} 6^2 + 1 - 37 \\ 6 - 1^2 - 5 \\ 6 + 1 + (-4) - 3 \end{bmatrix} = \begin{bmatrix} 0 \\ 0 \\ 0 \end{bmatrix}$$

This confirms that x_5 is the solution vector.

(b) using eqn. (8.1.5)

This is the form of eqn. 8.1.3 in which Gaussian elimination is used during the iteration procedure to solve the non-linear system.

Recall eqn. (8.1.5) $\rightarrow \qquad J(x_k)(x_{k+1} - x_k) = -F(x_k)$

Again we choose an appropriate vector x_0.

Let $\qquad\qquad x_0 = \begin{bmatrix} 1 \\ 1 \\ 1 \end{bmatrix}$ Hence $F(x_0) = \begin{bmatrix} -35 \\ -5 \\ 0 \end{bmatrix}$

The Jacobian matrix will have the same form as it did in the first part of the example.

$$J(x) = \begin{bmatrix} 2x_1 & 1 & 0 \\ 1 & -2x_2 & 0 \\ 1 & 1 & 1 \end{bmatrix}$$

Now

$$J(x_0) = \begin{bmatrix} 2(1) & 1 & 0 \\ 1 & -2(1) & 0 \\ 1 & 1 & 1 \end{bmatrix} = \begin{bmatrix} 2 & 1 & 0 \\ 1 & -2 & 0 \\ 1 & 1 & 1 \end{bmatrix}$$

Now the first iteration can be performed

$k = 0$

$$J(x_0)(x^{(1)} - x^{(0)}) = -F(x_0)$$

$$\begin{bmatrix} 2 & 1 & 0 \\ 1 & -2 & 0 \\ 1 & 1 & 1 \end{bmatrix} \begin{bmatrix} x_1^{(1)} - 1 \\ x_2^{(1)} - 1 \\ x_3^{(1)} - 1 \end{bmatrix} = \begin{bmatrix} 35 \\ 5 \\ 0 \end{bmatrix}$$

Solve for $x^{(1)}$ using Gaussian elimination with back substitution.

$$R_j - \frac{A_{ij}}{A_{ii}} R_i \rightarrow \text{New } R_j$$

$$\tilde{A}^1 = \begin{bmatrix} 2 & 1 & 0 & : & 35 \\ 1 & -2 & 0 & : & 5 \\ 1 & 1 & 1 & : & 0 \end{bmatrix}$$

$$\tilde{A}^2 = \begin{bmatrix} 2 & 1 & 0 & : & 35 \\ 0 & -2.5 & 0 & : & -12.5 \\ 0 & 0.5 & 1 & : & -17.5 \end{bmatrix}$$

Back substitution gives

$$-2.5y_2^{(1)} = -12.5$$
$$y_2^{(1)} = 5$$
$$2y_1^{(1)} + 5 = 35$$
$$y_1^{(1)} = 15$$
$$0.5(5) + y_1^{(1)} = -17.5$$
$$y_3^{(1)} = -20$$

Hence

$$y_1^{(1)} = (x^{(1)} - x^{(0)}) = \begin{bmatrix} 15 \\ 5 \\ -20 \end{bmatrix}$$

$$x^{(1)} = \begin{bmatrix} 15 \\ 5 \\ -20 \end{bmatrix} + \begin{bmatrix} 1 \\ 1 \\ 1 \end{bmatrix} = \begin{bmatrix} 16 \\ 6 \\ -19 \end{bmatrix}$$

This is the same $x^{(1)}$ found in 8.1.4.
The next iteration can now be performed
k = 1

$$J(x_1)y^{(1)} = J(x_1)(x^{(2)} - x^{(1)}) = -F(x_1)$$

$$\begin{bmatrix} 32 & 1 & 0 \\ 1 & -12 & 0 \\ 1 & 1 & 1 \end{bmatrix} \begin{bmatrix} x_1^{(2)} - 16 \\ x_2^{(2)} - 6 \\ x_3^{(2)} - (-19) \end{bmatrix} = \begin{bmatrix} -225 \\ 25 \\ 0 \end{bmatrix}$$

$$\tilde{A}^1 = \begin{bmatrix} 32 & 1 & 0 & : & -225 \\ 1 & -12 & 0 & : & 25 \\ 1 & 1 & 1 & : & 0 \end{bmatrix}$$

$$\widetilde{A}^2 = \begin{bmatrix} 32 & 1 & 0 & \vdots & 225 \\ 0 & -12.01 & 0 & \vdots & 32.03 \\ 0 & .96875 & 1 & \vdots & 7.03 \end{bmatrix}$$

Back substitution gives

$$-12.01y_2^{(2)} = 32.03$$
$$y_2^{(2)} = -2.667$$
$$.96875(-2.667) + y_3^{(2)} = 7.03$$
$$y_3^{(2)} = 9.61$$
$$32y_1^{(2)} + (-2.667) = -225$$
$$y_1^{(2)} = -6.95$$

$$\begin{bmatrix} x_1^{(2)} - 16 \\ x_2^{(2)} - 6 \\ x_3^{(2)} - (-19) \end{bmatrix} = \begin{bmatrix} -6.95 \\ -2.667 \\ 9.61 \end{bmatrix}$$

$$x^{(2)} = \begin{bmatrix} 9.05 \\ 3.33 \\ -9.39 \end{bmatrix}$$

$x^{(2)}$ is also the same as $x^{(2)}$ found from 8.1.4. This procedure can be continued to converge in the same manner as 8.1.4.

NEWTON'S METHOD FOR SYSTEM OF NON-LINEAR EQUATIONS

To solve system of n non-linear equations $\overline{F} = 0$ in an iterative fashion using an initial guess solution vector. Individual equations are $f_i(x) = 0, \quad i = 1, 2, \cdots, n$

STEP 1   Input: $\overline{F}(x) = 0$ system of equations

$x^{(0)}$ initial guess vector

ξ relative error

N maximum number of iterations

STEP 2   Set k = 1

STEP 3   Calculate $\overline{F}(x^{(0)}) = 0$ and $J(x^{(0)}) = \dfrac{\partial f_i(x^{(0)})}{\partial x}$ for $1 \le i, j \le n$

where $x^{(0)}$ denotes $y_i^{(0)}, i = 1, 2, \cdots, n$

STEP 4   Solve the nxn linear system of equations

$$J(x^{(0)})y = \overline{F}(x^{(0)})$$

STEP 5   Set $x^{(1)} = x^{(0)} + y$

STEP 6   If $\dfrac{\left\| x_{k+1} - x_k \right\|_p}{\left\| x_{k+1} \right\|_p} < \xi$ then go to 10

STEP 7   Set $k = k + 1$

STEP 8   If $k > N$ then go to 10

STEP 9   Set $x^{(0)} = x^{(1)}$ and go to 3

STEP 10   Output: $x^{(1)}$

PROBLEMS

1.   Obtain the solution of the nonlinear equation given below using Newton's method. Use 3 iterations.

$$-x_1 + 2x_1^2 - 2x_1x_2 + x_2^2 = 1$$

$$x_1^2 - 2x_1x_2 - x_2 + x_2^2 = 0$$

2.   Solve the following set of nonlinear equations using Newton's method. Use 3 iterations.

$$12x_1 - 3x_2^2 - 4x_3 = 7.17$$

$$x_1^2 + 10x_2 - x_3 = 11.54$$

$$x_2^3 + 7x_3 = 7.631$$

3.   Solve the following nonlinear system using Newton's method. Use 3 iterations.

$$x_1 + \cos(x_1x_2x_3) - 1 = 0$$

$$(1 - x_1)^{1/4} + x_2 + 0.05 x_3^2 - 0.15x_3 - 1 = 0$$

$$-x_1^2 - 0.1 x_2^2 + 0.01x_2 + x_3 - 1 = 0$$

Chapter 9

NUMERICAL SOLUTIONS OF ORDINARY DIFFERENTIAL EQUATIONS

9.1 INTRODUCTION

Differential equations describe the rate of change of a system's response with respect to some independent variable. The form of a nth order ordinary differential equation is given below.

$$F^n(x,y) = F\left(x, y, \frac{dy}{dx}, \frac{d^2y}{dx^2}, ..., \frac{d^ny}{dx^n}\right) = 0 \qquad (9.1.1)$$

This is a nth order differential equation because the largest derivative is of order n. The equation is termed as ordinary because there is only one independent variable. In the case of equation (9.1.1) the independent variable is x; time, t, is the independent variable for equations describing the response of a system as a function of time.

A solution to Equation (9.1.1) would be a function y (x), which is differentiable n times and results in the form of f^n (x, y). A unique solution can be obtained if values of y(x) or of its derivatives are known at a specified value of x. In order to insure a unique solution, n such values will be required. If these known values belong to the same value of x where x=x_0, then the problem is called an initial value problem. If the values are known at different values of x, then the problem is called a boundary-value problem.

Any n^{th} order ordinary differential equation can be broken down into n first order differential equations by introducing n-1 additional variables. For instance, consider the following equation.

$$t\frac{d^2y}{dt^2} + \frac{dy}{dt} + 3ty = 0$$

By letting $z = \dfrac{dy}{dt}$ we reduce this equation into two first order equations which are,

$$t\frac{dz}{dt} + z + 3ty = 0$$

and

$$\frac{dy}{dt} - z = 0$$

In this example, n=2; n-1=1 and hence one additional variable z is introduced.

For this reason, most of the methods described in this chapter will be devoted to solving first order differential equations.

9.2 EULER'S METHOD

This method is used to solve first order differential equations of the following form.

$$\frac{dy}{dx} = g(x, y) \tag{9.2.1}$$

The solution to such an equation is a function y(x), which satisfies one initial condition and whose derivative is $\frac{dy}{dx}$. To solve for y(x) in an analytical fashion is difficult if at all possible; therefore the function y(x) is solved numerically at discrete values of x. The accuracy of such a solution will depend on the size of the interval used between successive values.

In order to solve $\frac{dy}{dx}$ numerically, consider the following Taylor series expansion at a point x about the initial value x_0. Let the distance between x_0 and x be a discrete step size h where $h = x - x_0 = \Delta x$.

$$y(x_0+h) = y(x) = y(x_0) + hy'(x_0) + \ldots$$

In this equation we neglect the terms higher than $y'(x_0)$ so that a closed form solution can be obtained.

Of course this will introduce a truncation error as discussed in earlier chapters. The value $y'(x_0)$ can be obtained directly from eqn. (9.2.1), if the initial condition $y(x_0)$ is known. Then $y(x_0+h) = y(x) = y(x_0) + hg(x_0,y(x_0))$

Notice that $y'(x_0)$ is now replaced by $\frac{dy}{dx} = g(x,y)$

The following example will illustrate the method.

EXAMPEL 9.1

Solve the following differential equation at Δx increments of 0.1 using Euler's method for values of x in the range [0,1].

$$g(x,y) = \frac{dy}{dx} = -y^2$$
$$y(x_0) = y(0) = 1$$

Solution

$$y(x_{i+1}) = y(x_i) + hg\,(x_i,\,y(x_i))$$

where

$h = 0.1$

$y(.1) = y(0) + 0.1g(0,1) = 1 + 0.1(-(1)^2) = 0.9$
$y(.2) = y(.1) + 0.1g(.1,.9) = 0.9 + 0.1(-(0.9)^2) = 0.819$
$y(.3) = y(.2) + 0.1g(.2,.819) = 0.819 + .1(-(0.819)^2) = 0.752$
$y(.4) = y(.3) + 0.1g(.3,.752) = 0.752 + .1(-(0.752)^2) = 0.695$
$y(.5) = y(.4) + 0.1g(.4,.695) = 0.695 + .1(-(0.695)^2) = 0.647$
$y(.6) = y(.5) + 0.1g(.5,.647) = 0.647 + .1(-(0.647)^2) = 0.605$
$y(.7) = y(.6) + 0.1g(.6,.605) = 0.605 + .1(-(0.605)^2) = 0.568$
$y(.8) = y(.7) + 0.1g(.7,.569) = 0.568 + .1(-(0.568)^2) = 0.536$
$y(.9) = y(.8) + 0.1g(.8,.536) = 0.536 + .1(-(0.536)^2) = 0.507$
$y(1) = y(.9) + 0.1g(.9,.507) = 0.507 + .1(-(0.507)^2) = 0.481$

In order to check the accuracy of these results, the analytical solution is calculated using the separation of variables techniques.

The analytical solution is obtained in the following way.

$$\frac{dy}{dx} = -y^2$$

with
$$y(0) = 1$$

Separate the variables and perform the indefinite integral.

$$\int \frac{dy}{-y^2} = \int dx$$

This results in

$$\frac{1}{y} = x + c$$

The value of c is solved by applying the boundary condition.
$$y(0) = 1$$

Hence
$$\frac{1}{1} = 0 + c\;;\; c = 1$$

The solution is $y(x) = \dfrac{1}{x+1}$; the following table gives the analytical values and the corresponding relative error in steps of 0.1.

$y_{analytical}$	$\xi = \left\| \dfrac{y_{analytical} - y_{numerical}}{y_{analytical}} \right\|$
$y(0) = 1$	0
$y(.1) = .909$	.0099
$y(.2) = .833$	.0168
$y(.3) = .769$	.0221
$y(.4) = .714$	.0266
$y(.5) = .667$	.0299
$y(.6) = .625$	.032
$y(.7) = .588$	.034
$y(.8) = .556$	.036
$y(.9) = .526$	.037
$y(1.0) = .500$	.038

Notice ξ is increasing

EXAMPLE 9.2

Solve the following differential equation using Euler's method with step size h=.1 in the range 0 to 0.5.

$$\frac{dy}{dt} = g(t,y) = 3yt^2+2yt+1$$

with
$$y(0) = 1$$

Solution

$$Y(t_{i+1}) = y(t_i) + hg(t_i,y(t_i))$$
$$y(.1) = y(0) + .1g(0,1) = 1+.1(3(1)(0)^2+2(1)(0)+1) = 1.1$$

$$y(.2)= y(.1)+.1g(.1,1.1) =$$
$$9.1 + .1(3(1.1)(.1)^2+2(1.1)(.1)+1) = 1.23$$

$$y(.3) = y(.2) + .1g(.2,1.23) =$$
$$1.23+.1(3(1.23)(.2)^2+2+2(1.23)(.2)+1) = 1.39$$

$$y(.4) = y(.3)+ .1g(.3,1.39) =$$
$$1.39+.1(3(1.39)(.3)^2+2(1.39)(.3)+1) = 1.61$$

$$y(.5) = y(.4)+.1g(.4,1.61) =$$
$$1.61+.1(3(1.61)(.4)^2+2+2(1.61)(.4)+1) = 1.92$$

Table form;

i	0	1	2	3	4	5
t_i	0	0.1	0.2	0.3	0.4	0.5
y_i	1	1.1	1.23	1.39	1.61	1.92
y_i'	1	1.25	1.64	2.21	3.06	4.36

Note: This method is described as a self starting method because only one value is required to start solving the function values in the forward direction. Another name for this method is Taylor's method of order one.

EULER'S METHOD

To solve an ordinary differential equation $y' = g(x, y)$, $a \le x \le b$, $y(a) = \alpha$

STEP 1   Input: a, b endpoints

                    α initial condition

                    h step increment

STEP 2   Set $x_0 = a$, $y_0 = \alpha$, $N = \dfrac{b-a}{h}$

STEP 3 y_i, $y_{i-1} + h\, g(x_{i-1}, y_{i-1})$ for i = 1, 2, ... N

STEP 4   Output: y_i, i = 1, 2, ... N

CAUTION
The error in this method increases as the value of x increases. This is because successive values are formed by using the previous approximations. These previous approximations become less accurate each time due to the inherent truncation error. If the derivative terms higher than the first derivative are equal to zero, then no truncation error will exist.

9.3 HIGHER ORDER TAYLOR METHODS

In the previous method, only the first two terms were used to find a linearized approximation to the first order differential equation. This resulted in quite a large truncation error. In this method an extra term in the Taylor series will be used in order to solve equation (9.2.1) at discrete values of x more accurately. The Taylor expansion series will have the following form;

$$y(x_0+h) = y(x_0) + hy'(x_0) + \frac{h^2}{2} y''(x_0) + ...$$

The added complication of this method is that the second derivative must be evaluated. Since in the equation $y' = g(x, y)$ represents the first derivative expression, then $\left.\dfrac{dg}{dx}\right|_{x=x_0}$ will represent the $y''(x_0)$ term in the series expression. When forming $\dfrac{dg}{dx}$ it should be realized that since g(x,y) is a function of x and y, the chain rule must be used.

Chain rule to find $g'(x, y)$

$$g'(x,y) = \frac{dg}{dx} = \frac{\partial g}{\partial x} + \frac{\partial g}{\partial y}\frac{\partial y}{\partial x} = \frac{\partial g}{\partial x} + \frac{\partial g}{\partial y}g$$

The expansion expression becomes;

$$Y_{i+1} = y_i + hg(x_i, y_i) + \frac{h^2}{2}g'(x_i, y_i)$$

EXAMPLE 9.3

Solve the following differential equation using Taylor's method of order two. Use a step size h=0.1 in the range 0 to 0.5.

$$g(t,y) = \frac{dy}{dt} = -y^2 \qquad \text{with } y(0) = 1$$

Solution

The exact solution of this equation is found by separation of variables to be;

$$y = \frac{1}{1+t}$$

The numerical solution is found in the following manner

$$y_{i+1} = y_i + hg(t_i, y_i) + \frac{h^2}{2}g'(t_i, y_i)$$

with $h = 0.1$

and

$$g'(t,y) = \frac{\partial g}{\partial t} + \frac{\partial g}{\partial t}\frac{dy}{dt} = 0 + (-2y)(-y^2) = 2y^3$$

Hence, $y_{i+1} = y_i + 0.1(-y_i^2) + \frac{(0.1)^2}{2}(2y_i^3)$ with $y(0) = 1$

Approximation   Exact solution
                                                          $[y=(1+t)^{-1}]$

y(.1) = 1+.1(-1)+.005(2(1)) = .910   .909
y(.2) = .91+.1(-(.91)^2)+.005(2(.91)^3) = .835   .820
y(.3) = .835+.1(-(.835)^2)+.005(2(.835)^3) = .771   .769
y(.4) = .771+.1(-(.771)^2)+.005(2(.771)^3) = .716   .714
y(.5) = .716+.1(-(.716)^2)+.005(2(.716)^3) = .668   .667

Note: An error exists between the numerical solution and the exact solution because higher order derivatives exist. If the third and higher derivatives would be zero, then no truncation error will exist.

When using the Taylor method one should recognize whether the derivative g(x, y) expression can be differentiated a finite number of times so that eventually the derivative expression will be zero. This will be possible if the first derivative function g(x, y) is a function of the independent variable x alone. If this is the case, then the Taylor series should be expanded to the term that would result in zero, so that the truncation error is eliminated.

HIGER ORDER TAYLOR METHOD

To solve an ordinary differential equation $y' = g(x, y)$, $a \le x \le b$, $y(a) = \alpha$. Function g(x, y) is differentiable and has many derivatives g', g'', etc with respect to x.

STEP 1   Input: a, b endpoints

             α initial condition

             h step increment

STEP 2   Set $x_0 = a$, $y_0 = \alpha$, $N = \dfrac{b-a}{h}$

STEP 3 $y_i = y_{i-1} + hg(x_{i-1}, y_{i-1}) + \dfrac{h^2}{2!} g'(x_{i-1}, y_{i-1})$

            $+ \dfrac{h^3}{3!} g''(x_{i-1}, y_{i-1}) + ... \text{ for } i = 1, 2, ... N$

STEP 4   Output: y_i, i = 0, 1, 2, ... N

9.4 RUNGE-KUTTA METHODS

In the Taylor series expansion method, higher order derivatives of the primary function were required in order to evaluate the function value with reasonable accuracy. In Runge-kutta methods the tedious chore of evaluating higher order derivatives is avoided while the accuracy of the function approximation is retained.

In Euler's method, one derivative term in the Taylor series expansion was considered. In the Runge Kutta method of order two, we consider up to the second derivative term in the Taylor series expansion and then substitute the derivative terms with the appropriate function values in the interval.

Consider the Taylor series expansion of the function about y_i.

$$y_{i+1} = y_i + hy'(x_i, y_i) + \frac{h^2}{2} y''(x_i, y_i)$$

$$y_{i+1} = y_i + hg(x_i, y_i) + \frac{h^2}{2} g'(x_i, y_i)$$

$$y_{i+1} = y_i + h\left[g(x_i, y_i) + \frac{h}{2} g'(x_i, y_i) \right] \qquad (9.4.1)$$

Substituting for $g'(x_i, y_i) = \dfrac{\partial g}{\partial x} + \dfrac{\partial g}{\partial y} . g(x_i, y_i)$, where $\dfrac{dy}{dx} = g(x_i, y_i)$ from the differential equation,

we get

$$y_{i+1} = y_i + h\left[g(x_i, y_i) + \frac{h}{2} \frac{\partial g}{\partial x} + \frac{h}{2} \frac{\partial g}{\partial y} . g(x_i, y_i) \right] \qquad (9.4.2)$$

In equation (9.4.2) the factor inside the rectangular brackets involving the derivatives may be substituted with a function of the type. $g(x+\alpha, y+\beta)$ in a Taylor series expansion, so that from equation (9.4.1) we have

$$y_{i+1} = y_i + h[ag(x_i + \alpha, y_i + \beta)] \qquad (9.4.3)$$

Expanding the function $g(x_i + \alpha, y_i + \beta)$ in a Taylor series expansion involving two variables about (x_i, y_i) and retaining only the first derivative terms we get.

$$y_{i+1} = y_i + ha\left[g(x_i, y_i) + \alpha \frac{\partial g}{\partial x} + \beta . \frac{\partial g}{\partial y} \right] \qquad (9.4.4)$$

Equating coefficients of like terms on the right hand side of equations (9.4.2) and (9.4.4) we get.

$$a = 1$$
$$\alpha = \frac{h}{2}$$
$$\beta = \frac{h}{2} g(x_i, y_i)$$

Hence, equation (9.4.3) becomes

$$y_{i+1} = y_i + hg\left[x_i + \frac{h}{2}, y_i + \frac{h}{2} g(x_i, y_i) \right] \qquad (9.4.5)$$

This can also be written as

$$y_{i+1} = y_i + hk_2$$

with

$$k_2 = g\left[x_i + \frac{h}{2}, y_i + \frac{h}{2}k_1\right]$$

$$k_1 = g(x_i, y_i)$$

This method is also called the Midpoint method since the derivative is replaced by functions evaluated at the midpoint $x_i + \dfrac{h}{2}$.

For the fourth order Runge-Kutta formula the derivatives are evaluated at four points, once at each end and twice at the interval midpoint as given below.

$$y(x_{i+1}) = y(x_i) + \frac{h}{6}(k_1 + 2k_2 + 2k_3 + k_4)$$

$$k_1 = g(x_i, y_i(x_i))$$

$$k_2 = g\left(x_i + \frac{h}{2}, y(x_i) + \frac{1}{2}k_1 h\right)$$

$$k_3 = g\left(x_i + \frac{h}{2}, y(x_i) + \frac{1}{2}k_2 h\right)$$

$$k_4 = g(x_i + h, y(x_i) + k_3 h)$$

Note: The truncation error of this method is of the order h^5.

The fourth order formula is referred to as the Runge-Kutta formula since it is the most accurate formula available without extending outside the interval $[x_i, x_{i+1}]$.

EXAMPLE 9.4
Solve the following differential equation using the Runge-Kutta method of order two or Midpoint method for $0 \le t \le 1$ with an initial condition of y(0)=1. Let the step size be h=0.1.

$$\frac{dy}{dt} = g(t, y) = t^2 - y + 1$$

$$0 \le t \le 1$$
$$y(0)=1; \quad h=0.1$$

Solution
The exact analytical solution of the above equation is.

$$y(t) = -2e^{-t} + t^2 - 2t + 3$$

The midpoint formula is;

$$y(t_{i+1}) = y(t_i) + hk_2$$

$$y(t_{i+1}) = y(t_i) + hg\left(t_i + \frac{h}{2}, y(t_i) + \frac{h}{2}k_1\right)$$

$$y(t_{i+1}) = y(t_i) + hg\left(t_i + \frac{h}{2}, y(t_i) + \frac{h}{2}g(t_i, y(t_i))\right)$$

By directly substituting the derivative function $g(t_i,y(t_i))$ into this final form of the midpoint formula gives.

$$y(t_{i+1}) = y(t_i) + hg\left[\left(t_i + \frac{h}{2}\right)^2 - \left(y(t_i) + \frac{h}{2}(t_i^2 - y(t_i) + 1)\right) + 1\right]$$

with   h=0.1

$$y(.1) = 1 + .1\left[\left(0 + \frac{.1}{2}\right)^2 - \left(1 + \frac{.1}{2}\left((0)^2 - 1 + 1\right)\right) + 1\right] = 1.0025$$

By forming a table, the comparison between actual and numerical solutions can be made.

	Numerical solution	Actual solution
y(0.1)	1.00025	1.00033
y(0.2)	1.00243	1.00254
y(0.3)	1.00825	1.00836
y(0.4)	1.01926	1.01936
y(0.5)	1.03688	1.03694
y(0.6)	1.06238	1.0623
y(0.7)	1.09690	1.0968
y(0.8)	1.14150	1.1413
y(0.9)	1.19710	1.1969
y(1.0)	1.26458	1.2642

RUNGE-KUTTA METHOD OF ORDER 4

To solve an ordinary differential equation $y' = g(x, y)$, $a \le x \le b$, $y(a) = \alpha$.

STEP 1   Input: a ,b   endpoints

                         α initial condition

h step increment

STEP 2   Set $x_0 = a$, $y_0 = \alpha$, $N = \dfrac{b-a}{h}$

STEP 3   Set $\; k_1 = g(x_{i-1}, y_{i-1})$

$$k_2 = g(x_{i-1} + h/2, y_{i-1} + k_1 h/2)$$

$$k_3 = g(x_{i-1} + h/2, y_{i-1} + k_2 h/2)$$

$$k_4 = g(x_{i-1} + h, y_{i-1} + k_3 h)$$

$$y_i = y_{i-1} + \frac{h}{6}(k_1 + 2k_2 + 2k_3 + k_4)$$

for $i = 1, 2, \dots N$

STEP 4   Output: y_i, $i = 0, 1, 2, \dots N$

Note: Replace step 3 appropriately for Midpoint Method

Multistep Methods

All of the methods discussed up to now have been termed one step methods. In these methods the approximation of $y(x_{i+1})$ was evaluated by knowing the form of the differential equation and information at a single point x_i. In the multistep methods the differential equation must be known as well as information at several values of x, say x_{i-1}, x_{i-2}, etc. This means that the multistep methods require a one step method to provide this additional information to get started.

When the approximation to the value $y(x_{i+1})$ is expressed explicitly in terms of previously determined values, the expression is called an open method or explicit method.

For example, all the methods explained up to now have been explicit methods. If the approximation expression of $y(x_{i+1})$ contains a value of $y(x_{i+1})$ obtained from another expression, then the expression is known as an implicit or closed expression.

It will be seen in the following sections that a predictor expression is an explicit expression which gives a crude approximation to $y(x_{i+1})$ whereas a corrector expression is an implicit expression which starts with this crude approximation to get an improved value of $y(x_{i+1})$.

9.5 TWO STEP PREDICTOR-CORRECTOR METHODS

Simple two step predictor-corrector formulas can be derived as follows. Consider a Taylor series expansion of the function $y(x_{i+1})$ about x_i given symbolically by

$$y_{i+1} = y_i + h y_i' + \frac{h^2}{2} y_i'' \qquad (9.5.1)$$

where $y_i = y(x_i)$ and only terms with second derivatives are considered. The differential equation is of the form

$$y'(x) = g(x, y)$$

substituting this into equation 9.5.1 we get

$$y_{i+1} = y_i + hg_i + \frac{h^2}{2} g_i' \qquad (9.5.2)$$

where $g_i = g(x_i, y_i)$. Replacing g_i' by its backward difference form, we get

$$y_{i+1} = y_i + hg_i + \frac{h^2}{2} \left[(g_i - g_{i-1})/h \right]$$

Rearranging, we get

$$y_{i+1} = y_i + \frac{3h}{2} g_i - \frac{h}{2} g_{i-1}$$

This is called the Adams-Bashforth predictor formula. This is not self starting because initially y_i must be obtained using any one step method that were discussed earlier.

If g_i' is replaced by its forward difference form in equation (9.5.2) we get

$$y_{i+1} = y_i + hg_i + \frac{h^2}{2} \left[(g_{i+1} - g_i)/h \right]$$

Rearranging we get an iterative type of relation

$$y_{i+1} = y_i + \frac{h}{2} \left[(g_{i+1} + g_i) \right] \qquad (9.5.4)$$

Since g_{i+1} in the right hand side can be calculated only after y_{i+1} is obtained, this is an implicit formula. This is called the Adams-Moulton corrector formula. The right hand side can be calculated using a predicted value of y_{i+1} to start with. This enables us to calculate a better y_{i+1} value. A repeated application of equation (9.5.4) will improve the value of y_{i+1} which can be continued until the relative error in y_{i+1} is within a limit given by

$$\xi = \left| \frac{y_{i+1}^{(n+1)} - y_{i+1}^{(n)}}{y_{i+1}^{(n+1)}} \right| < \varepsilon$$

where $y_{i+1}^{(n+1)}$ corresponds to the nth iteration value of y_{i+1}.

EXAMPLE 9.5

Solve the following differential equation and obtain y(0.2) using Adams-Bashforth predictor and Adams-Moulton corrector:

$$\frac{dy}{dx} = x^2 - y + 1 \quad 0 \le x \le 1$$

with y(0)=1. Use h = 0.1 and obtain the answer to an accuracy of 6 digits.

Solution

We have y_0=y(0)=1. From the midpoint method we get

$$y_1 = y_0 + h\{(x_0+h/2)^2 - y_0 + 1 + (x_0^2 - y_0 + 1)h/2\}$$
$$= 1 + 0.1\{0.05^2 - 1 + 1 + (0-1+1)0.05\} = 1.00025$$

Using Adams-Bashforth predictor

$$y_2 = y_1 + \frac{3h}{2}g_1 - \frac{h}{2}g_0$$

where

$$g_0 = x_0^2 - y_0 + 1 = 0 - 1 + 1 = 0$$
$$g_1 = x_1^2 - y_1 + 1 = 0.1^2 - 1.00025 + 1 = 0.00975$$

Hence

$$y_2 = 1.00025 + 0.15 \times 0.00975 - 0.05 \times 0 = 1.00171$$

Now, using Adams-Moulton corrector we have

$$y_2 = y_1 + \frac{h}{2}(g_2 + g_1)$$

Denoting the iteration as a superscript for y_2 and g_2 we can write

$$y_2^{(1)} = y_1 + \frac{h}{2}\left(g_2^{(0)} + g_1\right)$$

where

$$g_2^{(0)} = g\left(x_2, y_2^{(0)}\right) = x_2^2 - y_2^{(0)} + 1$$

and

$$y_2^{(0)} = 1.00171 \text{ from the predictor method.}$$

Hence

$$g_2^{(0)} = 0.2^2 - 1.00171 + 1 = 0.03829$$
$$g_1 = x_1^2 - y_1 + 1 = 0.1^2 - 1.00025 + 1 = 0.00975$$

Consequently

$$y_2^{(1)} = 1.00025 + 0.05 \cdot (0.03829 + 0.00975) = 1.00265$$

$$\xi_1 = \left| \frac{y_2^{(1)} - y_2^{(0)}}{y_2^{(1)}} \right| = 9.38 \times 10^{-4}$$

Continuing the iteration

$$y_2^{(2)} = y_1 + \frac{h}{2}\left(g_2^{(1)} + g_1\right)$$

where

$$g_2^{(1)} = g(x_2, y_2^{(1)}) = x_2^2 - y_2^{(1)} + 1 = 0.03735$$

Hence

$$y_2^{(2)} = 1.00025 + 0.05(0.03735 + 0.00975) = 1.00261$$

$$\xi_2 = \left| \frac{y_2^{(2)} - y_2^{(1)}}{y_2^{(2)}} \right| = 3.99 \times 10^{-5}$$

Repeating the iterative process, we get

$$y_2^{(3)} = 1.00261$$

Here we stop the iterative process since

$$\xi_3 = \left| \frac{y_2^{(3)} - y_2^{(2)}}{y_2^{(3)}} \right| \approx 0 \le 5 \times 10^{-6}$$

Exact value $y_2 = 1.00254$ from example 9.4

9.6 MILNE'S PREDICTOR-SIMPSON'S CORRECTOR METHOD

This is a multistep method which uses an explicit and implicit expression to approximate $y(x_{i+1})$.
When such a combination is used the method is referred to as a Predictor-Corrector method. The predictor expression is a multistep formula which is derived by integrating Newton's backward formula between x_{i-3} and x_{i+1} to give the following formula:

$$y(x_{i+1}) = y(x_{i-3}) + \frac{4}{3}h\big(2g(x_i, y(x_i) - g(x_{i-1}, y(x_{i-1})) + 2g(x_{i-2}, y(x_{i-2}))\big) \qquad (9.6.1)$$

where

$$g(x, y) = \frac{dy}{dx} = y'(x)$$

The above equation is referred to as the Milne predictor expression. This expression requires function values at x_i, x_{i-1}, x_{i-2} and x_{i-3}. The initial condition $y(x_0)$ must be given. Then the Runge-Kutta fourth order formula or any of the other one step methods can be employed to get $y(x_1), y(x_2)$ and $y(x_3)$. Once these values are evaluated $y(x_{i+1})$ can be approximated using Milne's method. The value $y(x_{i+1})$ can then be improved by using an implicit corrector formula known as Simpson's formula. This expression is given below:

$$y(x_{i+1}) = y(x_{i-1}) + \frac{h}{3}\left(g(x_{i+1}, y(x_{i+1})) + 4g(x_i, y_i) + g(x_{i-1}, y_{i-1})\right)$$

where

$$g(x, y) = \frac{dy}{dx} = y'(x)$$

The above formula was derived by integrating a Newton's backward formula between x_{i-1} and x_{i+1}. Notice that this is a direct application of Simpson's integration rule in which the function value is approximated by replacing the function values on the right hand side by the derivative function values.

Once an improved value of $y(x_{i+1})$ is found it is replaced into Milne's equation, the value of i is incremented, and next crude approximation can be made. Note that the Runge-Kutta one step method can be used to start the method.

EXAMPLE 9.6
Solve the following differential equation for x=0.4 and x=0.5 using the Milne predictor-Simpson's corrector method.

$$\frac{dy}{dx} = g(x, y) = x^2 - y + 1$$
$$0 \le x \le 1$$
$$y(0) = 1$$

Use the Runge-Kutta midpoint formula to start the method.

Solution

The Runge-Kutta midpoint formula was used to solve this problem in the previous example. The first three values found by the midpoint formula are the following.

$$y(.1) = 1.00025$$
$$y(.2) = 1.00243$$
$$y(.3) = 1.00825$$

Now these values and the initial condition can be used to find a crude evaluation of y(.4) using Milne's equation in the following manner.

Milne's equation

$$y(.4) = 1 + \frac{4}{3}(.1)\left(2\left((.3)^2 - 1.00825 + 1\right) - \right.$$
$$\left.\left((.2)^2 - 1.00243 + 1\right) + 2\left((.1)^2 - 1.00025 + 1\right)\right)$$
$$y(.4) = 1.01939$$

Now this value is refined using the Simpson's equation.

Simpson's equation

$$y(.4) = 1.00243 + \frac{1}{3}(.1)\left(\left((.4)^2 - 1.01939 + 1\right) + \right.$$
$$\left. 4\left((.3)^2 - 1.00825 + 1\right) + \left((.2)^2 - 1.00243 + 1\right)\right)$$

$$y(.4) = 1.01927 \text{ (corrected value)}$$

Now this value can be used in the Milne's equation. Along with y(.3),y(.2) and y(.1) to approximate a value for y(.5).

Mine's equation

$$y(.5) = 1.00025 + \frac{4}{3}(.1)\left(2\left((.4)^2 - 1.01927 + 1\right) - \right.$$
$$\left.\left((.3)^2 - 1.00825 + 1\right) + 2\left((.2)^2 - 1.00243 + 1\right)\right)$$

$$y(.5) = 1.036790 \text{ (the predictor step)}$$

After application of corrector step y(0.5) = 1.03684
Hence the calculated values are

$$y(.4) = 1.01927$$
$$y(.5) = 1.04779$$

The values are more accurate than those found in the one step method.

MILNE-SIMPSON MULTISTEP PREDICTOR CORRECTOR METHOD

To solve an ordinary differential equation $y' = g(x, y)$, $a \leq x \leq b$, $y(a) = \alpha$.

STEP 1   Input: a, b endpoints

                   α initial increment

                   h step increment

                   ξ relative error

STEP 2   Set $x_0 = a$, $y_0 = \alpha$, $N = \dfrac{b-a}{h}$

STEP 3   For i =1, 2, and 3

$$k_1 = g(x_{i-1}, y_{i-1})$$
$$k_2 = g(x_{i-1} + h/2, y_{i-1} + k_1 h/2)$$
$$k_3 = g(x_{i-1} + h/2, y_{i-1} + k_2 h/2)$$
$$k_4 = g(x_{i-1} + h, y_{i-1} + k_3 h)$$
$$y_i = y_{i-1} + \frac{h}{6}(k_1 + 2k_2 + 2k_3 + k_4)$$

STEP 4   For i =3, 4, … N do steps until 8

STEP 5 $y_{i+1}^{(0)} = y_{i-3} + \dfrac{4}{3} h \left(2g(x_i, y_i) - g(x_{i-1}, y_{i-1}) + 2g(x_{i-2}, y_{i-2}) \right)$

STEP 6 $y_{i+1}^{(1)} = y_{i-1} + \dfrac{h}{3} \left(g(x_{i+1}, y_{i+1}^{(0)}) + 4g(x_i, y_i) + g((x_{i-1}, y_{i-1})) \right)$

STEP 7   If $\dfrac{\left| y_{i+1}^{(1)} - y_{i+1}^{(0)} \right|}{\left| y_{i+1}^{(1)} \right|} > \xi$ go to 6

STEP 8   Set $y_{i+1} = y_{i+1}^{(1)}$

STEP 9   Output: y_{i+1}, i = 0, 1, 2, … N

9.7 HAMMING'S METHOD

The Hamming formula is another corrector type formula which is used in conjunction with the Milne-predictor formula. This formula was derived in such a way as to increase the stability of the solution in comparison to the Milne-Simpson formula; however, this increase in stability was made at a slight cost of increasing the truncation error. For this reason, the formula is used only when the stability of the first order differential equation is questionable.

The Hamming   corrector equation is given below;

$$y(x_{i+1}) = \frac{1}{8}(9y(x_i) - y(x_{i-2})) + \frac{3}{8}h[g(x_{i+1}, y(x_{i+1})) +$$

$$2g(x_i, y(x_i)) \_ g(x_{i-1}, y(x_{i-1}))]$$

where;

$$g(x, y(x)) = \frac{dy}{dx}$$

EXAMPLE 9.7.1

Using the Hamming equation approximate y(.4) for the following equation.

$$\frac{dy}{dx} = g(x, y) = x^2 - y + 1; \text{ where } y(0) = 1$$

Use the Runge-kutta midpoint formula to start the method.

Solution

The first three values found from the Runge-Kutta one step method are the following.

y(.1) = 1.00025
y(.2) = 1.00243
y(.3) = 1.00825

Milne's predictor formula is then used to predict a value for y(.4).

Milne's equation

$$y(.4) = 1 + \frac{4}{3}(.1)[2((.3)^2 - 1.00825 + 1) -$$

$$((.2)^2 - 1.00243 + 1) + 2((.1)^2 - 1.00025 + 1)]$$

y(.4) = 1.01939

Now Hamming's equation can be used to correct this value.

Hamming's equation

$$y(.4) = \frac{1}{8}(9(1.00825 - 1.00025)) + \frac{3}{8}(.1)(0.14061 + 2(.08175 - 0.03757)) \qquad y(.4) = 1.01925$$

Notice that in this case y(.4) is the same as in the previous example. This will not always be the case.

MILNE-HAMMING-MULTISTEP PREDICTOR-CORRECTOR METHOD

Same as Milne-Simpson method except for step 6

STEP 6 $y_{i+1}^{(1)} = \frac{1}{8}(9y_i - y_{i-2}) + \frac{3}{8}h\left[g(x_{i+1}, y_{i+1}) + 2g(x_i, y_i) - g(x_{i-1}, y_{i-1})\right]$

9.8 HIGHER ORDER DIFFERENTIAL EQUATIONS AND SYSTEM OF DIFFERENTIAL EQUATIONS

An n-th order differential equation can be transformed into n number of first order differential equations. Let the n-th order differential equation be given as

$$y^{(n)}(t) = f\left(t, y, y^1, \dots y^{(n-1)}\right) \qquad (9.8.1)$$

with a $\le$ t $\le$ b and with initial conditions

$$y(a) = \alpha_1, \quad y^1(a) = \alpha_2, \dots, y^{(n-1)}(a) = \alpha_n$$

In this equation t is the independent parameter and y is the unknown dependent quantity which is to be solved. Introduce the following notations.

$$u_1(t) = y(t)$$
$$u_2(t) = y'(t)$$
$$\vdots \qquad\qquad (9.8.2)$$
$$u_n(t) = y^{(n-1)}(t)$$

Using the notation, we can write eqn. (9.8.1) as

$$\frac{du_1}{dt} = \frac{dy}{dt} = u_2$$
$$\frac{du_2}{dt} = \frac{dy'}{dt} = u_3$$
$$\vdots \qquad\qquad\qquad (9.8.3)$$
$$\frac{du_n}{dt} = \frac{dy^{(n-1)}}{dt} = y^{(n)} = f(t, y, y^1, \dots, y^{(n-1)})$$
$$= f_n(t, u_1, u_2, \dots, u_n)$$

In the above we have n number of first order differential equations. The last equation is the restatement of eqn. (9.8.1) in a different form, whereas the first (n-1) differential equations relate the additional (n-1) unknowns $u_1(t)$. The initial conditions are written as.

$$u_1(a) = y(a) \qquad = \alpha_1$$
$$u_2(a) = y'(a) \qquad = \alpha_2$$
$$\vdots$$
$$u_n(a) = y^{(n-1)}(a) = \alpha_n$$

$$(9.8.4)$$

In general, eqn. (9.8.3) can be written as a system of first order equations in the general form.

$$\frac{du_1}{dt} = f_1(t, u_1, u_2, ..., u_n)$$
$$\frac{du_2}{dt} = f_2(t, u_1, u_2, ..., u_n)$$
$$\vdots$$
$$\frac{du_n}{dt} = f_n(t, u_1, u_2, ..., u_n)$$

$$(9.8.5)$$

with $u_1(a) = \alpha_1$, i=1,2,....,n as initial condition.

Even though f_i (t, u_1, u_2, ... u_n) are shown as general functions of all unknowns and time, when we convert higher order differential equations into a system of first order differential equations, we note that.

$$f_i \text{ (t, } u_1, u_2, ... u_n) = u_{i+1} \qquad\qquad (9.8.6)$$
$$i = 1, 2, ..., n-1$$

Any method that can solve numerically a system of first order differential equations can solve a higher order differential equation also.

Methods developed to solve a single first order differential equation can be extended to solve a system of first order differential equations. We choose a suitable time step h, so that

$$t_j = a+jh \qquad\qquad j = 0, 1, 2, ... N \qquad\qquad (9.8.7)$$

where b = hN+a

Let us denote $u_{ij} = u_i(t_j)$. Then the initial conditions are given by

$$u_{10} = \alpha_1, u_{20} = \alpha_2, ...u_{no} = \alpha_n \qquad\qquad (9.8.8)$$

Using any method for solving first order differential equations, we obtain $u_{11}, u_{21}, u_{31}, \ldots. u_{n1}$ which are the solutions at the time $t_1 = a+h$. Once all the unknowns are obtained at t_1, we proceed to t_2 and find all the unknown there. Then we proceed to t_3 and so on. In the case of a higher order differential equation, if we want the value of y any y' at $b = t_N$, then at t_N we have to solve only for u_{1n} and u_{2n}. The expressions for finding u_{ij}, using different methods are given below:

i) <u>Euler's Method:</u>

$$u_{i,j+1} = u_{ij} + hf_i(t_j, u_{1j}, u_{2j}, \ldots, u_{nj}) \qquad (9.8.9)$$
$$i = 1, 2, \ldots n$$

ii) <u>Midpoint Method:</u>

$$u_{i,j+1} = u_{ij} + hf_i(t_j + \frac{h}{2}, u_{ij} + \frac{h}{2}k_{1j}, \ldots, u_{nj} + \frac{h}{2}k_{nj})$$

where $$k_{ij} = f_i(t_j, u_{1j}, u_{2j}, \ldots, u_{nj}) \qquad (9.8.10)$$

iii) <u>Fourth Order Runge-Kutta Method:</u>

$$u_{i,j+1} = u_{ij} + \frac{h}{6}\left[k_{1i} + 2k_{2i} + 2k_{3i} + k_{4i}\right]$$

where

$$k_{1i} = f_i(t_j, u_{1j}, u_{2j}, \ldots, u_{nj})$$

$$k_{2i} = f_i(t_j + \frac{h}{2}, u_{1j} + \frac{h}{2}k_{11}, u_{2j} + \frac{h}{2}k_{12}, \ldots, u_{nj} + \frac{h}{2}k_{1n})$$

$$k_{3i} = f_i(t_j + \frac{h}{2}, u_{1j} + \frac{h}{2}k_{21}, u_{2j} + \frac{h}{2}k_{22}, \ldots, u_{nj} + \frac{h}{2}k_{2n})$$

$$k_{4i} = f_i(t_j + h, u_{ij} + hk_{31}, u_{2j} + hk_{32}, \ldots, u_{nj} + hk_{3n})$$

$$i = 0, 1, \ldots, n$$
$$j = 0, 1, \ldots, N$$

EXAMPLE 9.8
Solve the following differential equation using Euler's method.

$$y'' - 2y' + 2y = e^{2t}\sin t$$

with $0 \le t \le 1$, $y(0) = -0.4$ and $y'(0) = 0.6$. Choose a time step of $h = 0.1$ and obtain $y(0.2)$.

Solution
Let $$y(t) = u_1, \ y'(t) = u_2$$

The two first order differential equations are

$$u_1' = u_2 = f_1(t, u_1, u_2)$$

$$u_2' = e^{2t} \sin t - 2u_1 + 2u_2 = f_2(t, u_1, u_2)$$

The initial conditions are

$$u_{10} = -0.4, \quad u_{20} = -0.6$$

Using eqn.(9.8.9) we have

$$u_{1,j+1} = u_{1j} + hf_1(t_j, u_{1j}, u_{2j}) = u_{1j} + 0.1u_{2j}$$

$$u_{2,j+1} = u_{2j} + hf_2(t_j, u_{1j}, u_{2j})$$

$$= u_{2j} + 0.1(e^{2t} \sin t_j - 2u_{1j} + 2u_{2j})$$

solving for j = 0, we have

$$u_{11} = -0.4 + 0.1(-0.6) = -0.46$$
$$u_{21} = -0.6 + 0.1[\, 0 - 2(-0.4) + 2(-0.6)\,] = -0.64$$

Solving for j = 1, we have

$$u_{12} = -0.46 + 0.1(-0.64) = -0.524$$

Hence, $y(0.2) = u_{12} = -0.524$

$$u_{22} = -0.64 + 0.1[e^{0.2}\sin(0.1) + 2 \times 0.46 - 2 \times 0.64] = -0.6638$$

Hence, $y(0.2) = u_{22} = -0.6638$

EXAMPLE 9.9

Solve the following differential equation using midpoint method.

$$y'' - 2y' + 2y = e^{2t} \sin t$$

with 0 $0 \le t \le 1$, $y(0) = -0.4$ and $y'(0) = -0.6$

use $h = 0.1$ and obtain $y(0.2)$ and $y'(0.2)$.

Solution

we have

$$u_1'(t) = u_2 = f_1(t, u_1, u_2)$$

$$u_2'(t) = e^{2t}\sin t - 2u_1(t) + 2u_2(t) = f_2(t, u_1, u_2)$$

and $\qquad u_{10} = -0.4, \; u_{20} = -0.6,$

Using eqn (9.8.10), we have

$$u_{1,j+1} = u_{1j} + hf_1\left(t_j + \frac{h}{2}, u_{1j} + \frac{h}{2}k_{1j}, u_{2j} + \frac{h}{2}k_{2j}\right)$$

$$= u_{1j} + 0.1\left[u_{2j} + \frac{h}{2}k_{2j}\right]$$

where

$$k_{2j} = f_2(t_j, u_{1j}, u_{2j}) = e^{2tj}\sin(t_j) - 2u_{1j} + 2u_{2j}$$

$$u_{2,j+1} = u_{2j} + hf_2\left(t_j + \frac{h}{2}, u_{1j} + \frac{h}{2}k_{1j}, u_{2j} + \frac{h}{2}k_{2j}\right)$$

$$= u_{2j} + 0.1\left[e^{2(t)+0.05}\sin(t_j + 0.05) - 2(u_{1j} + 0.05k_{1j} + 2(u_{2j} + 0.05k_{2j})\right]$$

where $\quad k_{1j} = f_1(t_j, u_{1j}, u_{2j}) = u_{2j}$
and k_{2j} is defined as previously.

Solving for j = 0,

$$u_{11} = -0.4 + 0.1(-0.6 + 0.05\,k_{20})$$
$$k_{20} = -2(-0.4) + 2(-0.6) = -0.4$$

Hence

$$u_{11} = -0.462$$

$$u_{21} = -0.6 + 0.1\left[e^{0.1}\sin 0.05 - 2(-0.4 + .05k_{10}) + 2(-0.6 + .05k_{20})\right]$$

$$k_{10} = u_{20} = -0.6$$

Hence, $\qquad u_{21} = -0.6325$

Solving at j = 1, we have

$$u_{12} = u_{11} + 0.1(u_{21} + 0.05k_{21})$$

$$k_{21} = e^{0.2}\sin(0.1) - 2u_{11} + 2u_{21} = -0.2191$$

Hence, $\qquad u_{12} = -0.5263 = y(0.2)$

Further,

$$u_{22} = u_{21} + 0.1\left[e^{0.3}\sin(0.15) - 2(u_{11} + 0.05k_{11}) + 2(u_{21} + 0.05k_{21})\right]$$

$$k_{11} = u_{21} = -0.6325$$

Hence,

$$u_{22} = -0.6423 = y'(0.2)$$

PROBLEMS

1.   Obtain the solution of the differential equation

$$\dot{y} = -0.1\,y + \sin(2\pi t) \text{ with } y(0) = 1.0$$

Tabulate the solution for $0 \le t \le 1$ at 0.1 interval. Use Euler's method with $h = 0.1$. Compare the solution with that obtained by Midpoint method.

2.   Given the following differential equation

$$\frac{dy}{dx} = 1 + x^2 y^2, \quad y(0) = 0$$

Solve by Euler's method in the range of 0 to 0.5 and $h = 0.1$

3.   Try Euler's method with step size $h = 0.5$ on the following problem.

$$y' = t \sin t + e^{-t}, \quad 0 \le t \le 1 \; y(0) = 0$$

4.   Apply Euler's method to the differential system

$$y' = ty^2 + y$$

$$0 \le t \le 2 \qquad h = 0.5 \qquad y(0) = 1$$

5.   An electrical circuit can be described by the following differential equation.

$$\frac{di}{dt} = \frac{1}{L} V_s - \frac{R}{L} i$$

where $V_s(t) = e^{-t/3} \sin(3t - \pi)$, $R = 1$, $L = 1$

with initial condition $i(0) = 1$.

Use Euler's method to solve for i on the interval [0.1] at time steps of $\Delta t = 0.2$

6.   Use Euler's method to solve the following initial value problem,

$$\frac{dy}{dx} = y^2, y(0) = 0.2 \quad h = 0.5 \; 0 \le x \le 1$$

7.   The function y(x) is defined by the following problem

$$\frac{dy}{dx} = x^2 - y^2 \qquad y(0) = 1$$

Find y(0.2) using

(a)   Euler's method: h = 0.1

(b)   Runge-Kutta method of order 4   with   h = 0.1

8.   Solve the differential equation to obtain y(0.4)

$$y = 1 + t \sin(ty) \qquad 0 \le t \le 1 \quad y(0) = 0$$

using Euler's method with h = 0.2

9.   Use Euler's method to approximate the solution to the initial value problem in the given range.

$$\frac{dy}{dt} = -y^2 + t + 1.1$$

$$0 \le t \le 1 \qquad y(0) = 1 \qquad h = 0.1$$

10.   Solve the differential equation

$$\ddot{y} + \dot{y} + y = \sin t$$

with y(0) = 1 and $\dot{y}$ (0) = 0, and obtain y(0.3) using Euler's method.

11.   Given the following equation

$$\dot{y} = \frac{y}{t} + te^t$$

with 1 ≤ t ≤ 2   y(1) = 0   h = 0.1

(a)   solve by Euler's method
(b)   solve by Runge-Kutta method of order 2.

12.   Solve the third order differential equation using Euler's method and obtain y(0.2).

$$y''' + xy' - 2y = x$$
$$y(0) = y''(0) = 0$$
$$y'(0) = 1$$

13.   Solve the differential equation

$$\dot{y} = t\sin t + e^{-t}$$

$0 \le t \le 1$ $\qquad$ $h = 0.1$ $\qquad$ $y(0) = 0$

(a)   using Euler's method
(b)   using Runge-Kutta method of order 4.

14.   An R-L circuit is described by the following second order differential equation

$$\frac{d^2i}{dt^2} + 8\frac{di}{dt} + 10\,i = 10, \text{ where } i(0) = 1, \; i'(0) = 1$$

Solve at $t = 0.1$ using

(a)   Midpoint method
(b)   Runge –Kutta method of order 4.

15.   Solve the following differential equation

$$y' = 2y + 2e^x$$

with $y(0) = 0$, $h = 0.1$ and $0 \le x \le 0.5$

using third order Taylor method.

16.   Solve by Midpoint method in the given range.

$$\dot{y} = -\frac{y^2}{2} + t + 1$$

$0 \le t \le 1, \; y(0) = 1, \; h = 0.1$

17.   Solve by Runge-Kutta method of order 4.

$$\frac{dy}{dt} = t^2 + 1$$

with the condition $y(0) = 1$, $h = 0.5$ and $0 \le t \le 2$

18.   The motion of an electron in an electrostatic field due to a positive charge can be described by the following second order differential equation.

$$m\,\ddot{x} + \frac{k}{x} = 0$$

with $m = 1$, $k = 2$, $\quad x(0) = 1$, $x'(0) = 1$

Solve the above equation by Runge-Kutta method of order 4 using $h = 1/3$ and $0 \le t \le 1$.

19.   Use Runge-Kutta 4 method with step length h = 0.25 and initial condition y(0) = 2 on the interval [0,1], given that

$$y' = \frac{\cos x}{1 - 4\sin x}$$

20.   Use midpoint method to solve the following differential equation in the given range.

$$y' = x + y$$

$$0 \le x \le 1, \ y(0) = 1, \ h = 0.1$$

21.   Solve the differential equation

$$y' = 3x^4 + x - 2$$

with step size h = 0.5 and y(0) = 0, in the interval [0,5] by Runge-Kutta method of order 2.

22.   Use Milne's predictor-Simpson's   corrector method with Runge-Kutta 4 to the following differential equation.

$$\frac{dy}{dx} = xy^{1/4}$$

$$y(1) = 1, \ 1 \le x \le 2, \ h = 0.25$$

find y(2).

23.   The inductor current i of a second order circuit is expressed by the following differential equation.

$$\frac{d^2 i}{dt^2} + 7\frac{di}{dt} + 6i = \frac{dv_s}{dt}$$

where $v_s(t) = e + e^{-0.1t} \sin(2\pi t)$, i(0) = 2 and $\frac{di(0)}{dt} = 1$

Use the Runge-Kutta method 4 to solve the above differential equation at t = 0.1 sec with h = 0.05

24.   The differential equation of motion of a falling object of mass m and speed v is given below.

$$\frac{dv}{dt} = 32 \ - \ \frac{k}{m}v^2, \frac{k}{m} = 3$$

Apply Milne's predictor-Simpson's corrector method to solve the equation at t = 0.2 and 0.4, given the initial condition v(0) = 1.

25.   Solve the third order differential equation

$$\frac{d^3y}{dx^3} - 3\frac{dy}{dx} + 2y = 2\,(\sin x - 2\cos x)$$

using a method of your choice to find y(0.1). Use a step size of 0.05. The initial conditions are

$$y(0) = 4,\ y'(0) = \text{-}2 \text{ and } y''(0) = 17$$

26.   Solve problem 19 using Runge-Kutta method of order two (Midpoint method). Use a step size of h = 0.1 and obtain the value of y(0.3).

27.   Obtain y(0.2) in the following initial value problem, using Runge-Kutta method of order two.

$$\frac{dy}{dx} = 3x^2y \qquad y(0) = 0 \text{ and } h = 0.1$$

28.   Solve problem 1 using Miline's Predictor – Hamming's Corrector Scheme with Runge-Kutta 4 and obtain y(0.5) to an accuracy of 5 significant digits.

29.   Write a computer program to implement Runge-Kutta method or order 4 and solve problem 1.

Chapter 10

INTRODUCTION TO PARTIAL DIFFERENTIAL EQUATIONS

10.1 INTRODUCTION

Partial differential equations are those which relate a function and its derivatives with respect to more than one independent variable with any other function of the independent variables. The first function may be the output of a system whereas the other functions may be the input to the system. The order of the partial differential equation is governed by the highest derivative. Many physical phenomena in, for example, heat transfer, quantum mechanics, electrodynamics, fluid/aero dynamics are described by partial differential equations. These equations may have boundary conditions in which case they are called steady state equations if the equations have initial conditions instead, they are called transient equations.

This chapter provides an introduction to partial differential equations and deals only with linear second order partial differential equations with two independent variables.

The most general from of a linear second order partial differential equation with independent variables x and y is:

$$a\frac{\partial^2 u}{\partial x^2} + 2b\frac{\partial^2 u}{\partial x \partial y} + c\frac{\partial^2 u}{\partial y^2} + d\frac{\partial u}{\partial x} + e\frac{\partial u}{\partial y} + fu = g$$

where a, b, c, d, e, f, and g may be functions of x and/or y. Classification of second order partial differential equations depends on the sign of b^2-ac and is as follows:

if $b^2-ac<0$ the equation is called elliptic:
if $b^2-ac=0$ the equation is called parabolic; and
if $b^2-ac>0$ the equations are called hyperbolic.

If the differential equation is transformed to a two dimensional frequency domain, the transform will have the form of an ellipse, or a parabola, or a hyperbola. Hence the names for the differential equations.

To solve partial differential equations analytically is often impossible. They are therefore solved numerically. Many of the methods discussed in earlier chapters can be expanded upon and applied to the solution of partial differential equations.

10.2 ELLIPTIC PARTIAL DIFFERENTIAL EQUATIONS

A common elliptic partial differential equation is the Laplace equation:

$$\frac{\partial^2 u}{\partial x^2} + \frac{\partial^2 u}{\partial y^2} = 0 \qquad (10.2.1)$$

which describes the steady state temperature distribution of a plate or the voltage distribution in a conductive plate, given known temperatures or temperature gradients at the boundaries; this is thus a boundary value problem.

The Laplace equation can be approximated using central difference expressions as discussed previously. Consider a grid with step size Δx in x and Δy in y defined over the region of interest (see Fig. 10.1).

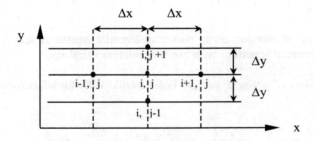

Figure 10.1: Elliptic partial differential equation approximated

The following finite difference expressions can be used to approximate the partial differentials;

$$\left(\frac{\partial^2 u}{\partial x^2}\right)_{i,j} = \frac{1}{(\Delta x)^2}(u_{i-1,j} - 2u_{i,j} + u_{i+1,j}) + 0(\Delta x)^2 \qquad (10.2.2)$$

$$\left(\frac{\partial^2 u}{\partial y^2}\right)_{i,j} = \frac{1}{(\Delta y)^2}(u_{i,j-1} - 2u_{i,j} + u_{i,j+1}) + 0(\Delta y)^2 \qquad (10.2.3)$$

Substituting these into eqn. (10.2.1) and setting $\Delta x = \Delta y = h$ we get;

$$\left(\frac{\partial^2 u}{\partial x^2} + \frac{\partial^2 u}{\partial y^2}\right)_{i,j} = \frac{1}{h^2}(u_{i-1,j} + u_{i,j+1} - 4u_{i,j} + u_{i+1,j} + u_{i,j-1}) + 0(h)^2 = 0 \quad (10.2.4)$$

EXAMPLE 10.1

Determine the temperatures at the inside nodes of the following L-shaped figure, made up of 12 equally sized squares, whereby the temperature along the boundaries are known.

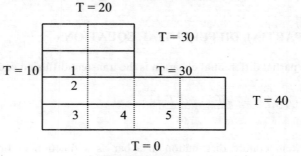

Figure 10.2

Solution

This is a boundary value problem for which eqn. (10.2.4) can be applied.

Using eqn. (10.2.2) for each of the nodes 1 to 5 we get;

Node 1:	$10+20-4T_1+30+T_2=0$
Node 2:	$10+T_1-4T_2+30+T_3=0$
Node 3:	$10+T_2-4T_3+T_4+ 0=0$
Node 4:	$T_3+30-4T_4+T_5 + 0=0$
Node 5:	$T_4 +30-4T_5+40+0=0$

These are five equations with five unknowns, which can also be written as;

$$\begin{bmatrix} 4 & -1 & 0 & 0 & 0 \\ -1 & 4 & -1 & 0 & 0 \\ 0 & -1 & 4 & -1 & 0 \\ 0 & 0 & -1 & 4 & -1 \\ 0 & 0 & 0 & -1 & 4 \end{bmatrix} \begin{bmatrix} T_1 \\ T_2 \\ T_3 \\ T_4 \\ T_5 \end{bmatrix} = \begin{bmatrix} 60 \\ 40 \\ 10 \\ 30 \\ 70 \end{bmatrix}$$

Solving above system of equations using Gauss-Seidel with starting values for $T_1,...,T_5$ respectively, as 20, 20,10,15, 20 (these are 'educated guesses', as can easily interpret the expected range), we can create the following table;

iteration	T_1	T_2	T_3	T_4	T_5
0	20.000	20.000	10.000	15.000	20.000
1	20.000	17.500	10.625	15.156	21.289
2	19.375	17.500	10.664	15.488	21.372
3	19.375	17.510	10.750	15.530	21.383
4	19.377	17.532	10.769	15.537	21.384
5	19.383	17.537	10.769	15.538	21.385
6	19.384	17.538	10.769	15.538	21.385
7	19.385	17.538	10.769	15.538	21.385

10.3 PARABOLIC PARTIAL DIFFERENTIAL EQUATIONS

A common parabolic partial differential equation is the transient diffusion equation or heat equation:

$$\frac{\partial T}{\partial t} = k \frac{\partial^2 T}{\partial x^2} \qquad (10.3.1)$$

which describes the temperature distribution in a bar as a function of time. The two independent variables are time t ($t \geq 0$) and distance x along the bar with end points a and b ($a \leq x \leq b$).

The transient diffusion equation can be solved numerically using a combination of central and forward difference expressions. Hereby a grid with step size Δx in x and Δt in t is defined over the region of interest (see fig. 10.3).

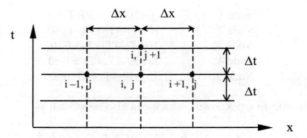

Figure 10.3: Parabolic partial differential equation approximated

These problems need both boundary conditions, and initial conditions. Temperatures specified at the end points of the bar are examples of boundary conditions, with the temperature distribution given at time t=0 as the initial condition. Typically, the decay in temperature is calculated as a function of time. It is therefore necessary to use forward difference expressions when dealing with t, and central difference expressions when dealing with x. Thus the following finite difference expressions can be used;

$$\left(\frac{\partial T}{\partial t}\right)_{i,j} = \frac{1}{\Delta t}(-T_{i,j} + T_{i,j+1}) + 0(\Delta t) \qquad (10.3.2)$$

$$\left(\frac{\partial^2 T}{\partial x^2}\right)_{i,j} = \frac{1}{(\Delta x)^2}(T_{i-1,j} - 2T_{i,j} + T_{i+1,j}) + 0(\Delta x)^2 \qquad (10.3.3)$$

Substituting these into eqn. (10.3.1) we get;

$$\frac{1}{\Delta t}(-T_{i,j} + T_{i,j+1}) = \frac{k}{(\Delta x)^2}(T_{i-1,j} - 2T_{i,j} + T_{i+1,j}) \qquad (10.3.4)$$

For given boundary values $T_{o,j}$ and $T_{n,j}$, and known initial conditions $T_{i,o}$ eqn. (10.3.4) can be solved directly for the only unknown $T_{i,j+1}$. This method is therefore called an explicit method. Unfortunately, the explicit method is prone to stability problems, and we can only find a solution as long as we satisfy the following stability criterion;

$$\Delta t \leq \frac{1}{2k}(\Delta x)^2$$

This means that the selection of Δt depends on the choice of Δx in order to find a stable solution.

In order to avoid above mentioned stability problems Crank-Nicholson proposed the following finite difference expression:

$$\left(\frac{\partial^2 T}{\partial x^2}\right)_{i,j} = \frac{1}{2}\left\{\frac{1}{(\Delta x)^2}(T_{i-1,j} - 2T_{i,j} + T_{i+1,j}) + \frac{1}{(\Delta x)^2}(T_{i-1,j+1} - 2T_{i,j+1} + T_{i+1,j+1})\right\} + 0(\Delta x)^2 \qquad (10.3.5)$$

Equation (10.3.5) is in effect the average central difference formula about the point (i,j+0.5)or the average of the central differences about the points (i,j) and (i,j+1). Substituting eqns. (10.3.2) and (10.3.5) into eqn. (10.3.1) we obtain,

$$\frac{1}{\Delta t}(-T_{i,j} + T_{i,j+1}) = \frac{k}{2(\Delta x)^2}\left\{T_{i-1,j} - 2T_{i,j} + T_{i+1,j} + T_{i-1,j+1} - 2T_{i,j+1} + T_{i+1,j+1}\right\} \qquad (10.3.6)$$

This equation has a stable solution for any value of Δt and Δx. However, it contains three unknowns $T_{i-1,j+1}$, $T_{i,j+1}$ and $T_{i+1,j+1}$. Therefore, we cannot directly solve for these, but we end up with a system of equations which can be solved implicitly. This method is thus an implicit method.

Rewriting eqn. (10.3.6) such that all unknowns are on the left, and all knowns on the right we obtain;

$$T_{i-1,j+1} - \left(\frac{2(\Delta x)^2}{k\Delta t} + 2\right)T_{i,j+1} + T_{i+1,j+1} = -T_{i-1,j} - \left(\frac{2(\Delta x)^2}{k\Delta t} - 2\right)T_{i,j} - T_{i+1,j} \quad (10.3.7)$$

Setting $\qquad\qquad F_i = -T_{i-1,j} - \left(\dfrac{2(\Delta x)^2}{k\Delta t} - 2\right)T_{i,j} - T_{i+1,j}$ and

$C = -\left(\dfrac{2(\Delta x)^2}{k\Delta t} + 2\right)$ we can rewrite eqn. (10.3.7) in the following matrix form:

$$
\begin{bmatrix}
c & 1 & 0 & . & . & . & . & 0 \\
1 & c & 1 & . & . & . & . & . \\
0 & 1 & c & 1 & . & . & . & . \\
. & . & . & . & . & . & . & . \\
. & . & . & . & . & . & 0 \\
. & . & . & 1 & c & 1 \\
0 & . & . & . & 0 & 1 & c
\end{bmatrix}
\begin{bmatrix}
T_{1,j+1} \\
T_{2,j+1} \\
T_{3,j+1} \\
. \\
. \\
T_{n-2,j+1} \\
T_{n-1,j+1}
\end{bmatrix}
=
\begin{bmatrix}
F_1 - T_0 \\
F_2 \\
F_3 \\
. \\
. \\
F_{n-2} \\
F_{n-1} - T_n
\end{bmatrix}
$$

where T_0 and T_n are known temperatures at the end points of the bar [a,b], and the length of the bar has been divided into n equal segments with $\Delta x = (b-a)/n$.

EXAMPLE 10.2

The temperature distribution in above bar can be described by the heat equation with k=1 and with boundary conditions T(0,t)=0 and T(1,t)=1 and initial condition T(x,0)=4x for 0≤x≤.5 and T(x,0)=3-2x for 0.5≤x≤1. Determine how the temperature changes in time using Δx=0.25 and Δt=0.1

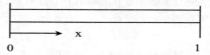

$$0 \qquad\qquad\qquad\qquad\qquad\qquad\qquad\qquad 1$$

Figure 10.4

Solution

To solve this problem we may apply eqn. 10.3.8 where n=(1-0)/0.25=4; $T(0) = T_0 = T_{0,j} = 0$; $T(1) = T_4 = T_{4,j} = 1$;

$$c = -\left(\frac{2*(0.25)^2}{1*(0.1)} + 2\right) = -3.25 \; ; \text{ and}$$

$$F_i = -T_{i-1,j} - \left(\frac{2*(0.25)^2}{1*(0.1)} - 2 \right) T_{i,j} - T_{i+1,j}$$

$$= -T_{i-1,j} + 0.75T_{i,j} - T_{i+1,j} \quad \text{for } i = 1,2,3 \text{ and } j = 0, 1\ldots\ldots$$

For j = 0 eqn. (10.3.8) becomes:

$$\begin{bmatrix} -3.25 & 1 & 0 \\ 1 & -3.25 & 1 \\ 0 & 1 & -3.25 \end{bmatrix} \begin{bmatrix} T_{1,1} \\ T_{2,1} \\ T_{3,1} \end{bmatrix} = \begin{bmatrix} F_1 - 0 \\ F_2 \\ F_3 - 1 \end{bmatrix}$$

where;

$$F_1 = -T_{0,0} + 0.75T_{1,0} - T_{2,0}$$

$$F_2 = -T_{1,0} + 0.75T_{2,0} - T_{3,0}$$

$$F_3 = -T_{2,0} + 0.75T_{3,0} - T_{4,0}$$

but;

$$T_{1,0} = 4(0.25) = 1$$

$$T_{2,0} = 4(0.5) = 3 - 2(0.5) = 2$$

$$T_{3,0} = 3 - 2(0.75) = 1.5$$

Therefore

$$F_1 = -0 + 0.75(1) - 2 = -1.25$$

$$F_2 = -1 + 0.75(2) - 1.5 = -1$$

$$F_3 = -1.2 + 0.75(1.5) - 1 = -1.875$$

We can now calculate $T_{1,1}$, $T_{2,1}$ and $T_{3,1}$ from

$$\begin{bmatrix} -3.25 & 1 & 0 \\ 1 & -3.25 & 1 \\ 0 & 1 & -3.25 \end{bmatrix} \begin{bmatrix} T_{1,1} \\ T_{2,1} \\ T_{3,1} \end{bmatrix} = \begin{bmatrix} -1.25 \\ -1 \\ -2.875 \end{bmatrix}$$

After applying Gaussian elimination, we obtain;

$$\begin{bmatrix} T_{11} \\ T_{2,1} \\ T_{3,1} \end{bmatrix} = \begin{bmatrix} .64964 \\ .86131 \\ 1.1496 \end{bmatrix}$$

We can now calculate a new right hand side for eqn. 10.3.8 and repeat the process for j=1, etc. as is summarized in the following table:

t	T_0	$T_{1,j+1}$	$T_{2,j+1}$	$T_{3,j+1}$	T_4	F_1-T_0	F_2	F_3-T_4
0	0	1.0000	2.0000	1.5000	1	-1.2500	-1.0000	-2.3750
.1	0	.6496	.8613	1.1496	1	- .3741	-1.1533	-1.9991
.2	0	.3351	.7149	.8351	1	- .4636	- .6340	-2.0886
.3	0	.3084	.5387	.8084	1	- .3074	- .7128	-1.9324
.4	0	.2583	.5321	.7583	1	- .3384	- .6175	-1.9634
.5	0	.2590	.5032	.7590	1	- .3090	- .6405	-1.9340
.6	0	.2505	.5051	.7505	1	- .3172	- .6222	-1.9422
.7	0	.2514	.5000	.7515	1			
.								
.								
∞	0	.2500	.5000	.7500	1			

10.4 HYPERBOLIC PARTIAL DIFFERENTIAL EQUATIONS

A common hyperbolic partial differential equation is the wave equation:

$$\frac{\partial^2 u}{\partial x^2} - \frac{\partial^2 u}{\partial t^2} = 0 \qquad\qquad (10.4.1)$$

which describes for example the vibration of a string.

We can express this hyperbolic partial differential equation using the same central difference expressions as for elliptic equations. Substituting eqns. (10.2.2) and (10.2.3) into eqn. (10.4.1) with t instead of y we get;

$$\frac{u_{i-1,j} - 2u_{i,j} + u_{i+1,j}}{(\Delta x)^2} = \frac{u_{i,j-1} \quad -2u_{i,j} + u_{i,j+1}}{(\Delta t)^2}$$

or:

$$u_{i,j+1} = 2u_{i,j} - u_{i,j-1} + \frac{(\Delta t)^2}{(\Delta x)^2}\left(u_{i-1,j} - 2u_{i,j} + u_{i+1,j}\right) \qquad (10.4.2)$$

In order to obtain a stable solution, we must satisfy the stability criterion

$$\Delta t \le \Delta x$$

EXAMPLE 10.3

The taut string in above figure is at time t=0 subjected to a transverse displacement of u=(1-x)x, at

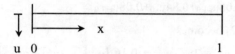

$$u \quad 0 \qquad\qquad\qquad\qquad\qquad 1$$

Figure 10.5

which time also $\dfrac{\partial u}{\partial t} = 0$ .Determine the displacement of the string using Δx=0.25 and Δt=0.1.

Solution

As the string is fixed at both ends, the boundary conditions are u(0,t)=0 and u(1,0)=0. With the string divided into n=(1-0)/0.5=4 segments, the initial condition u(x,0)=(1-x)x and boundary conditions can be written as

$$\begin{cases} u_{0,j} = 0 \\[4pt] u_{1,0} = (1-0.25)*0.25 = 0.1875 \\[4pt] u_{2,0} = (1-.50)*0.50 = 0.25 \\[4pt] u_{3,0} = (1-0.75)* 0.75 = 0.1875 \\[4pt] u_{4,j} = 0 \end{cases} \qquad (10.4.3)$$

where $u_{i,j}$ indicates the displacement at x=iΔx and t=jΔt.

In order to solve this problem, we express the partial derivative in the initial condition in terms of central finite differences;

$$\frac{\partial u}{\partial t}(x,0) = \left(\frac{\partial u}{\partial t}\right)_{i,1} = \frac{u_{i,1} - u_{i,-1}}{2\Delta t} + 0(\Delta t)^2 = 0$$

where $\Delta t = 0.1$

or

$$u_{i,1} - u_{i,-1} = 0 \qquad\qquad (10.4.4)$$

However, $u_{i,1}$ and $u_{i,-1}$ are unknown.

For $j=0$, and substituting the values for Δx and Δt eqn. (10.4.2) becomes:

$$u_{i,1} = 2u_{i,0} - u_{i,-1} + 0.16\left(u_{i-1,0} - 2u_{i,0} + u_{i+1,0}\right)$$

or

$$u_{i,1} = u_{i,-1} = 0.16u_{i-1,0} + 1.68u_{i,0} + 0.16u_{i+1,0}$$

Substituting eqn. (10.4.4) into above equation we obtain:

$$u_{i,1} = 0.08u_{i-1,0} + 0.84u_{i,0} + 0.08u_{i+1,0} \qquad\qquad (10.4.5)$$

For this example, eqn. (10.4.2) becomes:

$$u_{i,j+1} = 2u_{ij} - u_{i,j-1} + 0.16\left(u_{i-1,j} - 2u_{i,j} + u_{i+1,j}\right)$$

$$= 0.16u_{i-1,j} + 1.68u_{i,j} + 0.16u_{i+1,j} - u_{i,j-1} \qquad\qquad (10.4.6)$$

Equation (10.4.3), (10.4.5) and (10.4.6) allow us to calculate the following table:

j	$u_{0,j}$	$u_{1,j}$	$u_{2,j}$	$u_{3,j}$	$u_{4,j}$
0	0	0.1875	0.25	0.1875	0
1	0	0.1775	0.24	0.1775	0
2	0	0.1491	0.21	0.1491	0
3	0	0.1066	0.1605	0.1066	0
4	0	0.0556	0.0938	0.0556	0
5	0	0.0019	0.0148	0.0019	0
6	0	-0.0501	-0.0682	-0.0501	0
7	0	-0.0970	-0.1455	-0.0970	0
8	0	-0.1361	-0.2073	-0.1361	0

PROBLEMS

1.   For the following rectangular plate determine the temperature at A at (1,1).
      (a)   using the grid indicated
      (b)   using a grid made up of squares of 0.5 x 0.5

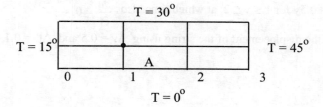

Use a computer program to solve augmented matrix

2.   For the following bridge determine the temperature at A. Use Gauss-Jordan with an accuracy
      of 2 decimals. Use computer program to solve augmented matrix.

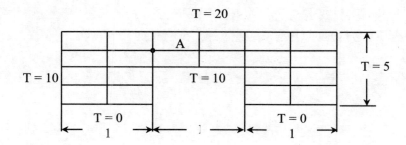

3.

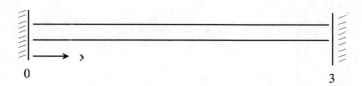

The temperature distribution in above bar follows the heat equation. Using $k = 1$, boundary
conditions $T(0,t)=0$, $T(3,t) = 10$, and initial condition $T(x,o) = x(3-x)$, determine how the
temperature changes in time using $\Delta x=1$, and $\Delta t = 1$

4.   Repeat problem 3, but with k = 0.5, T(3,t) = I and T(x,o) = 2x for $0 \leq x \leq 1.5$ and T(x,o) = 3 for $1.5 \leq x \leq 3$.

5.   A bent string is at time t = 0 subjected to a transverse displacement of u = 0.5 for $0 \leq y \leq 1$ and u = 1 – 0.5y for $1 \leq y \leq 2$, at which time also $\dfrac{\partial u}{\partial t} = 0$.

Determine the displacement of the string using $\Delta y = 0.5$ and $\Delta t = 0.1$.

Chapter 11

INTRODUCTION TO THE THEORY OF LINEAR VECTOR SPACES

11.1 PRELIMINARIES

An ordered set $\bar{x} = (x_1, x_2, ..., x_n)$ of n elements is called an n-tuple and is also called a point or vector in n-dimensional space. The scalars $x_1, x_2, ... x_n$ are said to be the coordinates of the vector $\bar{x}$.

Let two vectors be $\bar{x} = (x_1, x_2, ... x_n)$ and $\bar{y} = (y_1, y_2, ... y_n)$.

Equality of two vectors: Two vectors $\bar{x}$ *and* $\bar{y}$ are termed as equal iff (i.e. if and only if) $x_i = y_i$ for each i = 1, 2, ... n. The vector (0, 0, ... 0) is denoted by $\bar{0}$ and it is called the zero vector.

Addition of vectors: We define

$$\bar{x} + \bar{y} = (x_1 + y_1, x_2 + y_2, ... x_n + y_n)$$

Scalar multiplication: Let α be a scalar. Then
$$\alpha \bar{x} = (\alpha x_1, \alpha x_2, ... \alpha x_n)$$

Linear vector space: A collection of vectors $\bar{a}, \bar{b}, \bar{c}, ...$ is called a linear vector space or vector space or linear space V_n over the real number field R if the following rules are satisfied:

I. Vector addition: For every pair of vectors $\bar{a}$ and $\bar{b}$ there corresponds a unique vector $\bar{a} + \bar{b}$ (sum) in V_n such that:

(i) commutative: $\bar{a} + \bar{b} = \bar{b} + \bar{a}$

(ii) associative: $(\bar{a} + \bar{b}) + \bar{c} = \bar{a} + (\bar{b} + \bar{c})$

(iii) existence of additive identity element: there exists a unique vector $\bar{0}$, independent of $\bar{a}$ such that
$$\bar{a} + \bar{0} = \bar{a} = \bar{0} + \bar{a}$$

(iv) existence of the additive inverse element: for every $\bar{a}$, there exists a unique vector (depending on $\bar{a}$) denoted by $-\bar{a}$ such that $\bar{a} + (-\bar{a}) = \bar{0}$

II. Scalar multiplication: for every vector $\bar{a}$ and every real number $\alpha \in R$, there corresponds a unique vector $\alpha\bar{a}$ (product) in V_n with the properties -

(i) associative: $\alpha(\beta\bar{a}) = (\alpha\beta)\bar{a}, \ \alpha, \beta \in R$
(ii) distributive with respect to scalar addition: $(\alpha + \beta)\bar{a} = \alpha\bar{a} + \beta\bar{a}$
(iii) distributive with respect to vector addition: $\alpha(\bar{a} + \bar{b}) = \alpha\bar{a} + \alpha\bar{b}$
(iv) $1.\bar{a} = \bar{a}.1 = \bar{a}$

EXAMPLE 11.1:
Suppose we define a set of polynomials in an independent variable x over the field of real number R and define the vector addition as the addition of polynomials and the multiplication as the multiplication of a polynomial by an element of R. Then P is a linear vector space (Prove it).
[Hint: take P as $P(x) = c_0 + c_1 x + c_2 x^2 + \ldots\ldots$].

solution
Short cut method to prove a set of vectors is a vector space V_n :

(i) first define the rules of vector addition and scalar multiplication of a vector over the set.
(ii) closure property must be verified i.e. if $\bar{a}, \bar{b} \in V_n$ then $\alpha\bar{a} + \beta\bar{b} \in V_n$ for all scalars $\alpha, \beta \in R$.

Linear combination of vectors: If $\bar{x}^{(1)}, \bar{x}^{(2)}, \ldots\ldots\bar{x}^{(n)}$ are n vectors then any vector $\bar{x} = \alpha_1\bar{x}^{(1)} + \alpha_2\bar{x}^{(2)} + \ldots\ldots\ldots\alpha_n\bar{x}^{(n)}$ where $\alpha_1, \alpha_2, \ldots\ldots\ldots\alpha_n$ are scalars is called a linear combination of the vectors

$$\bar{x}^{(1)}, \bar{x}^{(2)}, \ldots\ldots\ldots\bar{x}^{(n)}.$$

Linear dependence and linear independence of vectors:

Linear Dependence: Let V_n be a vector space. A finite non-empty set $\{\bar{x}^{(1)}, \bar{x}^{(2)}, \ldots\ldots\ldots\bar{x}^{(m)}\}$ of vectors of V_n is said to be linearly dependent if there exist scalars $\alpha_1, \alpha_2 \ldots\ldots\ldots\alpha_m$ not all of them zero (some of them may be zero) such that

$$\alpha_1\bar{x}^{(1)} + \alpha_2\bar{x}^{(2)} + \ldots\ldots\ldots\ldots\alpha_n\bar{x}^{(m)} = 0 \qquad\qquad (1)$$

It implies that $\alpha_i = 0$ for each $1 \le i \le m$.

It is to be noted here that if $\alpha_m \ne 0$ (and other scalars are zero) then from equation. (1) we have

$$\bar{x}^{(m)} = \sum_{j=1}^{m-1} \beta_j \bar{x}^{(j)} \quad, \quad \beta_j = -\frac{\alpha_j}{\alpha_m}, \quad j = 1,2,\dots\dots(m-1)$$

So, the given vectors are <u>linearly dependent</u> if and only if one of them is a linear combination of the other vectors.

Next let us consider a system of equations -

$$\alpha_1 x_1^{(1)} + \alpha_2 x_1^{(2)} + \dots\dots + \alpha_m x_1^{(m)} = 0$$

$$\alpha_1 x_2^{(1)} + \alpha_2 x_2^{(2)} + \dots\dots + \alpha_m x_2^{(m)} = 0$$

.

.   (2)

.

$$\alpha_1 x_n^{(1)} + \alpha_2 x_n^{(2)} + \dots\dots + \alpha_m x_n^{(m)} = 0$$

A matrix M of the coordinates can be written in the form

$$M = \left[\begin{Bmatrix} x_1^{(1)} \\ x_2^{(1)} \\ . \\ . \\ x_n^{(1)} \end{Bmatrix} \begin{Bmatrix} x_1^{(2)} \\ x_2^{(2)} \\ . \\ x_n^{(2)} \end{Bmatrix} \quad . \quad . \quad \begin{Bmatrix} x_1^{(m)} \\ x_2^{(m)} \\ . \\ . \\ x_n^{(m)} \end{Bmatrix} \right]$$

(3)

$$= [\bar{x}^{(1)} \ \bar{x}^{(2)} \ \dots\dots \bar{x}^{(m)}]_{n \times m}$$

where $\bar{x}^{(j)}$, $j = 1,2,\dots\dots n$ are n vectors whose n components are

$$x_1^{(j)}, x_2^{(j)}, x_3^{(j)}, \dots\dots\dots x_n^{(j)}.$$

Let r be the rank of the matrix M.

An important result can be proved that the vectors $\bar{x}^{(1)}, \bar{x}^{(2)}, \dots\dots\dots \bar{x}^{(m)}$ are linearly dependent if r<m and linearly independent if r=m and r can not be greater than n for obvious reasons. The proof of this result is beyond the scope of the present treatment.

The set of all linear combinations of m vectors $\{\bar{x}^{(1)}, \bar{x}^{(2)}, \dots\dots\dots \bar{x}^{(m)}\}$ of a vector space V_n makes up a <u>subspace</u> of V_n. This subspace is spanned by $\{\bar{x}^{(1)}, \bar{x}^{(2)}, \dots\dots\dots \bar{x}^{(m)}\}$.

<u>Basis</u>: any set of n linearly independent vectors of an n-dimensional space is termed a basis of that space.

<u>Dimension</u>: a linear space V_n is called n-dimensional if it possesses a set of n linearly independent vectors, and every set of (n+1) vectors becomes a dependent set.

EXAMPLE 11.2:
A subset, $S=\{0,y,z\}$ of R^3 is a subspace of R^3

Solution:
  Let $\alpha, \beta \in R$ and $\bar{a}, \bar{b} \in S$ Then S will be a subspace of R^3 if we can prove that

$$\alpha\bar{a} + \beta\bar{b} \in S$$

Now, as per the definition of the subset S, the form of the vectors $\bar{a}, \bar{b}$ will be

$$\bar{a} = (0, a_2, a_3) \quad \text{and} \quad \bar{b} = (0, b_2, b_3)$$
$$\alpha\bar{a} + \beta\bar{b} = \alpha(0, a_2, a_3) + \beta(0, b_2, b_3)$$
$$= (0, \alpha a_2, \alpha a_3) + (0, \beta b_2, \beta b_3)$$
$$= (0, \alpha a_2 + \beta b_2, \alpha a_3 + \beta b_3)$$

which obviously belongs to S.
Hence S is a subspace of R.

<u>Linear transformation:</u>

  A transformation L from a linear vector space X into another linear vector space Y is a correspondence which assigns to each element x in X a unique element $y = Lx$ in Y.
A transformation L of a vector space X into a vector space Y, where X and Y have the same scalar field is termed as linear if

(a) homogeneous: $L(\alpha\bar{x}) = \alpha L(\bar{x})$ for all $\bar{x} \in X$, α is a scalar
(b) additive: $L(\bar{x}^{(1)} + \bar{x}^{(2)}) = L(\bar{x}^{(1)}) + L(\bar{x}^{(2)})$, for all $\bar{x}^{(1)}, \bar{x}^{(2)} \in X$
Otherwise it is termed as non-linear transformation.
The above conditions (a) and (b) can be written as:
L is a linear transformation if
$$L(\alpha\bar{x}^{(1)} + \beta\bar{x}^{(2)}) = \alpha L(\bar{x}^{(1)}) + \beta L(\bar{x}^{(2)}) \quad \text{for all} \quad \bar{x}^{(1)}, \bar{x}^{(2)} \in X$$
and α, β are scalars.

EXAMPLE 11.3:
The function L: $R^3 \to R^2$ defined by
$$L(x_1, x_2, x_3) = (x_1, x_2) \text{ for all } x_1, x_2, x_3 \in R$$
is a linear transformation from R^3 into R^2.

Solution:

Let $\bar{x} = (x_1, x_2, x_3)$, $\bar{y} = (y_1, y_2, y_3)$ both $\in R^3$.

If $\alpha, \beta \in R$ then

$$L(\alpha \bar{x} + \beta \bar{y}) = L[\alpha(x_1, x_2, x_3) + \beta(y_1, y_2, y_3)]$$
$$= L[(\alpha x_1 + \beta y_1), (\alpha x_2 + \beta y_2), (\alpha x_3 + \beta y_3)]$$
$$= [(\alpha x_1 + \beta y_1), (\alpha x_2 + \beta y_2)], \text{ by definition of } L$$
$$= \alpha(x_1, x_2) + \beta(y_1, y_2)$$
$$= \alpha L(x_1, x_2, x_3) + \beta L(y_1, y_2, y_3)$$
$$= \alpha L(\bar{x}) + \beta L(\bar{y})$$

Therefore L is a linear transformation from R^3 into R^2.

EXAMPLE 11.4:

The mapping L defined by $L(p) = \dfrac{dp}{dx}$ is a linear transformation on Q, where Q is the real linear space

of all polynomials p(x) with real coefficients defined in [0,1].

Solution:

If $p(x) \in Q \Rightarrow L(p) = \dfrac{dp}{dx}$

Let $p(x), q(x) \in Q$ and α, β are scalars $\in R$ then

$$L(\alpha p + \beta q) = \frac{d}{dx}(\alpha p + \beta q) = \alpha \frac{dp}{dx} + \beta \frac{dq}{dx} = \alpha L(p) + \beta L(q)$$

$\Rightarrow$ L is a linear transformation.

Inner Product and Norm:

To measure the length of a vector or the difference between two vectors of a linear vector space we introduce norm and to measure the angle between two vectors we introduce the concept of an inner product. Inner product is analogous to the scalar product of geometric vectors.

Norm of a vector:

Let V be a linear vector space over the real number field R. A norm on vector space V is a function that transforms every element $x \in V$ into a real number $\|x\|$ such that $\|x\|$ satisfies the following conditions:

a) $\|x\| \geq 0$ and $\|x\| = 0$ if $x = 0$

b) $\|\alpha x\| = |\alpha| \|x\|$, $\alpha \in R$

c) $\|x + y\| \leq \|x\| + \|y\|$, $x, y \in V$

A linear vector space on which a norm can be defined is called a <u>Normed Linear Space</u>.

<u>Inner Product:</u>

Let F be a field of real numbers. Then an inner product on a linear vector space V is defined as a mapping from VxV into F which assigns to each ordered pair of vectors x,y in V a scalar <x,y> in F in such a way that:

(*i*) $< x, y > = < y, x >$

(*ii*) $< \alpha x + \beta y, z > = \alpha < x, z > + \beta < y, z >$

(*iii*) $< x, x > \geq 0$ and $< x, x > = 0$ if $x = 0$

for any $x, y, z \in V$ and $\alpha, \beta \in F$.

The linear space V is then said to be an <u>Inner Product Space</u> with respect to that specified inner product defined on it.

We define a norm with every inner product by $\|x\| = \sqrt{< x, x >}$

<u>Orthogonal Systems of Vectors:</u>

Two vectors $\bar{x}$ and $\bar{y}$ in an inner product space I_n are called orthogonal if their inner product is zero i.e. if $< \bar{x}, \bar{y} > = 0$.

The orthogonality is in general tells about perpendicularity. If the vectors are non-zero, then orthogonality implies that the angle between the two vectors is $\pi / 2$.

A set of non-zero vectors $\bar{x}^{(1)}, \bar{x}^{(2)}, \ldots \ldots \bar{x}^{(k)}$ or in short $\{\bar{x}^{(i)}\}$ is called orthogonal set if any two vectors of the set are orthogonal to each other i.e.

$$< \bar{x}^{(i)}, \bar{x}^{(j)} > = 0 \quad \text{for} \quad i \neq j .$$

A set of vectors $\bar{x}^{(1)}, \bar{x}^{(2)}, \ldots \ldots \ldots \bar{x}^{(k)}$ is said to be orthonormal

if $< \bar{x}^{(i)}, \bar{x}^{(j)} > = \delta_{ij}$

where δ_{ij} is the Kronecker delta (i.e. $\delta_{ij} = 0$ *for* $i \neq j$ and $\delta_{ij} = 1$ for $i = j$).

Moreover any non-zero set of orthogonal vectors can be converted into an orthonormal set by replacing each vector $\bar{x}^{(i)}$ with $\bar{x}^{(i)} / \|\bar{x}^{(i)}\|$.

<u>Orthogonal system of functions:</u>

A set of functions $\{\phi_i(x)\}$ is said to be orthogonal over a set of points $\{x_i\}$ with respect to a weight function W(x) if

$$< \phi_j(x), \phi_k(x) >= 0, \; j \neq k \qquad\qquad (4)$$

where the inner product is defined as

$$< \phi_j(x), \phi_k(x) >= \sum_{i=0}^{N} W(x_i)\phi_j(x_i)\phi_k(x_i)$$

A set of functions $\{\phi_i(x)\}$ is said to be orthogonal on a closed interval [a,b] with respect to the weight function $W(x)$ if Eq. (11.4) holds and in this case the inner product is defined as,

$$< \phi_j(x), \phi_k(x) >= \int_{a}^{b} W(x)\phi_j(x)\phi_k(x)\,dx$$

The orthogonal functions $\{\phi_i(x)\}$ can be ortho-normalized by the relation

$$\frac{\phi_i(x)}{\|\phi_i(x)\|} \quad \text{and is denoted by } \hat{\phi}_i(x) \text{ and so } < \hat{\phi}_i(x), \hat{\phi}_j(x) >= \delta_{ij}.$$

Sequence of orthogonal polynomials:

A sequence of polynomials viz. $P_0(x), P_1(x), P_2(x), \ldots\ldots$ (finite or infinite) are orthogonal if $P_i(x)$ are all orthogonal to each other and each $P_i(x)$ is a polynomial of exact i^{th} degree.

This can be written as:
Sequence of polynomials $P_i(x)$, i =0, 1, 2,...... are orthogonal if,
(i) $< P_i(x), P_j(x) >= 0$ for $i \neq j$
(ii) $P_i(x) = C_i x^i + P_{i-1}$ for each i with $C_i \neq 0$

where P_{i-1} is a polynomial of degree less than i.

EXAMPLE 11.5:
The functions $P_0(x)=1$, $P_1(x)=(x-1)$ and $P_2(x)=(x^2 -2x+2/3)$ form a sequence of orthogonal polynomial which are orthogonal on [0,2] with respect to the weight function $W(x)=1$.

$$< P_0, P_1 >= \int_0^2 1.(x-1)dx = 0$$

$$< P_0, P_2 >= \int_0^2 1.(x^2 - 2x + 2/3)dx = 0 \text{ and}$$

$$< P_1, P_2 >= \int_0^2 (x-1)(x^2 - 2x + 2/3)dx = 0$$

$$< P_i(x), P_j(x) >= 0 \quad \text{for} \quad i \neq j$$

Hence P_0, P_1, P_2 are orthogonal polynomials.

EXAMPLE 11.6:

Prove that the set of functions $\{Sin\dfrac{n\pi x}{l}\}, n = 1, 2, ...$ is orthogonal in $[0, l]$ and find the orthonormal set.

Solution:

Let
$$\phi_i(x) = Sin\frac{i\pi x}{l}, \qquad \phi_j(x) = Sin\frac{j\pi x}{l}$$

then to prove these are orthogonal, we have to prove

$$< \phi_i, \phi_j >= 0 \quad \text{for} \quad i \neq j$$

i.e. we should have
$$\int_0^l \phi_i(x)\phi_j(x)dx = 0 \quad \text{for} \quad i \neq j$$

Now,

$$\int_0^l \phi_i \, \phi_j \, dx$$

$$= \int_0^l Sin(i\pi x/l)\sin(j\pi x/l)dx$$

$$= \frac{1}{2}\int_0^l [\cos((i-j)\pi x/l) - \cos((i+j)\pi x/l)]dx, \text{ where, } i \neq j$$

$$= \frac{1}{2}[\frac{l}{\pi(i-j)}\sin\frac{(i-j)\pi x}{l} - \frac{l}{\pi(i+j)}\sin\frac{(i+j)\pi x}{l}]_0^l$$

$$= 0$$

This shows that the given set of functions are orthogonal.
Now, norm of ϕ_i

$$= \left\| \sin\frac{i\pi x}{l} \right\|$$

$$= \left\{ \int_0^l \sin^2\left(\frac{i\pi x}{l}\right) dx \right\}^{1/2}$$

$$= \left[\frac{1}{2} \int_0^l \left(1 - \cos\frac{2i\pi x}{l}\right) dx \right]^{1/2}$$

$$= \left[\frac{1}{2}\left(x - \frac{l}{2i\pi}\sin\frac{2i\pi x}{l}\right)_0^l \right]^{1/2} = \left(\frac{l}{2}\right)^{1/2} = \sqrt{\frac{l}{2}}$$

Therefore, orthonormal set is given as,

$$\{\phi_i(x)\} = \left\{ \frac{\phi_i}{\|\phi_i\|} \right\} = \left\{ \sqrt{\frac{2}{l}}\sin\frac{i\pi x}{l} \right\}.$$

11.2 CONSTRUCTION OF ORTHOGONAL POLYNOMIALS

Three Term Recurrence Relation:

It is possible to construct a sequence of orthogonal polynomials using the following three terms recurrence relation:

$$\phi_{k+1}(x) = (x - e_k)\phi_k(x) - p_k\phi_{k-1}(x), \; k = 0,1,2,\ldots\ldots.$$
$$\text{and} \quad \phi_{-1} = 0$$

where, $e_k = \dfrac{<x\phi_k, \phi_k>}{<\phi_k, \phi_k>}$ and $p_k = \dfrac{<x\phi_k, \phi_{k-1}>}{<\phi_{k-1}, \phi_{k-1}>}$

Here it is to be noted that we get the sequence of orthogonal polynomials with leading coefficients as 1. If the leading coefficient of the sequence of orthogonal polynomials is other than unity then in general we can write the above three term recurrence relation as,

$$\phi_{k+1}(x) = (d_k x - e_k)\phi_k(x) - p_k\phi_{k-1}(x), \quad k = 0,1,2,\ldots\ldots..$$

where d_k, e_k, and p_k can be obtained from the orthogonality property.

Gram Schmidt Orthogonalisation Process:

The above recurrence relation is valid only for polynomials. Now let us suppose that we are given a set of function $(f_i(x),\ I=1,2,.....)$ in $[a,b]$. From these set of functions, we can construct appropriate orthogonal functions by using a well known procedure known as Gram-Schmidt orthogonalisation process as follows:

$$\phi_1 = f_1$$
$$\phi_2 = f_2 - \alpha_{21}\phi_1$$
$$\phi_3 = f_3 - \alpha_{31}\phi_1 - \alpha_{32}\phi_2$$
$$-------------$$

where, $\alpha_{21} = \dfrac{<f_2,\phi_1>}{<\phi_1,\phi_1>},\quad \alpha_{31} = \dfrac{<f_3,\phi_1>}{<\phi_1,\phi_1>},\quad \alpha_{32} = \dfrac{<f_3,\phi_2>}{<\phi_2,\phi_2>}$ etc.

In compact form we can write the above procedure as

$$\phi_1 = f_1$$
$$\phi_i = f_i - \sum_{j=1}^{i-1} \alpha_{ij}\phi_j$$

and $$\alpha_{ij} = \frac{<f_i,\phi_j>}{<\phi_j,\phi_j>} = \frac{\displaystyle\int_a^b W(x)f_i(x)\phi_j(x)dx}{\displaystyle\int_a^b W(x)\phi_j(x)\phi_j(x)dx}$$

The above procedure is valid only when the inner product as defined exists for the interval $[a,b]$ with respect to the weight function $W(x)$. Now we will show how the functions of example 11.5 can be obtained by Gram-Schmidt process:

EXAMPLE 11.7:
Using the Gram-Schmidt orthogonalisation process construct the three orthogonal polynomials in $[0,2]$. Take weight function $W(x)=1$ and given

$$f_1(x) = 1,\ f_2(x) = x\ \text{and}\ f_3(x) = x^2$$

Solution:
As per the Gram Schmidt procedure write:

$$\phi_1(x) = f_1(x) = 1 \tag{i}$$
$$\phi_2(x) = f_2 - \alpha_{21}\phi_1 = x - \alpha_{21}\phi_1$$
$$\phi_3(x) = f_3 - \alpha_{31}\phi_1 - \alpha_{32}\phi_2 = x^2 - \alpha_{31}\phi_1 - \alpha_{32}\phi_2$$

Now
$$\alpha_{21} = \frac{<x,\phi_1>}{<\phi_1,\phi_1>} = \frac{\int\limits_0^2 x\,dx}{\int\limits_0^2 dx} = 1$$

$$\phi_2(x) = (x-1) \qquad\qquad (ii)$$

and
$$\alpha_{31} = \frac{<x^2,\phi_1>}{<\phi_1,\phi_1>} = \frac{\int\limits_0^2 x^2\,dx}{\int\limits_0^2 dx} = \frac{4}{3}$$

$$\alpha_{32} = \frac{<x^2,\phi_2>}{<\phi_2,\phi_2>} = \frac{\int\limits_0^2 x^2(x-1)\,dx}{\int\limits_0^2 (x-1)^2\,dx} = 2$$

$$\phi_3 = x^2 - \frac{4}{3}.(1) - 2(x-1) = (x^2 - 2x + \frac{2}{3}) \qquad\qquad (iii)$$

It is seen in example 11.5 that ϕ_1, ϕ_2, ϕ_3 are orthogonal.

Thus the orthogonal polynomials $\{\phi_i\}$ can be generated from the set of functions $\{f_i\}$.

Some Standard Orthogonal Polynomials

Now we will give the well known orthogonal polynomials viz. Legendre and Chebyshev polynomials. It is to be noted here that although the Gram Schmidt orthogonalisation procedure may be applied to find a set of orthogonal polynomials, we often use the well-known orthogonal polynomials also.

Legendre Polynomials:

If we define the inner product of two functions u(x), v(x) in [-1,1] by

$$<u,v> = \int\limits_{-1}^{1} u(x)v(x)dx \qquad\qquad (i)$$

we generate orthogonal polynomials from the set of functions $f_0 = 1, f_1 = x, f_2 = x^2,\dots\dots$ by the Gram Schmidt procedure, which are known as Legendre polynomials (weight function w(x)=1).

Now by Gram Schmidt procedure we will write

$$\phi_0 = f_0 = 1 \qquad \qquad \text{(a)}$$

$$\phi_1 = f_1 - \alpha_{10}\phi_0$$

$$\phi_2 = f_2 - \alpha_{20}\phi_0 - \alpha_{21}\phi_1$$

$$\phi_3 = f_3 - \alpha_{30}\phi_0 - \alpha_{31}\phi_1 - \alpha_{32}\phi_2$$

where, $\alpha_{10}, \alpha_{20}, \alpha_{21}, \alpha_{30}, \alpha_{31}, \alpha_{32}$ etc. can be found out using orthogonality property and by (i),

$$\alpha_{10} = \frac{<f_1,\phi_0>}{<\phi_0,\phi_0>} = \frac{<x,1>}{<1,1>} = \frac{\int\limits_{-1}^{1} x\,dx}{\int\limits_{-1}^{1} dx} = 0$$

$$\phi_1 = f_1 = x \qquad \qquad \text{(b)}$$

$$\alpha_{2,0} = \frac{<x^2,1>}{<1,1>} = 1/3, \qquad \alpha_{21} = 0$$

$$\phi_2 = x^2 - 1/3 \qquad \qquad \text{(c)}$$

similarly we can find

$$\phi_3 = x^3 - \frac{3}{5}x \qquad \qquad \text{(d)}$$

etc.

The orthogonal polynomials described in (a) to (d) are known as Legendre polynomials.

Chebyshev polynomials:

If we define the inner product of two functions u(x), v(x) in [-1,1] by

$$<u,v> = \int\limits_{-1}^{1} \frac{u(x)v(x)}{\sqrt{1-x^2}}\,dx \qquad \qquad \text{(ii)}$$

we generate orthogonal polynomials from the set of functions $f_0 = 1, f_1 = x, f_2 = x^2, ...$ by the Gram Schmidt procedure and these orthogonal polynomials are known as Chebyshev polynomials (weight function $= \dfrac{1}{\sqrt{1-x^2}}$).

Here also as in Legendre polynomials we will find by Gram Schmidt procedure as follows:

$$\phi_0 = f_0 = 1 \qquad \qquad \text{(A)}$$

$$\phi_1 = f_1 - \alpha_{10}\phi_0$$
$$\phi_2 = f_2 - \alpha_{20}\phi_0 - \alpha_{21}\phi_1 \quad etc.$$

and again $\alpha_{10}, \alpha_{20}, \alpha_{21}, \ldots$ etc. can be found out by orthogonality and (ii) and after calculating these constants one can see easily that,

$$\phi_0 = 1$$

$$\phi_1 = x \tag{B}$$

$$\phi_2 = x^2 - \frac{1}{2} \tag{C}$$

The orthogonal polynomials described in (A) to (C) are known as Chebyshev polynomials.

11.3 BOUNDARY CHARACTERISTIC ORTHOGONAL POLYNOMIALS

In various engineering problems although the well known method of Rayleigh-Ritz (discussed in next chapter) is advantageous to apply, it is often difficult to obtain the meaningful deflection shape functions in the said method. So, a class of boundary characteristic orthogonal polynomials can be constructed using Gram Schmidt process and then these polynomials are employed as deflection functions in the Rayleigh-Ritz method. The orthogonal nature of the polynomials makes the analysis simple and straight forward. Moreover, ill-conditions of the problem may also be avoided.

Here the first member of the orthogonal polynomials set $\phi_1(x)$ is chosen as the simplest polynomial of the least order that satisfies both the geometrical and the natural boundary conditions. The other members of the orthogonal set in the interval $a \leq x \leq b$ are generated using Gram Schmidt process as follows -

$$\phi_2(x) = (x - B_2)\phi_1(x)$$
$$\phi_k(x) = (x - B_k)\phi_{k-1}(x) - C_k\phi_{k-2}(x)$$

where,

$$B_k = \frac{\int_a^b x(\phi_{k-1}(x))^2 W(x)dx}{\int_a^b (\phi_{k-1}(x))^2 W(x)dx}$$

$$C_k = \frac{\int_a^b x\phi_{k-1}(x)\phi_{k-2}(x)W(x)dx}{\int_a^b (\phi_{k-2}(x))^2 W(x)dx}$$

and W(x) is the weight function. The polynomials $\phi_k(x)$ satisfy the orthogonality condition

$$\int_a^b W(x)\phi_k(x)\phi_l(x)dx = 0 \quad \text{if} \quad k \neq l$$

$$\neq 0 \quad \text{if} \quad k = l$$

Here it is to be noted that, even though $\phi_1(x)$ satisfies all the boundary conditions both geometric and natural, the other members of the orthogonal set satisfy only geometric boundary conditions.

For example in vibration field let us consider a beam in [0,1] with both ends clamped:
So the boundary conditions are:

$$\phi_1(0) = \phi_1'(0) = \phi_1(1) = \phi_1'(1) = 0$$

Consider the deflection function,

$$\phi_1(x) = a_0 + a_1 x + a_2 x^2 + a_3 x^3 + a_4 x^4$$

Substituting the boundary conditions in the above equation, the coefficients a_i are determined to yield

$$\phi_1(x) = a_4(x^2 - 2x^3 + x^4)$$

The coefficient a_4 can be appropriately chosen so as to normalize $\phi_1(x)$ such that

$$\int_0^1 (\phi_1(x))^2 dx = 1$$

Then the Rayleigh-Ritz method may be applied as discussed in the next chapter.

Chapter 12

SOLUTION OF BOUNDARY VALUE PROBLEMS

12.1 PRELIMINARIES

Consider the following differential equation of second order

$$F(x, y, y', y'') = 0$$

Here the problem is to find y=y(x) in certain interval suppose in [a,b] which satisfy the above function F and also at the end points a and b (called the boundary points) of the interval [a,b]. These type of problems are known as Boundary Value Problems (BVP). Generally it is very difficult and in some cases not possible to get the exact solution. We will discuss here various methods for the approximate solution of these class of BVPs.

12.2 SHOOTING METHOD

Let the second order linear differential equation is

$$y'' = f(x, y, y') \text{ in } [a, b] \text{ such that } y(a) = y_0, \ y(b) = e \qquad (12.2.1)$$

It is well known that the above is a boundary value problem.
In the shooting method we solve the corresponding initial value problem

$$y'' = f(x, y, y') \ \ s/t \ \ y(a) = y_0, \ \ y'(a) = r_0 \text{ (say)} \qquad (12.2.2)$$

where r_0 is supposed to be the initial slope.

Using any method for solving intial value problem, we find the first approximation of y(b) and denote it by $y^{(1)}(b)$. Whereas the exact solution of y(b) is y(b)=e.

The above first approximation solution $y^{(1)}(b)$ is either above or below the required solution y(b)=e. Let us denote the error as
E(r)= $y^{(1)}(b) - y(b) = E(r_0)$ (say), where r_0 is the first approximation of e. It is to be noted that the exact solution is obtained when the error, E(r)=0.

If $E(r_0) \neq 0$ then   the   initial   value   problem   (12.2.2)   is   solved   again   with $y'(a) = r_1$ (say) where $r_1 = r_0 + h$, h is some increment. The following initial value problem now has to be solved:

$$y'' = f(x, y, y') \ s/t \ y(a) = y_0, \ y'(a) = r_1$$

Let the solution of this Initial Value Problem (IVP) gives the value of y(b) as $y^{(2)}(b)$. Again E(r) = $y^{(2)}(b) - y(b) = E(r_1)$ is computed. And this procedure is repeated untill we get E(r) = 0. But in actual computation we will have E(r) less than some specified accuracy. It is to be carefully noted that the computed solution for various slopes UNDERSHOOT or OVERSHOOT the exact (required) solution y(b) = e. This may clearly be understood from Fig 12.1.

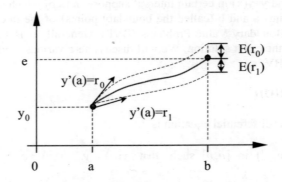

Figure 12.1: Initial value problem

Now if we plot a curve between $y'(a) = r$ and $y^{(i)}(b)$, $i = 1,2,\dots$ as in Fig. 12.2, then by applying Secant Method for E(r) = 0 we have

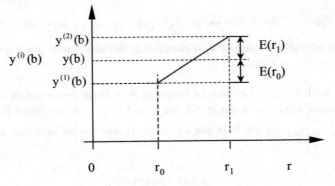

Figure 12.2: Secant Method for E(r)

$$r_2 = r_1 - \left[\frac{(r_1 - r_0)}{(E(r_1) - E(r_0))} \right] E(r_1)$$

$$= r_1 - \left[\frac{r_1 - r_0}{(y^{(2)}(b) - y(b)) - (y^{(1)}(b) - y(b))} \right] E(r_1)$$

$$\therefore r_2 = r_1 - \frac{(r_1 - r_0)}{\{y^{(2)}(b) - y^{(1)}(b)\}} * E(r_1) \tag{12.2.3}$$

Similarly, we can write,

$$r_3 = r_2 - \frac{(r_2 - r_1)}{(y^{(3)}(b) - y^{(2)}(b))} * E(r_2) \tag{12.2.4}$$

and in general this can be written as:

$$r_{n+1} = r_n - \frac{(r_n - r_{n-1})}{\{y^{(n+1)}(b) - y^{(n)}(b)\}} * E(r_n) \tag{12.2.5}$$

which is actually a linear interpolation.

Thus we get r_i that is the guess of the initial slope at each step and we solve the initial value problem:

$$y'' = f(x, y, y') \quad s/t \quad y(a) = y_0, \quad y'(a) = r_i \tag{12.2.6}$$

Solution of (12.2.6) gives $y^{(i)}(b)$. When we get $\left| y^{(i)}(b) - y(b)(= e) \right|$ less than a specified accuracy the computation is stopped and the value of y so obtained is the solution of the Boundary Value Problem (BVP) (12.2.1).

The solution of BVP (12.2.1) can also be obtained by a linear combination of two assumed initial slopes, if the differential equation is linear. So, let y_1 and y_2 are two solutions for two different initial slopes $y_1'(a)$ and $y_2'(a)$, where both the solutions y_1 and y_2 are obtained using the same initial value,

$$y_1(a) = y_2(a) = y_0 \tag{12.2.7}$$

Now we write linear combination of y_1 and y_2 as the solution

$$y(x) = c_1 y_1(x) + c_2 y_2(x) \tag{12.2.8}$$

At left end x = a;

$$y(a) = c_1 y_1(a) + c_2 y_2(a)$$

$$\Rightarrow y_0 = c_1 y_0 + c_2 y_0, \quad \text{by} \quad (12.2.7) \Rightarrow c_1 + c_2 = 1 \tag{12.2.9}$$

Again from (12.2.8) at right end x = b we can write:

$$y(b) = c_1 y_1(b) + c_2 y_2(b)$$
$$\Rightarrow e = c_1 y_1(b) + c_2 y_2(b) \tag{12.2.10}$$

Solving (12.2.9) and (12.2.10) we get,

$$c_2 = \frac{e - y_1(b)}{y_2(b) - y_1(b)} \tag{12.2.11}$$

and

$$c_1 = 1 - c_2 \tag{12.2.12}$$

Substituting (12.2.11) and (12.2.12) in (12.2.8) we may get the approximate solution of the BVP (12.2.1).

Similar procedures can be used to solve higher order linear BVP using the known values of first few ordinates y(x) and computing the errors in the boundary conditions at the other end. This method of computation becomes considerably more complicated and inaccurate as the order of the differential equation is higher than third or fourth.

Nonlinear Differential Equations:

Let us first consider a system of two differential equations in two unknowns:

$$\frac{dy}{dx} = y' = \phi(y, z; x)$$
$$\frac{dz}{dx} = z' = \varphi(y, z; x)$$

(12.2.13)

with two conditions given as

$$y(x = 0) = y_0 \quad \text{and}$$
$$z(x = L) = z_L$$

(12.2.14)

Suppose that $z(x = 0) = \delta_0$ is the initial guess, whereas $z(x = 0) = \delta$ is the actual value of z at x=0.

Now solve (12.2.13) with conditions $y(x = 0) = y_0$ and $z(x = 0) = \delta_0$ and denote the value of z at x=L by $z(\delta_0; L)$. Expanding $z(\delta; L)$ into a Taylor series upto two terms (linear) we have,

$$z(\delta; L) = z(\delta_0; L) + (\delta - \delta_0) z'(\delta_0; L)$$

(12.2.15)

Now

(i)   solve (12.2.13) with initial conditions
$$y(x = 0) = y_0 \quad \text{and} \quad z(x = 0) = \delta_0 \quad \text{and}$$

(ii)   solve (12.2.13) with initial conditions
$$y(x = 0) = y_0 \quad \text{and} \quad z(x = 0) = \delta_0 + \Delta\delta_0$$

(12.2.16)

where $\Delta\delta_0$ is small increment.

Then we can write,

$$\frac{z(\delta_0 + \Delta\delta_0; L) - z(\delta_0; L)}{\Delta\delta_0} \approx z'(\delta_0; L)$$

(12.2.17)

where $z(\delta_0 + \Delta\delta_0; L)$ and $z(\delta_0; L)$ are known from (12.2.16).

So, $z'(\delta_0; L)$ may be known. Now from (12.2.15), writing $z(\delta; L) = z_L$ from (12.2.14) (given condition) we have,

$$(\delta - \delta_0) = \frac{z_L - z(\delta_0; L)}{z'(\delta_0; L)}$$

(12.2.18)

all the quantities in the right hand side of (12.2.18) are known. Denoting $\delta - \delta_0 = \bar{\delta}_0$ we write

$$\bar{\delta}_0 = \frac{z_L - z(\delta_0;L)}{z'(\delta_0;L)}$$

Next we write the new estimates of the guessed initial values, as, $\delta_1 = \delta_0 + \bar{\delta}_0$. Again the process is repeated starting with the initial conditions $y(x = 0) = y_0$ and $z(x = 0) = \delta_1$. The procedure is to be stopped when

$$\left| z(\delta_1;L) - z(\delta_0;L) \right| \leq \varepsilon$$

where ε is the desired accuracy. satisfaction of the above condition gives the true value of the initial slope. after getting the true (approximated) value of the initial slope, say $z(x = 0) = \delta$ and the given initial condition $y(x = 0) = y_0$, the IVP (12.2.13) can be solved to get the required solution.

   Next we consider general situation of a nonlinear system of four equations in four unknowns given by

$$\frac{dp}{dx} = p' = f_1(p,q,r,s,x)$$
$$\frac{dq}{dx} = q' = f_2(p,q,r,s,x) \qquad (12.2.19)$$
$$\frac{dr}{dx} = r' = f_3(p,q,r,s,x)$$
$$\frac{ds}{dx} = s' = f_4(p,q,r,s,x)$$

The two conditions at x = 0, say, are

$$p(0) = p_0$$
$$q(0) = q_0 \qquad (12.2.20a)$$

and the two conditions at x = L, say, are

$$r(L) = r_L$$
$$s(L) = s_L \qquad (12.2.20b)$$

Let us suppose that the true initial values of r and s at x = 0 are r(0)=δ and s(0) =μ.
To start with we again suppose that δ_0 and μ_0 are the initial guesses for r and s at x = 0. Now solve (12.2.19) taking the initial values δ_0, μ_0 and denote the values of r and s at x=L by $r(\delta_0, \mu_0; L)$ and $s(\delta_0, \mu_0; L)$.

Next we expand r(δ,μ;L) and s(δ,μ;L) into a Taylor series for two variables up to the linear terms. This gives,

$$r(\delta, \mu,; L) = r(\delta_0, \mu_0; L) + (\delta - \delta_0)\frac{\partial r}{\partial \delta}(\delta_0, \mu_0; L)$$

$$+ (\mu - \mu_0)\frac{\partial r}{\partial \mu}(\delta_0, \mu_0; L)$$

(12.2.21)

and

$$s(\delta, \mu,; L) = s(\delta_0, \mu_0; L) + (\delta - \delta_0)\frac{\partial s}{\partial \delta}(\delta_0, \mu_0; L)$$

$$+ (\mu - \mu_0)\frac{\partial s}{\partial \mu}(\delta_0, \mu_0; L)$$

(12.2.22)

Now (i) solve (12.2.19) with initial conditions $p_0, q_0, \delta_0, \mu_0$

(ii) solve (12.2.19) with initial conditions $p_0, q_0, \delta_0 + \Delta\delta_0, \mu_0$

and (iii) solve (12.2.19) with initial condition $p_0, q_0, \delta_0, \mu_0 + \Delta\mu_0$

(12.2.23)

where $\Delta\delta_0$ and $\Delta\mu_0$ are small increments.

We then write the first derivative in the form of difference equations as follows:

$$\frac{r(\delta_0 + \Delta\delta_0, \mu_0; L) - r(\delta_0, \mu_0; L)}{\Delta\delta_0} \approx \frac{\partial r(\delta_0, \mu_0; L)}{\partial \delta}$$

$$\frac{s(\delta_0 + \Delta\delta_0, \mu_0; L) - s(\delta_0, \mu_0; L)}{\Delta\delta_0} \approx \frac{\partial s(\delta_0, \mu_0; L)}{\partial \delta}$$

$$\frac{r(\delta_0, \mu_0 + \Delta\mu_0; L) - r(\delta_0, \mu_0; L)}{\Delta\mu_0} \approx \frac{\partial r(\delta_0, \mu_0; L)}{\partial \mu}$$

(12.2.24)

$$\frac{s(\delta_0, \mu_0 + \Delta\mu_0; L) - s(\delta_0, \mu_0; L)}{\Delta\mu_0} \approx \frac{\partial s(\delta_0, \mu_0; L)}{\partial \mu}$$

Left hand side of (12.2.24) are known from the solution (12.2.23). So, the partial derivatives of (12.2.24) may be known. Now replace $r(\delta, \mu; L)$ and $s(\delta, \mu; L)$ by the original given conditions r_L and s_L respectively and thus we can solve (12.2.21) and (12.2.22) for $(\delta - \delta_0)$ and $(\mu - \mu_0)$.

Let us denote $(\delta - \delta_0)$ by $\bar{\delta}_0$ and $(\mu - \mu_0)$ by $\bar{\mu}_0$. Then proceed with the new estimates of the guessed initial values as, $\delta_1 = \delta_0 + \bar{\delta}_0$ and $\mu_1 = \mu_0 + \bar{\mu}_0$ for the true initial values of $r(0) = \delta$ and $s(0) = \mu$..

The above procedure is repeated as discussed in the system of two differential equations, taking the initial values as $p_0, q_0, \delta_1, \mu_1$.

EXAMPLE 12.1:
Solve the following BVP by shooting method:

$$y'' = y \quad s/t \quad y(0) = 0.0, \ y(1.0) = 1.1752$$

Use 4th order Taylor series method.

Solution
   We will write the differential equation

$$y'' = y \tag{12.2.25}$$

into two systems of first order differential equations as,

$$\frac{dy}{dx} = z \quad s/t \quad y(0) = 0.0 \quad \text{and}$$
$$\frac{dz}{dx} = y \quad s/t \quad z(0) = y'(0) = 0.5 \tag{12.2.26}$$

where we have supposed the slope $y'(0) = 0.5$.

 Now we will solve (12.2.26) considering the initial slope as 0.5 and find the value of $y(1.0)$, as discussed earlier.

To solve (12.2.26) we will use 4th order Taylor series method:

$$y(x) = y(0) + xy'(0) + \frac{x^2}{2!}y''(0) + \frac{x^3}{3!}y'''(0) + \frac{x^4}{4!}y^{iv}(0) \tag{12.2.27}$$

$$z(x) = z(0) + xz'(0) + \frac{x^2}{2!}z''(0) + \frac{x^3}{3!}z'''(0) + \frac{x^4}{4!}z^{iv}(0) \tag{12.2.28}$$

Now we have

$$y' = z \Rightarrow y'(0) = 0.5$$
$$z' = y \Rightarrow z'(0) = 0.0$$
$$y'' = z' \Rightarrow y''(0) = 0.0$$
$$z'' = y' \Rightarrow z''(0) = 0.5$$
$$y''' = z'' \Rightarrow y'''(0) = 0.5$$
$$z''' = y'' \Rightarrow z'''(0) = 0.0$$
$$y^{iv} = z''' \Rightarrow y^{iv}(0) = 0.0$$
$$z^{iv} = y''' \Rightarrow z^{iv}(0) = 0.5$$

and so on.

Putting these in (12.2.27) and (12.2.28) give

$$y(x) = 0.5x + \frac{0.5x^3}{6} \tag{12.2.29}$$

and

$$z(x) = 0.5 + \frac{0.5x^2}{2} + \frac{0.5x^4}{24} \tag{12.2.30}$$

For various values of x we can have values of y(x) and z(x). Then we match the value of y(1.0) with the given value of y(1.0)=1.1752. In Table 12.2.1 we give the results of the computations. It is seen from this Table that the result for y(1.0)=0.58333 is lower than the given value of 1.1752. So, next we suppose a different (larger) value of the slope. Let us assume now that the initial slope is $y'(0) = 0.6$. Again solving the present IVP as above we get y(1.0)=0.70000 as shown in Table 12.2.1.

After the above two trial values of the initial slopes we will use the relation (12.2.5) to have the more accurate intial slope given by $y'(0) = 1.00731$. Finally this value of the initial slope gives the correct value of y(1.0)=1.1752 (see, Table 12.2.1). It is important to note that, in other problems it may need more number of iterations to get the correct result. Required number of iterations will however largely depend upon the choice of the initial approximations of the slope.

Table 12.2.1

x	y'(0)=0.5		y'(0)=0.6		y'(0)=1.00731	
	y	z	y	z	y	z
0.0	0.0	0.5	0.0	0.6	0.0	1.00731
0.1	0.05008	0.50250	0.06010	0.60300	0.10090	1.01236
0.2	0.10067	0.51003	0.12080	0.61204	0.20281	1.02753
0.3	0.15225	0.52267	0.18270	0.62720	0.30673	1.05298
0.4	0.20533	0.54053	0.24640	0.64864	0.41367	1.08897
0.5	0.26042	0.56380	0.31250	0.67656	0.52464	1.13585
0.6	0.31800	0.59270	0.38160	0.71124	0.64065	1.19407
0.7	0.37858	0.62750	0.45430	0.75300	0.76270	1.26418
0.8	0.44267	0.66853	0.53120	0.80224	0.89181	1.34685

| 0.9 | 0.51075 | 0.71617 | 0.61290 | 0.85940 | 1.02897 | 1.44281 |
| 1.0 | 0.58333 | 0.77083 | 0.70000 | 0.92500 | 1.17520 | 1.55294 |

As discussed earlier the solution of the BVP (12.2.25) can be obtained by a linear combination of the two assumed initial slopes as long as the BVP is linear. For the above example we will find c_1 and c_2 by (12.2.11) and (12.2.12) using the two different initial slopes $y'(0) = 0.5$ and $y'(0) = 0.6$ and these are $c_1 = -4.07314$ and $c_2 = 5.07314$. For various values of x we can have the solution

$$y(x) = c_1 y_1(x) + c_2 y_2(x) \tag{12.2.31}$$

where $y_1(x)$ and $y_2(x)$ are the solutions obtained for the slopes 0.5 and 0.6 respectively. Finally it can be seen that the value of y(1.0) from (12.2.31) exactly tally the given value of y(1.0) = 1.17520.

12.3 RAYLEIGH-RITZ METHOD

Let us consider the following second order BVP in [a,b] :

$$y'' + p(x)y = q(x) \quad s/t \quad y(a) = \alpha, \ y(b) = \beta \tag{12.3.1}$$

We can find a functional (i.e. a function of functions)

$$I = \int_a^b F(x, y, y') dx \tag{12.3.2}$$

where

$$F(x, y, y') = \left(\frac{dy}{dx}\right)^2 - py^2 + 2qy \tag{12.3.3}$$

From the functional (12.3.3) we can have the differential equation (12.3.1) by applying the well-known Euler-Lagrange equation of calculus of variations viz. by employing

$$\frac{\partial F}{\partial y} - \frac{d}{dx}\left(\frac{\partial F}{\partial y'}\right) = 0 \tag{12.3.4}$$

$$\Rightarrow -2py + 2q - \frac{d}{dx}\left(2\frac{dy}{dx}\right) = 0 \quad \text{(by (12.3.3))}$$

$$\Rightarrow y'' + py = q$$

which is the differential equation (12.3.1).

In the Rayleigh-Ritz or commonly known as Ritz method, we assume an approximate solution satisfying the boundary conditions and involving unknown constants $c_0, c_1, c_2, \ldots\ldots\ldots, c_n$,

$$y(x) = \sum_{i=0}^{n} c_i \phi_i(x) \tag{12.3.5}$$

where ϕ_i's are linearly independent suitably chosen functions and are known as the shape functions. Substitute this in the integrand (12.3.2) and thus evaluate I as a function of c_i's by minimizing I. The necessary conditions for the extremum value of I from ordinary calculus gives:

$$\frac{\partial I}{\partial c_j} = 0, \quad j = 0,........n \tag{12.3.6}$$

These lead to (n+1) equations in (n+1) unknowns. The system of equations (12.3.6) can be solved for the c_i's to get the approximate solution from (12.3.5). It is interesting to note here that if we solve a system in terms of Rayleigh quotient then we lead to the similar solution particularly in vibration problems where one is interested in to the Eigen values and Eigen vectors of the system thereby giving system natural frequencies and mode shapes.

Starting with the approximate solution of the deflection as in (12.3.5) in the Rayleigh-Ritz method we first write the maximum kinetic energy and maximum potential energy of the system and denote these by T_{max} and L_{max}, respectively. Now we define the Rayleigh quotient as,

$$\text{Rayleigh Quotient} = \frac{L_{max}}{T^*_{max}} \tag{12.3.7}$$

Where $T_{max} = \omega^2 T^*_{max}$ for harmonic vibrations
Now putting the series (12.3.5) in (12.3.7) and employing the stationarity of (12.3.7) we turn into an eigenvalue problem which can be solved for the system characteristics.

The above two procedures will be clear by the following vibration problem.

EXAMPLE 12.2
Consider the problem of transverse vibration of a string. The corresponding differential equation is given by

$$S \frac{d}{dx}\left(\frac{dw(x,t)}{dx} \right) = m \frac{d^2 w(x,t)}{dt^2} \quad \text{in} \ [0,1] \tag{12.3.8}$$

s/t $w(0,t)=0$ and $w(1,t)=0$.
where w: transverse deflection, S: string tension and m: mass per unit length of the string.

By introducing $w(x,t)=W(x)\cos(\omega t)$ for simple harmonic motion where ω is system frequency , (12.3.8) becomes,

$$\frac{d^2W}{dx^2} + \beta^2 W = 0 \tag{12.3.9}$$

$$\text{s/t } W(0)=0 \text{ and } W(1)=0 \tag{12.3.10}$$

where

$$\beta^2 = \frac{\omega^2 m}{S}$$

Now we can find a functional of (12.3.9) as,

$$F = \left(\frac{dW}{dx}\right)^2 - \beta^2 W^2 = (W')^2 - \beta^2 W^2 \tag{12.3.11}$$

and by using Euler Lagrange equation

$$\frac{\partial F}{\partial W} - \frac{d}{dx}\left(\frac{\partial F}{\partial W'}\right) = 0 \tag{12.3.12}$$

we get

$$-\beta^2 2W - \frac{d}{dx}(2W') = 0$$

$$\Rightarrow W'' + \beta^2 W = 0 \text{ which is equation (12.3.9).}$$

Therefore, we can use the Ritz procedure for the functional F in (12.3.11).

Next we assume an approximate solution satisfying the boundary conditions (12.3.10) and involving unknown constants c_i's, as

$$W = \sum_{i=0}^{n} c_i \phi_i(x) \tag{12.3.13}$$

where for the present problem let us choose

$$\phi_i = f\, x^i, \quad i = 0,1,\ldots\ldots\ldots n \tag{12.3.14}$$

and

$$f = x(1-x) \tag{12.3.15}$$

In order to satisfy the boundary conditions (12.3.10), f is chosen as in (12.3.15). This makes W zero at x=0 and at x=1. Now as in (12.3.2) we first write from (12.3.12),

$$I = \int_0^1 [(W')^2 - \beta^2 W^2] dx \qquad (12.3.16)$$

Substituting the approximate solution of W from (12.3.13) in (12.3.16) we have,

$$I = \int_0^1 [(\sum_{i=0}^n c_i \phi_i')^2 - \beta^2 (\sum_{i=0}^n c_i \phi_i)^2] dx \qquad (12.3.17)$$

From (12.3. 6) $\dfrac{\partial I}{\partial c_j} = 0, \; j = 0,1,........n$ gives

$$\int_0^1 [2(\sum_{i=0}^n c_i \phi_i')\phi_j' - \beta^2 2(\sum_{i=0}^n c_i \phi_i)\phi_j] dx = 0$$

$$\Rightarrow \sum_{i=0}^n c_i [\int_0^1 \phi_i' \phi_j' \, dx - \beta^2 \int_0^1 \phi_i \phi_j dx] = 0, \quad j = 0,........n$$

This gives the eigenvalue problem,

$$[A] - \lambda[B] = 0 \qquad (12.3.18)$$

where,

$$[A] = \int_0^1 \phi_i' \phi_j' \, dx, \;\; [B] = \int_0^1 \phi_i \phi_j dx \;\; \text{and} \;\; \lambda = \beta^2 \qquad (12.3.19)$$

Solution of the Eigen value problem (12.3.18) gives the system's Eigen values (natural frequencies) and the corresponding Eigen vectors (mode shapes).

The above example will now be solved using the Rayleigh quotient method.

The expressions for kinetic energy T and potential energy for the system of the present example can be written as,

$$T = \frac{1}{2} \int_0^1 m \dot{w}^2 (x,t) dx \qquad (12.3.20)$$

$$L = \frac{1}{2} \int_0^1 S \left(\frac{dw(x,t)}{dx} \right)^2 dx \qquad (12.3.21)$$

again putting w(x,t)=W(x) cos(ωt) for simple harmonic motion we can write the maximum kinetic energy and potential energy as

$$T_{max} = \frac{\omega^2}{2} \int_0^1 m W^2 dx$$

and

$$L_{max} = S \int_0^1 (W')^2 dx$$

Now, equating $T_{max} = L_{max}$ we have, the Rayleigh quotient

$$\beta^2 = \frac{\int_0^1 W'^2 dx}{\int_0^1 W^2 dx} \qquad (12.3.22)$$

where

$$\beta^2 = \frac{\omega^2 m}{S}$$

Again define W as in (12.3.13) and substituting in (12.3.22) leads to,

$$\beta^2 = \frac{\int_0^1 (\sum_{i=0}^n c_i \phi_i')^2 dx}{\int_0^1 (\sum_{i=0}^n c_i \phi_i)^2 dx} \qquad (12.3.23)$$

Now for the stationarity of the Rayleigh quotient we will differentiate (12.3.23) with respect to c_j's and equate it to zero. i.e.

$$\frac{\partial \beta^2}{\partial c_j} = 0, \quad j = 0,1,\ldots\ldots\ldots n$$

which gives,

$$\sum_{i=0}^n [\int_0^1 \phi_i' \phi_j' dx - \beta^2 \int_0^1 \phi_i \phi_j dx] = 0, \quad j = 0,1,\ldots\ldots\ldots n$$

This turns out to be the same eigenvalue problem as in (12.3.18)

$$[A] - \lambda[B] = 0$$

This again may be solved for the system's characteristics.

12.4 COLLOCATION METHOD

Consider again the following second order BVP in [a,b] as

$$y'' + p(x)y = q(x) \quad s/t \quad y(a) = \alpha, \ y(b) = \beta \tag{12.4.1}$$

As in the Ritz method we assume an approximate solution satisfying the boundary conditions and involving unknown constants $c_0, c_1, \ldots \ldots c_n$;

$$y(x) = \sum_{i=0}^{n} c_i \phi_i(x) \tag{12.4.2}$$

where ϕ_i's are linearly independent suitably chosen functions that satisfy the geometrical boundary conditions.

Substitute (12.4.2) in (12.4.1) and find the residual R as

$$R(x; c_0, c_1, \ldots \ldots c_n) = \sum_{i=0}^{n} c_i \phi_i''(x) + p(x) \sum_{i=0}^{n} c_i \phi_i''(x) - q(x) \tag{12.4.3}$$

Now force the residual R to become zero at (n+1) points $x_0, x_1, \ldots \ldots x_n$ in [a,b] i.e.

$$\sum_{i=0}^{n} c_i \phi_i''(x_j) + p(x) \sum_{i=0}^{n} c_i \phi_i(x_j) - q(x_j) = 0 \tag{12.4.4}$$

$$j = 0, 1, \ldots n$$

Eqn. (12.4.4) gives (n+1) simultaneous equations in (n+1) unknowns. By solving above system by any of the methods discussed earlier one can find the unknowns $c_0, c_1, \ldots \ldots c_n$. The final solution can be obtained by putting these constants in (12.4.2).

EXAMPLE 12.3:

Using collocation method, approximate the solution of the equation: $y'' + y + x = 0$ subject to the boundary conditions $y(0) = y(1) = 0$.

Solution:

Let us take two terms of equation (12.4.2),

$$y(x) = c_0 \phi_0 + c_1 \phi_1$$

Take $\phi_0 = x(1-x)$ and $\phi_1 = x^2(1-x)$

So, y(x) can be written as,

$$y(x) = x(1-x)(c_0 + c_1 x)$$
$$= c_0(x - x^2) + c_1(x^2 - x^3) \tag{12.4.5}$$

It is to be noted here that ϕ_0 *and* ϕ_1 are so chosen that they satisfy the given boundary conditions. So, $y(x)$ satisfies the boundary conditions. Now find the following from (12.4.5):

$$y'(x) = c_0(1 - 2x) + c_1(2x - 3x^2)$$
$$y''(x) = -2c_0 + c_1(2 - 6x)$$

Therefore, residual, R can be written as

$$R(x; c_0, c_1) = -2c_0 + c_1(2 - 6x) + c_0(x - x^3) + c_1(x^2 - x^3) + x$$
$$= c_0(x - x^2 - 2) + c_1(2 - 6x + x^2 - x^3) + x$$

Let us take the collocating point as
$$x_0 = 1/3 \quad \text{and} \quad x_1 = 2/3$$

$$R(x_0; c_0, c_1) = 0$$
$$\Rightarrow c_0(-\tfrac{1}{9} + \tfrac{1}{3} - 2) + c_1(2 - \tfrac{6}{3} + \tfrac{1}{9} - \tfrac{1}{27}) + \tfrac{1}{3} = 0 \text{.}$$

$$\Rightarrow -\frac{5}{3}c_0 + \frac{2}{27}c_1 = -\frac{1}{3} \tag{12.4.6}$$

$$R(x_1; c_0, c_1) = 0$$
$$\Rightarrow c_0(-\tfrac{4}{9} + \tfrac{2}{3} - 2) + c_1(2 - \tfrac{12}{3} + \tfrac{4}{9} - \tfrac{8}{27}) + \tfrac{2}{3} = 0$$

$$\Rightarrow \frac{16}{9}c_0 + \frac{50}{27}c_1 = \frac{2}{3} \tag{12.4.7}$$

Solving for c_0 and c_1 from (12.4.6) and (12.4.7) we get

$$c_0 = \frac{81}{391} \quad \text{and} \quad c_1 = \frac{63}{391}$$

$$y = x(1-x)\left(\frac{81}{391} + \frac{63}{391}x\right) \tag{12.4.8}$$

This is the required approximate solution.
Clearly we have the exact solution of the given differential equation as,

$$y = \frac{\sin x}{\sin 1} - x \tag{12.4.9}$$

Now we can compare the approximate (by collocation) y_c and exact y_e solutions for particular values of x by the following table:

x	y_c	y_e
0.25	0.04640	0.04401
0.5	0.07193	0.06975
0.75	0.06150	0.06006

12.5 GALERKIN'S METHOD

Considering the same second order BVP in [a, b] as

$$y'' + p(x)y = q(x) \quad s/t \quad y(a) = \alpha, \ y(b) = \beta \tag{12.5.1}$$

an approximate solution as in earlier methods is assumed

$$y(x) = \sum_{i=0}^{n} c_i \phi_i(x) \tag{12.5.2}$$

which satisfies the given boundary conditions. Next we find the residual R as in collocation method,

$$R(x; c_0, c_1, \ldots c_n) = \sum_{i=0}^{n} c_i \phi_i''(x) + p(x) \sum_{i=0}^{n} c_i \phi_i(x) - q(x) \tag{12.5.3}$$

Here the residual R is orthogonalized to the (n+1) functions $\phi_0, \phi_1, \ldots \phi_n$. This gives

$$\int_a^b R(x; c_0, c_1, \ldots c_n) \phi_j(x) dx = 0, \quad j = 0, \ldots n \tag{12.5.4}$$

$$\Rightarrow \sum_{i=0}^{n} \left[\int_a^b \left\{ \phi_i''(x) \phi_j(x) + p(x) \phi_i(x) \phi_j(x) - q(x) \phi_j(x) \right\} dx \right] = 0 \tag{12.5.5}$$

$$j = 0, 1 \ldots n$$

Again eqn. (12.5.5) is (n+1) simultaneous equations in (n+1) unknowns, which can be solved by any standard method discussed earlier. Finally putting the evaluated constants $c_0, c_1, \ldots c_n$ in (12.5.2) we get the approximate solution for the BVP (12.5.1).

EXAMPLE 12.4:

Using the Galerkin's method, approximate the solution of equation $y'' + y + x = 0$ subject to the boundary conditions y(0) = y(1)=0.

Solution:

Let us take again the two terms of the series in (12.5.2) for the approximate solution as,

$$y(x) = x(1-x)(c_0 + c_1 x) \qquad (12.5.6)$$

where

$$\phi_0 = x(1-x) \quad \text{and} \quad \phi_1 = x^2(1-x) \qquad (12.5.7)$$

Clearly y(x) satisfies the given boundary conditions. As in collocation method substituting (12.5.6) in the given differential equation the residual R can be written as,

$$R(x; c_0, c_1) = c_0(x - x^2 - 2) + c_1(2 - 6x + x^2 - x^3) + x \qquad (12.5.8)$$

Now using eqn. (12.5.4) for ϕ_0 and ϕ_1 we have two equations,

$$\int_0^1 \phi_0 R(x; c_0, c_1) dx = 0 \qquad (12.5.9)$$

and

$$\int_0^1 \phi_1 R(x; c_0, c_1) dx = 0 \qquad (12.5.10)$$

Equation (12.5.9) gives

$$\int_0^1 x(1-x)[c_0(x - x^2 - 2) + c_1(2 - 6x + x^2 - x^3) + x]dx = 0$$

$$\Rightarrow \frac{3}{10} c_0 + \frac{3}{20} c_1 = \frac{1}{12} \qquad (12.5.11)$$

and the eqn. (12.5.10) gives,

$$\int_0^1 x^2(1-x)[c_0(x - x^2 - 2) + c_1(2 - 6x + x^2 - x^3) + x]dx = 0$$

$$\Rightarrow \frac{3}{20} c_0 + \frac{13}{105} c_2 = \frac{1}{20} \qquad (12.5.12)$$

The constants c_0 and c_1 can be obtained by solving (12.5.11) and (12.5.12) and are

$$c_0 = \frac{71}{369} \quad \text{and} \quad c_1 = \frac{7}{41} \tag{12.5.13}$$

So, the approximate solution by the Galerkin's method can be written as,

$$y(x) = x(1-x)[\frac{71}{369} + \frac{7}{41}x] \tag{12.5.14}$$

Now we can compare the solutions from Galerkin's y_g, collocation y_c (from the previous example 12.4.1), and the exact y_e for the given BVP for particular values of x by the following Table:

x	y_c	y_g	y_e
0	0	0	0
0.25	0.04640	0.04408	0.04401
0.5	0.07193	0.06944	0.06975
0.75	0.06150	0.06009	0.06006
1.0	0	0	0

It is clearly seen that Galerkin's method gives more accurate solution than the collocation. Moreover, the accuracy can be increased if we take more terms in (12.5.2). The solution also largely depends upon the suitable choice of the functions ϕ_i's, (which are called shape functions) so that (12.5.2) satisfies the boundary conditions also.

Chapter 2

2-1. Bisection method

```
function p = bis(f,a,b,n,er)

if f(a)*f(b)>0
    disp('same sign at points a and b')
else
    for i=1:n;
    p = (a + b)/2;
    err = abs(b-a)/2;
    while err > er
    if f(a)*f(p)<0
        b = p;
    else
        a = p;
    end
     p = (a + b)/2;
    err = abs(b-a)/2;
    end
    end
end
```

2-2. False Position Method

```
function p = fpm(f,a,b,n,er)

if f(a)*f(b)>0
```

```
        disp('same sign at points a and b')
else
    for i=1:n;
    p = (a*f(b)-b*f(a))/(f(b)-f(a));
    err=abs((p-a)/p);
    if err < er
        break
    end
    if f(a)*f(p)<0
        b = p;
    else
        a = p;
    end


    end
    end

end
```

2-3. Newton-Raphson Method

```
function p = nrm(f,a,n,er)

for i=1:n
    syms x;
    d=diff(f(x));
    dy=vpa(subs(d,x,a));
    y=f(a);
    p =a-(y/dy);
    err=abs((p-a)/p);
    a = p;
```

```
      if er>err
          break
end
end
```

2-4. Modified Newton-Raphson Method

```
function p = nrm(f,a,n,er)

for i=1:n
    syms x;
    d=diff(f(x));
    dd=diff(d);
    dy=vpa(subs(d,x,a));
    ddy=vpa(subs(dd,x,a));
    y=f(a);
    p =a-((y*dy)/(dy^2-y*ddy));
    err=abs((p-a)/p);
    a = p;
    if er>err
          break
end
end
```

2-5. Secant Method

```
function x1 = secant(f,x0,x1,n,er)

for i=1:n
    dy= (f(x1)-f(x0)/(x1-x0)));
```

```
        y=f(x1);
        x2 =x1-(y/dy);
        err=abs((x2-x1)/x2);
        x0 = x1;
        x1 = x2;
        if er>err
            break
    end
end
```

Chapter 3

3-1. Gaussian Elimination

```
function x=GE(A,B)

[n, n] = size(A);
[n, k] = size(B);
x = zeros(n,k);
for i = 1:n-1
    m = -A(i+1:n,i)/A(i,i);
    A(i+1:n,:) = A(i+1:n,:) + m*A(i,:);
    B(i+1:n,:) = B(i+1:n,:) + m*B(i,:);
end
x(n,:) = B(n,:)/A(n,n);
for i = n-1:-1:1
    x(i,:) = (B(i,:) - A(i,i+1:n)*x(i+1:n,:))/A(i,i);
end
```

3-2. Gauss-Jordan Method

```
function  X = GJ(A,B)
```

```
[n n] = size(A);
A = [A';B']';
X = zeros(n,1);
for p = 1:n,
  for k = [1:p-1,p+1:n],
    if A(p,p)==0, break, end
    mult = A(k,p)/A(p,p);
    A(k,:) = A(k,:) - mult*A(p,:);
  end
end
X = A(:,n+1)./diag(A);
```

3-3. Cholesky Method

```
function [F]=Chol(A,option)
[m,n]=size(A);
        L=zeros(m,m);%Initialize to all zeros
        row=1;
        col=1;
        j=1;
        for i=1:m,
            a11=sqrt(A(1,1));
            L(row,col)=a11;
            if(m~=1)  %Reached the last partition
                L21=A(j+1:m,1)/a11;
                L(row+1:end,col)=L21;
                A=(A(j+1:m,j+1:m)-L21*L21');
                [m,n]=size(A);
                row=row+1;
                col=col+1;
            end
```

```
            end
        if strcmpi(option,'upper'),F=L';
        else if strcmpi(option,'lower'),F=L;
            else error('Invalid option');end
        end
```

3-3. Jacobi Method

```
function [X] = jacobi(A,B,X,er,N)


Z = X';
n = length(B);
for i = 1:n
    j = 1:n;
    j(i) = [];
    S = abs(A(i,j));
    T(i) = abs(A(i,i)) - sum(S);
    if T(i) < 0
    fprintf('The matrix is not strictly diagonally dominant
therefore convergence might not occur')
    end
end

Xnew = X;
for k=1:N,
  for r = 1:n,
    Sum1 = B(r) - A(r,[1:r-1,r+1:n])*X([1:r-1,r+1:n]);
    Xnew(r) = Sum1/A(r,r);
  end
  dX = abs(Xnew-X);
  err = norm(dX);
  relerr = err/(norm(Xnew)+eps);
```

```
    X = Xnew;
    Z = [Z;X'];
    if (err<er)|(relerr<er),
        break,
end
end
```

3-4. Gauss-Seidel Method

```
function [X] = GS(A,B,X,er,N)
Z = X';
n = length(B);
for i = 1:n
    j = 1:n;
    j(i) = [];
    S = abs(A(i,j));
    T(i) = abs(A(i,i)) - sum(S);
    if T(i) < 0
    fprintf('The matrix is not strictly diagonally dominant
therefore convergence might not occur')
    end
end
Xold = X;
for k=1:N,
  for r = 1:n,
    Sum1 = B(r) - A(r,[1:r-1,r+1:n])*X([1:r-1,r+1:n]);
    X(r) = Sum1/A(r,r);
  end
  dX = abs(Xold-X);
  err = norm(dX);
  relerr = err/(norm(X)+eps);
  Xold = X;
```

```
    Z = [Z;X'];
    if (err<er)|(relerr<er),
    break,
end
end
```

Chapter 4

4-1. Least Square Method

```
function [p]= lsq(x,y,m)
n=length(x);
for i=1:2*m
    xsum(i)=sum(x.^(i));
end
a(1,1)=n;
b(1,1)=sum(y);
for j=2:m+1
    a(1,j)=xsum(j-1);
end
for i=2:m+1
    for j=1:m+1
        a(i,j)=xsum(j+i-2);
    end
    b(i,1)=sum(x.^(i-1).*y);
end
p=(a\b)
```

4-2. Lagrange polynomial Interpolation

```
function yi= lagI(x,y,xi)
```

```
n=length(x);
for i=1:n
    L(i)=1;
    for j=1:n
        if j~=i
            L(i)=L(i)*(xi-x(j))/(x(i)-x(j));
        end
    end
end
yi=sum(y.*L);
```

4-3. Interpolation with divided differences

```
function yi= DDI(x,y,xi)

n=length(x);
a(1)=y(1);

for i=1:n-1
    D(i,1)=(y(i+1)-y(i))/(x(i+1)-x(i));
end
for j=2:n-1
    for i=1:n-j
        D(i,j)=(D(i+1,j-1)-D(i,j-1))/(x(j+i)+x(i));
    end
end
for j=2:n
    a(j)=D(1,j-1);
end
yi=a(1);
xn=1;
```

```
for k=2:n
    xn=xn*(xi-x(k-1));
    yi=yi+a(k)*xn;
end .
```

Chapter 5

5-1. Trapezoidal Rule

```
function I = trap(f,a,b,N,er)

for n=1:N
h=(b-a)/n;
x=a:h:b;
for i=1:n+1
    d(i)= f(x(i));
end
I(n)=(h/2)*(d(1)+2*sum(d(2:n))+d(n+1));
if n>=2
err=abs((I(n)-I(n-1))/I(n));
else err=1;
end
if err<er
break
end
end
```

5-2. Simpson Rule

```
function I = simp_er(f,a,b,N,er)
```

```
for n=2:2:N

h=(b-a)/n;
x=a:h:b;
for i=1:n+1
    d(i)= f(x(i));
end
I(n)=(h/3)*(d(1)+4*sum(d(2:2:n))+2*sum(d(3:2:n-1))+d(n+1));
if n>2
err=abs((I(n)-I(n-2))/I(n));
else err=1;
end
if err<er
break
end
end
```

5-3. Romberg Method

```
function r = romberg(f,a,b,m)

h = (b - a) ./ (2.^(0:m-1));
r(1,1) = (b - a) * (f(a) + f(b)) / 2;
for j = 2:m
    subtotal = 0;
    for i = 1:2^(j-2)
        subtotal = subtotal + f(a + (2 * i - 1) * h(j));
    end
    r(j,1) = r(j-1,1) / 2 + h(j) * subtotal;
    for k = 2:j
        r(j,k) = (4^(k-1) * r(j,k-1) - r(j-1,k-1)) / (4^(k-
1) - 1);
```

```
      end
  end;
```

Chapter 7

7-1. Power Method

```
function [vec,value]=power_m(x,A,n,er)
 for i = 1:n;
     y = A*x;
     [t r] = max(abs(y));
     mu = y(r);
     x = y/y(r);
     y = A*x;
     [t r] = max(abs(y));
     mu1 = y(r);
     err=abs((mu1-mu)/mu1);
     if err<er
         break
     end
 end
vec=x;
value=mu
```

7-2. Inverse Power Method

```
function [vec,value]=Inpow(x,A,n,er)
 for i = 1:n;
     y = inv(A)*x;
     [t r] = max(abs(y));
     mu = y(r);
```

```
        l=(1/mu);
        x = y/y(r);
        y = inv(A)*x;
        [t r] = max(abs(y));
        mu1 = y(r);
        l1=(1/mu1);
        err=abs((l1-l)/l1);
        if err<er
            break
        end
    end
vec=x;
value=l
```

7-3. Q R factorization Method

```
function [e]=QRf(A,N)
for k=1:N
nmatrix = size(A);
n=nmatrix(1);
I=eye(n);
Q=I;
for j=2:n-1
c=A(:,j)
c(1,j-1)=0;
e(1:n,1)=0;
if c(j)>0
e(j)=1;
else
e(j)=-1;
end
clength=sqrt(c'*c);
```

```
v=c+clength*e;
H=I-2/(v'*v)*v*v';
Q=Q*H;
R=H*A;
end
A=R*Q;
end
A
e=diag(A)
```

Chapter 9

9-1. Euler Method

```
function [y] = eul(g,x0,xn,h,y0)
x(1)=x0;
y(1)=y0;
n=(xn-x0)/h;
for i=1:n
x(i+1)=x(i)+h;
y(i+1)=y(i)+h*g(x(i),y(i));
end
```

9-2. Runge-Kutta order 2

```
function [y] = rk2(g,x0,xn,h,y0)

x(1)=x0;
y(1)=y0;
n=(xn-x0)/h;
for i=1:n
```

```
x(i+1)=x(i)+h;
k1=g(x(i),y(i));
xh=x(i)+h/2;
yh=y(i)+h*k1;
k2=g(xh,yh)
y(i+1)=y(i)+h*k2;
end
```

9-3. Runge-Kutta order 4

```
function [y] = rk4(g,x0,xn,h,y0)

x(1)=x0;
y(1)=y0;
n=(xn-x0)/h;
for i=1:n
x(i+1)=x(i)+h;
k1=g(x(i),y(i));
xh=x(i)+h/2;
yh1=y(i)+k1*h/2;
k2=g(xh,yh1);
yh2=y(i)+k2*h/2;
k3=g(xh,yh2);
yh3=y(i)+k3*h;
k4=g(x(i+1),yh3);
y(i+1)=y(i)+(k1+2*k2+2*k3+k4)*h/6;
end
```

REFERENCES

(1)   AI-Khafaji, A.W. and Tooley, J.R., "Numerical methods in engineering practice", Holt, Rinehart and Winston, Inc., New York, NY, 1986.

(2)   Atkinson, K. 'elementary numerical analysis", John Willey and Sons, New York, NY, 1985.

(3)   Atkinson, K.e., "An introduction to numerical analysis", second edition, John Wiley and Sons, New York, NY, 1989.

(4)   Atkinson, L.V., Harley, P.J. ad Hudson, J.d., "Numerical methods with FORTRAN-77 a practical introduction", Addison-Wesley Publishing co.., New York, NY, 1989.

(5)   Burden, R.L, Faires, J.D. and Reynolds, A.C., 'Numerical analysis", second edition, Prindle, Weber and Schmidt Boston, MA, 1981.

(6)   Chabra, S.C. and Canale, R.P., 'Numerical methods for engineers", second edition, McGraw-Hill, New York, NY, 1988.

(7)   Dahlquist, G. and Bjork, A., "Numerical Methods", translated by Anderson, N., Prentice Hall, Englewood Cliffs, NJ, 1974.

(8)   Gerald, C.F., "Applied numerical analysis", second edition, Addison-Wesley Publishing Co., Reading, MA, 1980.

(9)   James, M.L., Smith, G.M. and Wolford, J.C., "Applied numerical methods for digital computation", third edition, Harper and Row Publishers, New York, NY, 1985.

(10)   Johnson, L.W. and Riess, R.D., "Numerical analysis", second edition, Addison-Wesley Publishing Co., Reading, MA, 1982.

(11)   Johnston, R.L., "Numerical methods, a software approach", John Wiley and Sons, New York, NY, 1982.

(12)   Kahaneer, D., Moher, C. and Nash, S., "Numerical methods and software", Prentice Hall, Englewood Cliffs, NJ, 1989.

(13)   Lastman, G.J. and Sinha, N.K., "Microcomputer Based numerical methods for science and engineering", Saunders College Publishing, New York, NY, 1989.

(14)   Ralston, A, ad Rabinowitz, P., "A first course in numerical analysis", second edition, McGraw-Hill New York, NY, 1978.

(15)   Ralston, A. and Wilf, H.S., Eds., "Mathematical methods for digital computers", Vols. 1 and 2, John Wiley and Sons, New York, NY, 1967.

(16)   Smith, W.A., "Elementary numerical analysis", Prentice Hall, Englewood Cliffs, NJ, 1986.

(17)   Wendroff, B., "Theoretical numerical analysis", Academic Press, New York, NY, 1966.

(18)   Yakowitz, S. and Szidarovszky, F., "An introduction to numerical computations", Macmillan Publishing Co., New York, NY, 1986.

INDEX